Rural protest groups and populist political parties

Rural protest groups and populist political parties

edited by:
D. Strijker
G. Voerman
I.J. Terluin

MONTESQUIEU INSTITUTE
FROM SCIENCE TO SOCIETY

Wageningen Academic
Publishers

This publication is supported by the Montesquieu Institute (www.montesquieu-institute.eu): a multifunctional centre for comparative European parliamentary history and constitutional development. The institute aims at disclosing relevant information in a user-friendly and manageable way – through publications and the electronic knowledge exchange network – for students and specific target groups such as politicians, civil servants, educators, journalists and those employed by interest groups.

EAN: 9789086862597
e-EAN: 9789086868070
ISBN: 978-90-8686-259-7
e-ISBN: 978-90-8686-807-0
DOI: 10.3920/978-90-8686-807-0

Photo cover:
Nationaal Archief/Collectie Spaarnestad/ANP
Hendrik Koekoek, leader of the Dutch Boerenpartij, addresses 'free farmers' in 1965, commemorating the 1963 Hollandscheveld riots

First published, 2015

Preface

Projects have a beginning and an end. The start of this project lies in the summer of 2010, when Ida Terluin announced she would spend her sabbatical leave at the Faculty of Spatial Sciences of the University of Groningen (the Netherlands) and that she was willing to join forces to do 'something' associated with the 100th birthday of Hendrik Koekoek in 2012. Koekoek was the leader of a farmers' protest group in the Netherlands and of the Farmers' Party (Boerenpartij) in the period from the late 1940s until the early 1980s. The Farmers' Party was a right-wing protest party with a populist style, mobilising small traditional farmers in the poorer parts of the Netherlands, but in the 1960s it was also able to attract protesting urban voters. The political dimension is not straightforward for agricultural economists, and therefore Gerrit Voerman, director of the Documentation Centre for Dutch Political Parties at the University of Groningen, was invited to add his expertise.

One of the reasons to remember Koekoek and his political ideas was that a few years earlier the 100th birthday of Sicco Mansholt was exuberantly celebrated, with symposia, exhibitions and a statue. In the context of agriculture in the Netherlands, Mansholt and Koekoek were antipodes, one being the icon of modernisation and state intervention in agriculture, and a strong supporter of European and international approaches, the other the voice of agricultural anti-modernism and anti-state intervention.

In addition to a book about Koekoek in Dutch and a meeting for the general public in Hollandscheveld – Koekoek's birthplace – on 19 May 2012, we decided to organise an academic seminar on the relationships between rural protest groups and right-wing populist parties. The seminar was held in Groningen on 24-25 May 2012, with invited participants from seven countries. Its aim, even at that time, was to publish a book on the same theme.

Now, three years later, we have arrived at the end of the project. We would like to thank the authors of the chapters of this book for their efforts and their patience, the University of Groningen Language Centre for editing the English, and the secretariat of the Faculty of Spatial Sciences for their contribution throughout all phases of the project. Finally, a special word of thanks to Eliza van der Ploeg, a member of the secretariat, for her work on the layout of this volume.

Groningen, the Netherlands, March 2015

Dirk Strijker, Gerrit Voerman, Ida Terluin

List of contributors

Ivan Bruneau, Department of Political Science, Triangle (Research centre), University of Lyon 2 (France)
Ivan Bruneau is a political sociologist who does field research in France and the United States on a variety of issues related to rural areas, from agricultural policy and farmers' unions to the relationship between social configurations and practices of local political power.

Joël Gombin, CURAPP-ESS, Université de Picardie-Jules Verne/CNRS (France)
Joël Gombin is completing a PhD in political science on the vote for the French Front National. He is also interested in ecological analysis and the electoral behaviour of farmers.

Darren R. Halpin, Research School of Social Sciences, Australian National University (Australia)
Darren Halpin is Professor of Political Science and Sociology at the Australian National University. He publishes on interest groups and democracy, comparative public policy and rural and agricultural policy.

Gilles Ivaldi, URMIS-CNRS, Université de Nice Sophia Antipolis (France)
Gilles Ivaldi is a CNRS researcher in political science at the University of Nice. His main fields of research include French politics, elections, party system change and the study of radical right parties in France and Europe.

Ann-Cathrine Jungar, School of Social Sciences, Södertörn University, Stockholm (Sweden)
Ann-Cathrine Jungar is a Senior Lecturer in political science. She has published on parliamentarism, governmental coalitions, gender equality, europeanisation and populism.

Sarah L. de Lange, Department of Political Science, University of Amsterdam (the Netherlands)
Sarah L. de Lange is Associate Professor in the Department of Political Science at the University of Amsterdam and co-convenor of the ECPR Standing Group on Extremism and Democracy. She has published extensively on the populism and the radical right in the Netherlands and Europe.

Paul Lucardie, Documentation Centre Dutch Political Parties (DNPP), Groningen University (the Netherlands)
Paul Lucardie is a (semi-retired) political scientist at the Documentation Centre Dutch Political Parties at Groningen University. He publishes on populism and democratic radicalism, among others. His latest book is Democratic Extremism in Theory and Practice. All power to the people (Routledge, Abingdon, UK, New York, USA, 2014).

Matthew Reed, The Countryside and Community Research Institute, University of Gloucestershire (England, UK)
Matthew Reed is a political sociologist who has published on a range of topics including the organic movement, rural protests, rural policy, populism and urban agriculture.

Matthijs Rooduijn, Department of Political Science, University of Amsterdam (the Netherlands)
Matthijs Rooduijn is a political sociologist and works as lecturer and postdoctoral researcher at the University of Amsterdam. He is interested in topics such as populism, left- and right-wing radicalism and voting behavior.

Franz Seifert, Independent Researcher (Austria)
Franz Seifert is biologist, political scientist, and university lecturer in Vienna. His major fields of interest are currently social movements engaging in technology controversies such as the controversy over agricultural biotechnology or nanotechnology; comparative research; the linkage of local and global political dynamics; the democratization of technology policies.

Ben Stanley, School of Law, Politics and Sociology, The University of Sussex (England, UK), and University of Social Sciences and Humanities, Warsaw (Poland)
Ben Stanley is a political scientist, and is currently Marie Curie Intra-European Fellow at the School of Law, Politics and Sociology, University of Sussex. He specialises in party politics and voting behaviour in Central and Eastern Europe, with a particular emphasis on Poland.

Dirk Strijker, Faculty Spatial Sciences, Groningen University (the Netherlands)
Dirk Strijker is an agricultural economist, and holds a special chair in Rural Development (the Mansholt chair) at Groningen University. He published on various subjects, such as agricultural policy, rural development, and the spatial aspects of agriculture.

Ida J. Terluin, LEI Wageningen UR, The Hague (the Netherlands)
Ida Terluin is rural economist. Her research topics include economic development of rural regions in the EU and a broad range of aspects related to the Common Agricultural Policy with the emphasis on structural and rural development.

Gerrit Voerman, Documentation Centre Dutch Political Parties (DNPP), Groningen University (the Netherlands)
Gerrit Voerman is Director of the DNPP and Professor of Development and functioning of the Dutch and European party system at Groningen University. He publishes widely on Dutch political parties (party history, organisation, identity, membership, candidate selection, relations with europarties).

Koen Vossen, Political Science, Radboud Universiteit Nijmegen (the Netherlands)
Koen Vossen is lecturer in Political Science Radboud Universiteit Nijmegen. He has published various books and articles on populism and new political parties in the Netherlands.

Michael Woods, Department of Geography and Earth Sciences, Aberystwyth University (Wales, UK)
Michael Woods is Professor of Human Geography at Aberystwyth University, specialising in rural geography and political geography, and Co-director of the ESRC WISERD/Civil Society Research Centre. His major research interests include globalization and rural areas, rural protest and politics, and community governance.

Table of contents

1 Introduction

D. Strijker[1], G. Voerman[2] and I.J. Terluin[3]*

[1]Faculty Spatial Sciences, Groningen University, Groningen, the Netherlands; [2]Documentation Centre Dutch Political Parties (DNPP), Groningen University, Groningen, the Netherlands; [3]LEI Wageningen UR, The Hague, the Netherlands; d.strijker@rug.nl

1.1 Introduction

This volume examines rural protest groups, populist political parties and the possible relationship between both in Europe following the Second World War. We find that protest groups sometimes acted as a breeding ground for populist parties, such as Self-Defence in Poland in the 1990s and to a certain extent the Farmers' Party in the Netherlands in the 1960s. However, there are far more examples of modern populist parties that do not have a rural origin, and of rural protest groups that did not develop into populist parties. These different trajectories are analysed in this volume in order to gain an understanding of the specific conditions behind the variety of scenarios.

In recent decades many studies have been published on the emergence and development of populist parties in the Western world, and in Europe in particular. Far less academic attention has been paid to rural protest groups in the developed world following the Second World War, with the notable exception of a special issue of the *Journal of Rural Studies*, guest edited by Woods (2008). The literature generally neglects the relationship between rural protest groups and populist parties, probably because it touches upon various disciplines, such as political science, history, agricultural economy, rural geography and sociology. In this volume, researchers drawn from these disciplines are brought together in order to reflect upon the ties between rural protest groups and political parties in the modern Western world.

Social protest groups or movements use a number of strategies to pursue their goals. They may attempt to influence public policy by attracting media attention and/or mobilising support within society through direct collective action, by connecting to partisan politics, making use of ties with congenial political parties, or by founding a party of their own and participating in elections. One basic question addressed in

D. Strijker, G. Voerman and I.J. Terluin (eds.) **Rural protest groups and populist political parties**
DOI 10.3920/978-90-8686-807-0_1, © Wageningen Academic Publishers 2015

this volume concerns why some rural protest groups spawned a political party, or even transformed themselves into a party, while others did not. Various answers arise, depending on the type of protest, the relationship with existing political parties, the leadership's perception that a new strategy is required for the achievement of its aims, the political opportunity structure, or the institutional context. The inverse question, also dealt with in this volume, concerns how political parties of a populist nature relate to 'the rural'. If grounded in rural protest, do they still focus on agricultural and countryside issues in their platforms? If their genesis is non-rural, do they target rural voters? What conditions influence these strategic choices?

Thus, on the basis of these two questions this book approaches the relationship between protest group and political party from two directions. In the first part of the volume (Chapters 2-7), the links between protest groups and political parties are examined, with populist parties in particular being analysed; in the second part (Chapters 8-13) populist political parties are taken as the object of study, exploring their relationships with the rural realm and rural protest groups. Both parts of the book start with a contribution of a comparative nature (Chapters 2 and 8), offering a framework for the case studies of that part of the book. The following countries are studied: Australia, Austria, Finland, France, the Netherlands, Poland and the United Kingdom. These countries were selected because geographically they represent a cross-section of Western countries, mostly inside but also outside Europe, mostly from the West but also from the East of Europe. In terms of agrarian characteristics, we examine countries with highly productive and export-oriented agricultural production systems, as well as those with more modest agricultural sectors.

1.2 Rural protest groups

Our starting point was the observation that rural protest groups have been or still are active in many countries in the Western world. In some cases, rural protests merely concern demonstrations and actions undertaken by farmers and others who feel connected to the agricultural sector. This is the case for many protests and actions in France, for example, but can also be said of Austria and the Netherlands, where there have been more or less purely agriculture-oriented actions. However, there are also cases where agriculture plays only a minor role in rural protest groups, such as the rallies of the Countryside Alliance in England. In some cases it may even play no role at all, for example, the many groups protesting against wind farms, or who more generally aim to protect the countryside. Woods (2008: p. 129) labels these groups which emerged in the 1970s 'new grassroots movements of rural residents' and 'rural identity movements', as distinct from 'peasant farmers' movements'. Thus, in this volume 'rural' has a wider meaning than 'agricultural', in line with the approach to the 'rural' in rural studies (see also Woods, 2005 and Chapter 2: Woods, 2015).

Rural protest groups can take the form of fluid or more formalised protest groups, whose members actively participate in many kinds of protest activities, some playful and noisy, some grim or even violent, such as the demonstrations and campaigns against a perceived external threat to rural interests or values (Woods, 2003). These protests might be directed against modern farming practices, agricultural policy, the globalisation of the food trade, land-ownership patterns, or in defence of purportedly historical, natural and agrarian-centred rural 'ways of life'. This volume analyses examples of all of these.

According to Shanin (1971: pp. 257-258), three types of rural peasant protest groups can be distinguished. While they may be mutually exclusive categories in theory, they possibly overlap in practice: (1) independent class action; (2) guided political action; and (3) spontaneous political action. In the first category, a previously passive social class becomes active during a conflict. Shanin mentions the Zapatista movement in Mexico as an example. More recently, the success of the Countryside Alliance in England could possibly be placed in the same category. The marches in London in 1998 and 2002, originally directed against a ban on hunting with hounds, developed into a mass movement for the preservation of rural life in England, creating some sort of rural self-awareness. In Chapter 4, Reed (2015) analyses the transformation of this movement and connects it to the political field.

The second category – guided political action – concerns groups that unite around an external leader or political party, such as Mao's People's Army and rural protest groups in Latin America. These movements often originate in urban areas, and to paraphrase Shanin (1971: p. 257), the peasantry may be used in such circumstances for quite distinct goals. The boundary between the first and second categories is often not very clear. However, in relation to our present aims, neither of these types of rural protest are particularly significant to the post-war modern Western world, since both cases presuppose a large, passive and predominantly ignorant, silent or neglected rural population.

The third category refers to spontaneous action, a type of protest which has been predominant in the Western world in recent times, not only in relation to peasants but also to other rural protest groups as specified above. According to Woods (2008), these actions always occur in response to social and economic restructuring that negatively impacts on 'the settled rural order'. The variety of such spontaneous rural actions is large, reflecting distinct circumstances, traditions and experiences. As Chapters 2-7 will show, fear of modernisation and societal and rural change, and concerns about the economic position of the agricultural population or elements of it, have been important reasons for spontaneous rural protest.

The nature of rural protest groups may vary from violent and revolutionary to peaceful and passive (Wolf, 1969). In the cases studied in this volume, almost all of the protests were relatively peaceful, although there were some victims, as in Brussels in 1971 in massive protests against Mansholt's Common Agricultural Policy (Van Merriënboer, 2006), or serious material damage, such as the 'dismantling' of a McDonald's restaurant in Millau in 1999.

Rural protest can have a national dimension or a take on a more regional character. In most of the cases selected here, the protests are regionally focused, but in general the issues they raise have a national or even international background. Protests related to low agricultural prices, or high production costs (e.g. taxes on fuel, the cost of hired labour, land leases), are always more acute in some sectors or sub-sectors than in others (e.g. in dairy, pork or fruit). As these sectors are spatially concentrated, a regional focus to the protests can be expected. Examples include protests in France against low prices and competing imports, from Spain in particular, in 2009 and 2014, and protests in the Netherlands against the low prices being offered for arable products, especially wheat and barley, in 1990. The anti-modernist protest movement, from which the Dutch Farmers' Party developed in the 1960s, had a national background, despite being concentrated in regions with relatively low-income, small-scale farming.

Agrarian and rural protest

There are different reasons for agrarian and rural protests. Today in the Western world France especially is known for its militant agrarian protests, which generally concern agricultural crises related to low prices and the resulting bad economic circumstances. However, protests of economic origin are of course not restricted to France. In Australia, as described by Halpin (2015) in Chapter 7, despite a traditionally fragmented structure of representation of farmers' interests, economic reasons for protests have also been prevalent, while economic discontent in the Netherlands was also the trigger for mass demonstrations by farmers in the mid-1970s and the late 1980s.

However, it is not only economic pressure that is a reason for protest. Aversion to the consequences of agricultural modernisation and to competition from foreign products also often plays a role. Moreover, as Bruneau (2015) shows in Chapter 5, rural protests in France are not limited to agricultural issues, as farmers today are a relatively small minority in rural areas. In this respect, British (as described in Chapter 2 by Woods and in Chapter 4 by Reed (2015)) and French rural protests exhibit many similarities. The protection of traditional hunting activities has been an issue in both countries. The protection of the countryside in relation to, for example, wind farms in Britain and the Netherlands (Chapters 3 and 4), are also clear cases in which rural coalitions of non-farmers took the lead in the protests rather than farmers or the more traditional population of rural areas.

The French Confédération Paysanne (CP), as analysed by Bruneau, is a typical peasant farmers' movement that reveals the link between traditional farmers' protest groups and anti-globalisation issues, with the farmer and activist José Bové as its icon (he was one of the two leading candidates for the European Green Party in the European elections in May 2014). Bové was one of the founders of the CP and one of the leaders of the campaign that resulted in the destruction of a MacDonalds building in Millau. Interestingly, this link can also be found in the agenda of some protest groups in the Netherlands, as shown in Chapter 3 (Strijker and Terluin, 2015). However, more than in the case of the CP in France, some of the Dutch groups had strong links with Third World issues.

Protests against the introduction of genetically modified (GM) crops, as an element of unwanted modernisation, have occurred in many countries in Western Europe (Scatasta *et al.*, 2007). These protests do not have a specifically rural nature; in general, they could be considered as consumer-based protests. However, there is also a rural component when farmers perceive GM crops as a threat to their own business and to the environment (Chapter 6: Seifert, 2015). Modernisation tendencies in farming in general – increasing size and scale – and state intervention are also reasons for protest, as was the case in Austria (Chapter 6). Protests of an anti-modernist nature also occurred in the Netherlands in the 1950s-1970s (the Farmers' Party and related organisations, Chapter 3 by Strijker and Terluin and Chapter 13 by Vossen), in France, as described by Bruneau, and in Poland, as discussed by Stanley in Chapter 9.

The relationship between rural protest groups and political parties

The Farmers' Party in the Netherlands and Self-Defence in Poland serve as examples of rural – primarily agricultural – protest groups which gradually transformed into political parties of a populist nature. In Austria and France such protest movements did not evolve into political parties, but maintained relationships (sometimes close) with new or existing political parties. This was also the case with rural protest groups such as the Countryside Alliance in England and Australian and other Dutch protest groups that did not undergo such an organisational transformation into a populist political party, also partially because of a relationship with existing political parties. Today, the electoral prospects of genuine agrarian parties have deteriorated, as the size of the agrarian sector is now relatively small. These observations lead to the question of the relationship between protest groups and political parties, and more specifically, why some rural protest groups do evolve into political parties of the populist type, while others do not.

As Woods shows in Chapter 2, a comparative study of Britain, France and Australia, the reasons for rural protests are universal, but the impact of these general processes and the responses are very much determined by the national context. In particular, the national political context shapes both the timing and the form of protests. The political context includes the spatial distribution of the electorate, the political culture, and the structure and position of the national government. Woods also found that while the rural protests in the three countries were quite different, in all cases they resembled each other in their results: all were unsuccessful in changing policies. Thus, he argues that rural activists would have been more effective had they participated in existing political parties and mobilised urban voters as well, as their own electoral share is too small to have any real impact on national rural and agricultural policies.

In Chapter 3, Strijker and Terluin distinguish four different types of rural protest groups in the Netherlands. The first type concerns the anti-modernist farmers' movement, of which the Farmers' Party was the political wing. This is one of the few examples of a protest group that developed into a political party. The vicissitudes of the Farmers' Party also reveals that the numerical share of the relevant agricultural population, and thus the proportion of the vote, was too small to influence national policies. The Farmers' Party was born of the specific Dutch agricultural and political constellation at that time, in which the existing political parties did not represent the opinions and concerns of the small, traditional farmers. The second type includes the direct activist groups that challenged the existing farmers' institutional-representative bodies, with generally quite weak ties to political parties. The third type covers the more progressive farmers' coalitions that have existed since the beginning of the 1970s. Some of these, especially those advocating environmental and nature-friendly policies, were quite successful in building coalitions with non-farmer interest groups, and were thus able to influence national policies. The last type consists of the non-farmer coalitions, the defenders of the traditional countryside. Most of these groups remained fragmented in the Netherlands and did not become influential at the national level.

Reed focuses on the British situation in Chapter 4, covering the last two decades. He shows that in the 1990s rural movements in the UK were often intertwined with politics and government programmes, but that more recently the linkages have become less explicit. He also demonstrates that the anti-wind farm movement and the Transition Town movement represent two different visions of rural life: the rural idyll versus the rural as the centre of social innovation. Both movements make use of populist arguments and exhibit a sense of urgency that is lacking in the political parties. The anti-wind farm movement is very active at the local level, but is not able to raise a common voice at the national level. Thus, their influence on national politics is low and comparable to the situation in the Netherlands. The main reasons for this are that the physical impact of wind farms is quite local, and that the protests have a strong not-in-my-backyard (NIMBY) nature.

In Chapter 5, Bruneau analyses the relationship between the left-wing militant Confédération Paysannne (CP) and the political field, sketching the complex world of agricultural representation in France. The subtle relationships with the political realm are drawn in a detailed case study of the CP in the Aveyron region, where two different groups of farmers joined forces: the traditional farmers – born into farming – and the *néo-ruraux*, farmers with a more urban background. Their different backgrounds may explain differences in their understanding of the political realm and different approaches and skills when it comes to activism.

The Austrian situation, as described by Seifert in Chapter 6, has some similarities with the French case, in terms of the institutional arrangements (elected Chambers of Agriculture), and with the Dutch case, in terms of a strong mainstream agricultural organisation representing the interest of the majority of the farmers but challenged by smaller groups with specific interests. The strong Farmers Association (Bauernbund), which represents a large share of Austrian farmers and maintains close ties with the conservative Austrian People's Party (ÖVP), is challenged by smaller groups of farmers, often with more progressive ideas. These groups correspond to the direct activist farmers' groups in the Netherlands and their disputes with the national farmers' union, LTO Nederland. However, the response of the Farmers Association has been quite different: it has chosen adaptation. While it may have initially neglected or ridiculed critical ideas, in a later phase it incorporated them into its own programme. A specific aspect of the Austrian situation concerns the position of the populist Freedom Party of the late Jörg Haider, the agricultural branch of which reflected Haider's right-wing populism. The existence of the Freedom Party has limited the space for a new populist party to emerge from these small agrarian protest groups.

In Chapter 7, Halpin deals with farmers' protests and militancy in Australia. From the 1960s onwards militancy and farmers' protests were common; however, the extent of fragmentation of the representation of agricultural interests is much greater than that found in most European countries. This has to do with the size of Australia, which means communities are more disparate and there are large 'regional' differences in resources and production systems. Halpin shows that the various organisations and movements were gradually amalgamated into the National Farmers' Federation (NFF), first in the state of New South Wales, and later at the national level. This process was enhanced with the recognition by its leaders of the great political cost of disunity. The Federation is now the dominant body representing farmers' interests at a political level. As the NFF is an umbrella organisation, it allows for divergence and in this way militancy is controlled.

The conclusion of this part of the book is that the national context and the political opportunity structure are of primary importance in shaping the political dimension of rural social movements. In the Netherlands, the Farmers' Party developed because existing political parties neglected the concerns of a large group of small-hold, traditional farmers. In the UK, the electoral system makes it almost impossible for minority groups to develop their own political party with any chance of political success. In France, as in many other countries, there is a dominant farmers' organisation which has strong ties to mainstream political parties and is able to thus influence the political agenda. Protest groups in France are generally too small to have any chance of developing into new, relevant political parties. In Austria, the populist Freedom Party voices rural and agricultural discontent to a certain extent, leaving limited space for other protest parties. As in France, the united NFF in Australia is able to influence the mainstream political parties when it comes to rural and agricultural concerns.

1.3 Populist parties

In the past, agrarian protest parties have quite often had a populist orientation, but certainly not all populist parties had associations with the agrarian realm. In the second part of this volume, a number of populist parties in various European countries are reviewed, analysing their agrarian – or more broadly – rural characteristics in terms of genesis, platform and electorate.

Populism as a concept

'Populism is a difficult, slippery concept', according to the British political scientist Paul Taggart (2000: p. 2). Because of its ambiguity and negative political connotation, it is also an 'essentially contested concept'. Some political scientists consider populism a political style or form of rhetoric; others regard it as a strategy or practice (Lucardie and Voerman, 2012: pp. 12-13). In this volume, we take populism to be an ideology, but not a 'thick', fully fledged one, such as socialism or liberalism, with universal core values and a vision on many aspects of life and society. Instead, we consider populism to be a 'thin', partial ideology, consisting of a set of more or less coherent concepts regarding the relationship between the people and the elite (see also Chapter 8: De Lange and Rooduijn, 2015). Populism seldom emerges on its own; it usually attaches itself to or is combined with one of the more comprehensive ideologies – today this is often conservative nationalism.

Thus, populism focuses not so much on society as a whole but on the political domain, and more specifically on representative politics. Its point of departure is the Manichean dichotomy between the people, considered to be essentially virtuous, noble and pure; and the selfish, corrupt and evil elite (Albertazzi and McDonnell, 2008; Mudde, 2004). Both entities are regarded as homogeneous and their antithesis defined in moral terms. The powerful establishment is alienated from the ordinary people and ignores their will. The political class serves its own interests, a parasite on the back of the common people (the 'folk'), exposing them to external threats, such as those perceived to arise from multiculturalism (Mudde, 2007). Populism thrives especially in times of transition and perceived or real crises that are blamed on globalisation, immigration and/or European integration. It quite often identifies with an idealised heartland which has to be defended against corrupt, alien or hostile elements (Taggart, 2000: pp. 2-4). When populists refer to 'the people', they may be focused on different groups, depending on the context: the people may consist of peasants or farmers, the ordinary person in the street, the 'natives' or 'the Dutch', for example. The substantiation of the term, however, is clearly exclusionist: there are groups within society which do not belong to 'the people'.

For populists it is evident that the people and the elite are separated by a wide gap. In order to resolve the antagonistic relationship between both, the political influence of the people must be increased. Representative, indirect democracy, in which political parties play a dominant role as intermediaries between society and the state, must be replaced with or at least supplemented by direct democracy, using instruments such as referenda, people's initiatives, recall processes, or direct elections of mayors and other officials, which give the power back to ordinary men and women. In other words, the sovereignty of the people has to be restored; their undiluted will should be reflected in government.

Thus, there is a 'lowest common denominator' in populism: people-centrism, anti-elitism, homogeneity of the people and the exploitation of a perceived or real crisis (Rooduijn, 2014). All of the authors in the second part of this volume, which focuses on the possible agrarian roots of populist parties, take these core elements of populism as their point of departure, in one way or another. Apart from these primary characteristics, however, populism also has some 'secondary features' or 'additives' (Lucardie and Voerman, 2012: pp. 17-19; Taggart, 2000; Vossen, 2010), about which there is greater disagreement. The most important 'additive' is charismatic leadership, which implies a personalised bond between the leader and their followers. Such leaders are able to convince the people that their needs are understood and that they are the only one who can represent and protect their interests. They use simplistic, direct, even vulgar language, mirroring that of the ordinary people, and sometimes melodramatic rhetoric – another secondary characteristic. In doing so they identify with the people, both distancing themselves from the establishment and its nuanced manner of expression, and ignoring the complexity of modern politics. The central role of the leader does not fit well with an organisational model that includes checks, balances and accountability: such leaders prefer to make the decisions themselves. This third feature implies that along with the populist preference for direct democracy and its corresponding dislike of intermediary actors between the leader and the people, populist parties are usually not well organised. It is no coincidence that a number refer to themselves as a 'movement' rather than a party: a weak or loose informal structure fits them better than a bureaucratic organisation with formal membership rights, which require the consultation of all the relevant party bodies before decisions are taken (Rooduijn, 2014).

Populism in its historical manifestations

Populism is rather fluid – not only conceptually, but also in its historical manifestations. Here we focus on agrarian populism, which is 'a kind of rural radicalism' (Canovan, 1981: p. 8, 98), a grassroots movement opposed to the political or economic elite and narrowing the notion of 'the people' to refer to its agrarian subgroup: the peasants or the farmers who are idealised as 'the source of morality, and agricultural life the foundation of society' (Mudde, 2001: p. 35). Canovan (1981) has distinguished three types of agrarian populism, briefly presented here: peasant populism, farmer populism and a populism of the intelligentsia (see also Chapter 8). All three are based on the antithetical idealisation of the people (i.e. peasants or farmers) and the demonisation of the elites.

Agrarian populism has generally made its appearance in times of profound social transformation. When tradition and modernisation collided, resulting in friction and polarisation 'between the rural and the urban complex', agrarian populism flourished, especially 'where this contrast is great and not tempered or forced into the background by other conflicts between social groups or subsystems' (Van Niekerk, 1974: p. 25). The organisational form of this kind of populism is mainly determined by the social and political context in which it originated.

The term 'populism' was used for the first time in the second half of the nineteenth century – or to be more precise, the Russian word '*narodnichestvo*' (usually translated as 'populism') was used for the first time around 1870 (Ionescu and Gellner, 1969: p. 3; Venturi, 1960: pp. xxxii-xxxiii). It was derived from the Russian word '*narod*', which means 'people', and it referred to a radical movement emerging in predominantly rural Russia in the 1860s and 1870s which was directed against appalling social conditions in the still feudal countryside.[1] During these years, the *narodniki*, as they were called, tried to mobilise the deprived and exploited 'innocent' and 'pure' peasants against the autocratic Tsarist regime (Taggart, 2000: p. 50). Romanticising peasant life, the *narodniki* wanted to preserve the traditional agrarian village commune (*obshchina*) – which they considered the embryo of a future socialist society because of its supposedly egalitarian traditions – against the rise of capitalism (Canovan, 1981: pp. 66-70). In 1874, the *narodniki* actually went 'to the people', leaving the cities for the countryside to live among the peasants who they idealised 'as martyrs..., and as embodiments of simple uncorrupted virtue' (Berlin, 1960: p. viii). The *narodniki* tried, rather unsuccessfully, to instigate a peasant rebellion against the absolutist system and to liberate them by realising a direct transition to socialism, bypassing capitalism (Walicki, 1969: p. 68). As urban intellectuals originating from higher social classes and advocating socialism, they were often treated with distrust by the peasants, who were deeply dedicated to the Tsar

[1] Some authors claim, however, that the translation of 'narodniki' as populists is incorrect, see Walicki, 1969.

and the Orthodox Church. In the context of harsh repression by the Tsarist authorities, some of the *narodniki* turned into terrorists, assassinating Tsar Alexander II in 1881.

The *narodniki* fit into the populist tradition, with their hostility to Western liberal, representative parliamentary democracy and their perception of the Russian peasants and their community as their heartland (Taggart, 2000: pp. 54-58). In a way, however, they themselves belonged to an elite – for this reason, Canovan qualifies *narodnichestvo* as a 'populism of the intelligentsia'. After the Second World War, a similar brand of populism, combined with agrarian socialist views, could be observed among intellectuals in Third World countries (Canovan, 1981: pp. 105-110).

In a centralised and autocratic political system such as that of Tsarist Russia in the late nineteenth century it was impossible for an agrarian protest movement to develop into a political party. Such a transformation was possible, however, in a democracy such as the United States of America. In the early 1890s, the People's Party was founded, emerging from the Farmers' Alliance (consisting of grassroots regional agrarian organisations), which campaigned for the needs of the honest and hard-working small farmer in the South and the West against the social infringement of organised capitalism. The 'Populist Party', as the People's Party was widely known, promoted the interests of cotton and wheat tenant farmers who had been impoverished during the economic depression. It directed their grievances against the economic and political elite of the North and East, who 'lived off the labor of others' (Canovan, 1981: p. 55). The People's Party wanted to end the hegemony of the Republicans and the Democrats and to return power to the ordinary and honest American worker, to whom it had originally belonged: 'the people, not the plutocrats, must control government' (Hicks, 1961: p. 406).

As its name indicates, the People's Party was committed to 'direct legislation by the people', which in practice meant bottom-up instruments such as referenda, citizens' initiatives and the recall of elected officials. In a way, its nickname, 'Populist Party' almost seems a *contradictio in terminis*: after all, advocating direct democratic reforms implied eliminating or at least weakening intermediary, representative structures such as political parties. After some initial electoral success, the People's Party disappeared from the political scene at the beginning of the twentieth century. Its political aims, however, found their way into American politics, for example, through the use of primaries for party nominations (Hicks, 1961: pp. 404-412; Hofstadter, 1969, Taggart, 2000: pp. 25-45).

Canovan labels the People's Party a case of 'farmers' populism' (1981: pp. 100-105). This concept refers to commercial farmers being hostile to big companies and demanding the regulation of the economy by a people's government. This seems to be a bit of a paradox, with farmers who possessed property advocating state intervention. However, there have been other cases of this type of agrarian populism, such as the farmers' movements in Canada in the 1930s: the Cooperative Commonwealth Federation and the Social Credit Party.

In its initial phase, populism had an indisputable agrarian character. In the last decades of the nineteenth century, it manifested itself in radical rural movements in both Russia and the US, however different they may have been. They articulated the interests of peasants and farmers (the 'people') against a selfish urban political elite, aiming at a 'grassroots control of society' (Canovan, 1981: p. 96). In a second wave, populism became a dominant political force in Latin America (Taggart, 2000), in a way moving from the countryside to the city. While the Russian and American 'classic' populists responded to rural poverty and shared 'a common belief in rural values', Latin American populists 'turned their backs on the peasantry: they draw no spiritual sustenance from visions of pastoral or rural innocence: they receive no more than fringe support from peasant organisations: they are rural only in so far as peasants exert pressure on towns by becoming migrants and thus ceasing to be peasants' (Hennessy, 1969: pp. 28, 35; Taggart, 2000: p. 19). As in Russia and the US, populism in Latin America manifested itself as a form of protest against the modernisation of society, but contrary to its predecessors, it was first and foremost an urban movement, although also focused on the migrants of rural origin living in the cities who were being left by the wayside (Van Niekerk, 1974: p. 180). The dominant role that strong charismatic leaders played was also different. While in the case of the People's Party, mobilisation was bottom-up, in Latin America it was top-down: here populist personalistic and charismatic leaders were able to inspire and rally the 'disposable' masses, utilising patronage (Hennessy, 1969).

The failure of the first populist movement in Russia in the early 1880s and the first populist party in the US around 1900 did not mean the end of agrarian populism in the Western world. In the interwar period, 'peasant populism', the third type of agrarian populism according to Canovan (1981), became 'the dominant ideology' in most Central and Eastern European countries, in which the majority of the population were peasants (Gollwitzer, 1977; Mudde, 2001: p. 39). Despite all national differences, grassroots peasant movements in these countries were all opposed to capitalism and socialism, opting for a third, morally superior way, 'the vision of a democratic society based on small family property, widespread cooperation and respect for traditional rural values', and all advocating direct democratic instruments (Canovan, 1981: pp. 112, 126). They also shared their suspicion of urban interference. These peasant movements were called the 'Green Uprising', and its leaders were to various degrees influenced by the Russian *narodniki*.

Peasant populist parties emerged in Eastern Europe after the First World War, mainly due to the fact that the countryside had suffered greatly during the war, as well as the creation of new nation-states after the dissolution of the Habsburg Empire and the introduction of universal suffrage. A number of these parties also came to power, but without being able to form stable governments, with most disappearing from the scene in the 1930s when authoritarian and fascist regimes were established (Gollwitzer, 1977: p. 66). After the Second World War, when the Soviet Union installed communist regimes in what came to be known as the 'Eastern Bloc', their time was definitely over. While peasant parties re-emerged in Eastern Europe after the fall of the Berlin Wall in 1989 and the collapse of communism, they were not very successful, especially not the populist ones. The process of industrialisation had been accompanied by the decline of the peasantry, with the traditional family farm also destroyed by the collectivisation of the agrarian sector by the Communists. As a result, the breeding ground for agrarian populism has almost disappeared. According to Mudde, it 'only survived in those postcommunist societies where collectivisation was either successfully resisted by the rural population or was moderated by 'goulash socialism'' (2001: p. 41). The latter refers to Hungary and the former to Poland, where the agrarian populist party Self-Defence did relatively well in the post-communist period – at least temporarily (as Stanley (2015) demonstrates in Chapter 9).

In Western Europe, nationally based, well-organised and durable agrarian parties were in general a marginal phenomenon in the twentieth century (Gollwitzer, 1977; Puhle, 1977; Urwin, 1980). Among other factors, this also had to do with the fact that the cleavage between the urban and the rural was usually less pronounced than in Eastern Europe, even in Mediterranean agricultural regions. Only in Scandinavia and Finland, where there was a sharp political and cultural contrast between the farmers and the urban elite, did agrarian parties flourish electorally. They were founded in the late nineteenth and early twentieth centuries defending the countryside against 'the city' (Urwin, 1980). Around 1960, such parties in Sweden, Norway and Finland all changed their names to include 'Centre'. Due to the shrinking agricultural sector, they attempted to expand their electoral potential by targeting non-agrarian groups within society and broadening their platform. The same holds true for the agrarian party in Switzerland. The Party of Farmers, Traders and Citizens, a merger of several cantonal agrarian parties, changed its name to the Swiss People's Party (SWP) in 1971 after merging with another small party. The SWP adopted a catch-all strategy in order to reach more voters. In the 1990s it turned into a radical-right populist party, rejecting Swiss EU membership, calling for strict limitations on immigration and taking an explicitly anti-Islam stance.

Not only was agrarian populism quite exotic in Western Europe for a long time, populism more generally was also basically absent there. It made a marked appearance in Western Europe in the 1980s (although in some countries such as Belgium and Denmark it manifested itself earlier), and in Eastern Europe after the fall of the Berlin Wall in 1989 and the collapse of the communist states. 'Coming to Europe, the new populism of the far right becomes fundamentally urban', according to Taggart (2000: p. 6). While populism in the past often had rural roots, it is now relatively strong in advanced, industrialised democracies, as is demonstrated by De Lange and Rooduijn in Chapter 8, who provide a comprehensive overview of contemporary populist parties in Europe. They first examine the characteristics and subtypes of populism (e.g. agrarian populism, conservative or neo-liberal populism, national populism and social populism) and subsequently assess its empirical manifestations. Since the early 1980s, at least 29 populist parties in Western Europe have gained parliamentary representation, and 40 in Central and Eastern Europe. It can be seen that populism has firmly integrated itself within the various national-level party systems in Europe. After presenting possible explanations for the electoral successes of populist parties, the authors focus on the relationship between agrarian populism, the populist vote and rural support. They conclude that although agrarian populism is no longer a widespread phenomenon in Europe, most populist parties have a clearly defined (nationalist and protectionist) stance on agrarian issues and do attract considerable electoral support from the countryside.

Of the 69 populist parties that De Lange and Rooduijn found to have existed in Europe in the past 35 years, they identified only two as agrarian populist: the Independent Smallholders, Agrarian Workers and Civic Party (Független Kisgazda, Földmunkás és Polgári Párt) in Hungary and Self-Defence of the Republic of Poland (Samoobrona Rzeczypospolitej Polski) in Poland – the two countries also referred to by Mudde. Both parties emerged in Central and Eastern European countries which still had a relatively large agricultural sector. In Chapter 9, Self-Defence in Poland is analysed by Stanley as a textbook example of a populist party arising from agrarian protest. Farmers were severely affected in the period of post-communist economic reform. In the early 1990s, one of them, the charismatic Andrzej Lepper, organised a protest movement 'for the Self-Defence of Farmers', directed against the politics of liberal-democratic transition. In 1992 it also registered as a political party. Self-Defence was hostile towards the domestic elites, which in its eyes had bartered away the interests of the Polish people. From the start, its programme did not confine itself to agrarian demands, but also included broader issues so as to widen its electoral appeal to other sections of society that had suffered from the effects of the transition after the collapse of communism. In the early 2000s, the party was electorally successful. In 2006 it took part in a government coalition, but this success was the beginning of the end of the party. After the suicide of Lepper in 2011 his party practically vanished.

Compared to the presence of agrarian populism in Central and Eastern Europe in the first half of the twentieth century, today it has been marginalised in the former communist countries. In contemporary Western Europe, however, there are no longer agrarian populist parties in parliament at all; their natural electoral bases were diminished by the contraction of the agrarian sector in these post-industrialist societies, threatening to make them irrelevant. Some existing agrarian parties have responded by transforming themselves into national-populist parties, such as the SWP referred to above. Another example is provided by Finland, where a tradition of radical agrarian movements and agrarian populist parties have existed since the early 1920s. In 1959 a party was founded which, after some time, came to be called the Finnish Rural Party (FRP). It aimed to defend the interests of small farmers and the rural population in the process of urbanisation and modernisation. It was not the product of an agrarian movement but had broken away from an established party. Its participation in government ultimately resulted in electoral decline, and in 1995 the FRP made way for a new party, the True Finns (Perussuomalaiset). According to Jungar (2015) in Chapter 10, the True Finns are 'a milder version of agrarian populism', representing the disadvantaged ordinary people and the rural population in general, who were all affected negatively by the reconstruction of the Finnish welfare state, underway since the 1990s. The party was more successful than its predecessor partly because the True Finns were able to broaden their electoral base: in the national elections of 2011 it gained nearly 20% of the vote. The traditional populist appeals to 'rural virtues' are still reflected in the rhetoric of the True Finns, although their representation has adapted to the need to reach a wider public.

The case of Poland, and especially that of Finland, show agrarian populist parties transforming themselves into more or less catch-all populist parties. They incorporate their agrarian or rural focus into a wider platform that is explicitly designed to attract voters from other sectors of society. To a certain extent, a reverse process is also possible, in which nationalist-populist parties deliberately try to target rural voters whom they previously had neglected. A perfect example is the Front National (FN) in France, founded by Jean-Marie Le Pen in the early 1970s. In Chapter 11, Ivaldi and Gombin (2015) describe the process of 'ruralisation' of the FN. Initially, the FN was predominantly an urban party: it did not originate in an agrarian movement, and its voters were found in the industrialised and urban regions of France. Especially since the mid-1990s, however, the party had progressively formulated a more comprehensive rural agenda, idealising agrarian values and specifically addressing farmers' interests and concerns. Le Pen's successor, his daughter Marine, has contributed to the further ruralisation of the FN, appealing to the 'forgotten rural France' – addressing not only farmers but also voters with lower incomes in periurban areas.

While the strategy of focusing explicitly on the interests of the farmers has been electorally fruitful for the FN, most of the non-agrarian populist parties elsewhere in Western Europe have not followed its example and refrain from 'ruralising' their platforms. The reason is, of course, the relatively small size of the agricultural sector workforce. The Netherlands is a good example. Although this country is one of the world's largest exporters of agricultural products, in 2013 less than 3% of the Dutch working population was engaged in the agricultural sector (17% in 1950). Thus, there is little wonder that contemporary populist parties in the Netherlands are not prioritising the interests and concerns of the farmers – the group is negligible within the electorate as a whole. This helps to explain why modern Dutch populist parties do not have agrarian roots, and have not 'ruralised' their programmes, as Lucardie and Voerman (2015) demonstrate in an overview of populism in the Netherlands in Chapter 12. Ideologically, the populist parties they describe are quite diverse. The Dutch case, according to these authors, is therefore a good example of the multiple forms populism can take, combining other ideologies such as socialism, liberalism or nationalism. Despite all this diversity, however, most Dutch populist parties have one thing in common, as already mentioned: they do not have agricultural characteristics, either in relation to their genesis, or in their programmatic priorities.

The only exception, to a certain extent, is the Farmers' Party, which cannot be labelled a fully fledged populist party due to its rejection of the idea of the people's sovereignty. It clearly evolved from an organised agrarian protest movement directed against the interference of the Dutch Agricultural Board in the decade after the Second World War. The Farmers' Party, and its pre-war forerunner, the Countryside League, are examined in more detail from a historical point of view by Vossen (2015) in Chapter 13, who reveals that both parties were led by farmers who voiced the concerns of their adherents about excessive state intervention in agriculture and attempted to cultivate the cultural contrast between the countryside and the city in a vulgar, anti-elitist, popular style. These agrarian protest parties remained quite marginal, being represented in Dutch parliament from 1918 to 1937 and from 1963 to 1981, generally with only a few seats.

1.4 Concluding remarks

In the past, populist political parties have emerged from a number of rural protest movements that rejected an urban elite which they said ignored the interests of farmers and the rural community. One of the main points in this introduction was the question of under which circumstances a rural protest movement could lead to the foundation of a populist party or transform itself into such a party (Lucardie 2000; Puhle 1977: pp. 605-608). The basis for this successful party formation was deep dissatisfaction in groups in the agrarian sector of society or in the countryside as a whole, which is to say a widely held feeling of being disadvantaged or neglected by the established political parties and political elite. These groups hoped to bring about change with their own political party.

The existence of a protest movement as such obviously indicates social unrest. Whether a political party develops, a party that stands the test of time, depends also on the political opportunity structure. Alongside socio-economic and cultural factors (such as urban-rural cleavage), further influential factors include the position of existing political parties and the institutional context. In an autocratic system such as Tsarist Russia during the second half of the nineteenth century, a movement like the *narodniki* could not become a political party, because there were neither political freedoms nor universal suffrage. Within a democracy such as the United States, however, such an evolution was possible, as the emergence of the People's Party illustrates. The emergence of agrarian populist parties after the First World War in Central and Eastern Europe was linked to the introduction of universal suffrage, which enabled political expression to be given to the dissatisfaction of the rural population after the hard years of war.

Alongside accessibility, the responsiveness of the political system is relevant. If a rural protest movement has the opportunity to influence the political agenda through the established parties, this removes an important incentive for the movement to form its own party. The emergence of the People's Party in the United States was stimulated by the existing Democratic and Republican parties ignoring the agrarian protest movement. The fact that the People's Party exited the stage soon afterwards was primarily because the Democrats had taken on some of its demands. This responsiveness also relates to institutionalized agrarian interest groups, which generally have close ties with the established political order. How responsive these are to the demands of the rural protest movement, as well as their willingness to use their influence in established politics, also plays a role in the decision to establish a new political party. For instance, the Farmers' Party in the Netherlands and Self-Defence in Poland emerged when protest movements felt that the established political order and the agrarian interest groups were not listening to them. The political constellation is also relevant here. If a populist party that aims to mobilize the voice of agrarian protest is already active, such as the Freiheitliche Partei Österreichs (FPÖ) in Austria, this reduces the opportunities for an agrarian-populist newcomer. Finally, the size of the country can also play a role. In Australia, regional protest movements initially proved unable to exercise influence at the national level, partly due to the large regional differences and physical distances.

A niche in the electoral market is not the only condition the newcomer has to fulfill – it must also possess resources, 'members, money, management and mass media exposure' (Lucardie 2000: p. 175). A protest movement can provide this to some extent; it is the job of a charismatic leader to attract indispensable media attention. The success, or at least temporary success, of agrarian-populist parties in Finland and the Netherlands was partly due to their mediagenic leaders, who emphasized the cultural contrast between the countryside and the city.

If it wants to be durable and successful, the newcomer must possess substantial electoral potential. The agrarian sector across Europe initially represented a significant part of society, but the modernization process that took place after the Second World War caused this to decrease considerably. This made agrarian-populist parties much less viable. In response, the Swedish, Norwegian, Finnish and Swiss parties tried to halt their decline with a catch-all strategy; in order to tap into new electoral groups, they toned down their agrarian character, broadened their electoral programmes and changed their names.

The result was that agrarian populism as such disappeared from the political stage. This marginalization took place before and during the emergence of populism in Western Europe in the final decades of the twentieth century. This leads to the second question that was posed at the start of this introduction: how do these relatively new populist parties relate to the rural? The answer is initially determined by how these parties rate the voter market: do they stand to gain an electoral advantage by focusing on agricultural and countryside issues in their manifesto? The French FN and the Dutch populist parties illustrate two contrasting scenarios. The FN, originally urban by nature, sought to engage with the countryside by 'ruralizing' its manifesto, thus managing to strengthen its position in the countryside. The Dutch populist parties, however, do not focus on the agrarian or rural voter; their main emphasis is urban concerns about animal welfare and the industrialization of agriculture. The agrarian electorate does not carry sufficient weight for them.

It is safe to say that in Europe agrarian protest will continue to exist as long as there is an agricultural sector, and rural protest as long as there is a countryside. But the chances that these protests will develop into an agrarian populist party with representation in a national parliament, as has happened in the past, seem extremely small – the agricultural sector has become too small and the countryside too heterogeneous. All in all, the age of agrarian populism seems to be over.

References

Albertazzi, D. and McDonnell, D., 2008. Twenty-first century populism: the spectre of Western European democracy. Palgrave Macmillan, Basingstoke, UK.

Berlin, I., 1960. Introduction. In: Venturi, F. (ed.) Roots of revolution. A history of the populist and socialist movements in nineteenth-century Russia. Weidenfeld & Nicolson, London, UK, pp. vii-xxx.

Bruneau, I., 2015. The Confédération Paysanne and the political field: a conflicted history. Chapter 5. In: Strijker, D., Voerman, G. and Terluin, I.J. (eds.) Rural protest groups and populist political parties. Wageningen Academic Publishers, Wageningen, the Netherlands, pp 101-126.

Canovan, M., 1981. Populism. Harcourt Brace Jovanovich, New York, NY, USA.

De Lange, S.L. and Rooduijn, M., 2015. Contemporary populism, the agrarian and the rural in Central Eastern and Western Europe. Chapter 8. In: Strijker, D., Voerman, G. and Terluin, I.J. (eds.) Rural protest groups and populist political parties. Wageningen Academic Publishers, Wageningen, the Netherlands, pp 163-190.

Gollwitzer, H. (ed.), 1977. Europäische Bauernparteien im 20. Jahrhundert. Gustav Fischer Verlag, Stuttgart, Germany.

Halpin, D.R., 2015. Farm protest and militancy in Australia: supporting or under-mining interest-group politics? Chapter 7. In: Strijker, D., Voerman, G. and Terluin, I.J. (eds.) Rural protest groups and populist political parties. Wageningen Academic Publishers, Wageningen, the Netherlands, pp 145-161.

Hennessy, A., 1969. Latin America. In: Ionescu, G. and Gellner, E. (eds.) Populism: its meaning and national characteristics. Weidenfeld & Nicolson, London, UK, pp. 28-61.

Hicks, J.D., 1961. The populist revolt: a history of the farmers' alliance and the people's party. University of Nebraska Press, Lincoln, NE, USA.

Hofstadter, R., 1969. North America. In: Ionescu, G. and Gellner, E. (eds.) Populism: its meaning and national characteristics. Weidenfeld & Nicolson, London, UK, pp. 9-27.

Ionescu, G. and Gellner, E. (eds.), 1969. Populism: its meaning and national characteristics. Weidenfeld & Nicolson, London, UK.

Ivaldi, G. and Gombin, J., 2015. The Front National and the new politics of the rural in France. Chapter 11. In: Strijker, D., Voerman, G. and Terluin, I.J. (eds.) Rural protest groups and populist political parties. Wageningen Academic Publishers, Wageningen, the Netherlands, pp 243-263.

Jungar, A.C., 2015. Agrarian populism in Finland: continuity and change. Chapter 10. In: Strijker, D., Voerman, G. and Terluin, I.J. (eds.) Rural protest groups and populist political parties. Wageningen Academic Publishers, Wageningen, the Netherlands, pp 217-241.

Lucardie, P. and Voerman, G., 2012. Populisten in de polder. Boom, Amsterdam, the Netherlands.

Lucardie, P. and Voerman, G., 2015. Rootless populists? The Dutch Pim Fortuyn List, the Freedom Party and others. Chapter 12. In: Strijker, D., Voerman, G. and Terluin, I.J. (eds.) Rural protest groups and populist political parties. Wageningen Academic Publishers, Wageningen, the Netherlands, pp 265-290.

Lucardie, P., 2000. Prophets, purifiers and prolocutors. Towards a Theory on the Emergence of New Parties. Party Politics 6(2): 175-185.

Mudde, C., 2001. In the name of the peasantry, the proletariat, and the people: populisms in Eastern Europe. East European Politics and Societies 14(2): 33-53.

Mudde, C., 2004. The populist zeitgeist. Government and Opposition 39(4): 542-563.

Mudde, C., 2007. Populist radical right parties in Europe. Cambridge University Press, Cambridge, UK.

Puhle, H.-J., 1977. Warum gibt es in Westeuropa keine Bauernparteien? Zum politischen Potential des Agrarsektors in Frankreich und Spanien. In: Gollwitzer, H. (ed.) Europäische Bauernparteien im 20. Jahrhundert. Gustav Fischer Verlag, Stuttgart, Germany, pp. 603-667.

Reed, M., 2015. Actors in the distance: rural protests in the UK and the parliamentary parties. Chapter 4. In: Strijker, D., Voerman, G. and Terluin, I.J. (eds.) Rural protest groups and populist political parties. Wageningen Academic Publishers, Wageningen, the Netherlands, pp 79-99.

Rooduijn, M., 2014. The nucleus of populism: In search of the lowest common denominator. Government and Opposition 49(4): 573-599.

Scatasta, S., Wesseler, J. and Hobbs, J., 2007. Differentiating the consumer benefits from labeling of GM food products. Agricultural Economics 37(2-3): 237-242.

Seifert, F., 2015. Austrian pastoral: rural protest at the margins of a hegemonic policy landscape. Chapter 6. In: Strijker, D., Voerman, G. and Terluin, I.J. (eds.) Rural protest groups and populist political parties. Wageningen Academic Publishers, Wageningen, the Netherlands, pp 127-144.

Shanin, T., 1971. Peasantry as a political factor. In: Shanin, T. (ed.) Peasants and peasants societies. Penguin Books, Harmondsworth, UK, pp. 238-263.

Stanley, B., 2015. The Polish Self-Defence party: from agrarian protest to the politics of populism, 1991-2007. Chapter 9. In: Strijker, D., Voerman, G. and Terluin, I.J. (eds.) Rural protest groups and populist political parties. Wageningen Academic Publishers, Wageningen, the Netherlands, pp 191-215.

Strijker, D. and Terluin, I.J., 2015. Rural protest groups in the Netherlands. Chapter 3. In: Strijker, D., Voerman, G. and Terluin, I.J. (eds.) Rural protest groups and populist political parties. Wageningen Academic Publishers, Wageningen, the Netherlands, pp 63-77.

Taggart, P., 2000. Populism. Open University Press, Buckingham, UK.

Urwin, D.W., 1980. From ploughshare to ballotbox. The politics of agrarian defence in Europe. Tromsø: Universitetsforlaget, Oslo, Norway.

Van Merriënboer, J., 2006. Mansholt. Een biografie. Boom, Amsterdam, the Netherlands.

Van Niekerk, A.E., 1974. Populism and political development in Latin America. University Press Rotterdam, Rotterdam, the Netherlands.

Venturi, F., 1960. Roots of revolution. A history of the populist and socialist movements in nineteenth-century Russia. Weidenfeld & Nicolson, London, UK.

Vossen, K., 2010. Populism in the Netherlands after Fortuyn: Rita Verdonk and Geert Wilders compared. Perspectives on European Politics and Society 11(1): 22-38.

Vossen, K., 2015. Agrarian parties in the Netherlands. The Plattelandersbond and the Boerenpartij. Chapter 13. In: Strijker, D., Voerman, G. and Terluin, I.J. (eds.) Rural protest groups and populist political parties. Wageningen Academic Publishers, Wageningen, the Netherlands, pp 291-301.

Walicki, A., 1969. Russia. In: Ionescu, G. and Gellner, E. (eds.) Populism: its meaning and national characteristics. Weidenfeld & Nicolson, London, UK, pp. 62-96.

Wolf, E., 1969. Peasant wars of the twentieth century. Harper & Row, New York, NY, USA.

Woods, M., 2003. Deconstructing rural protest: the emergence of a new social movement. Journal of Rural Studies 19: 300-325.

Woods, M., 2005. Rural Geography. SAGE, London, UK.

Woods, M., 2008. Social movements and rural politics. Journal of Rural Studies 24: 129-137.

Woods, M., 2015. Explaining rural protest: a comparative analysis. Chapter 2. In: Strijker, D., Voerman, G. and Terluin, I.J. (eds.) Rural protest groups and populist political parties. Wageningen Academic Publishers, Wageningen, the Netherlands, pp 35-62.

2 Explaining rural protest: a comparative analysis

M. Woods

Department of Geography and Earth Sciences, Aberystwyth University, Wales, United Kingdom; zzp@aber.ac.uk

Abstract

This chapter seeks to understand and explain the divergent trajectories of rural protest in different countries of the global north in the late twentieth and early twenty-first centuries. Drawing on a series of research projects undertaken since 2001, the chapter proposes that the determination of 'trigger issues', the organisational forms adopted by protest movements, and the forms and tactics of mobilisation chosen all reflect a combination of factors, including historic policy regimes and patterns of rural representation, the dynamics of urban-rural relations and influence, and domestic political systems and the opportunities that they provide for different forms of intervention. This model is developed further and the historical context discussed in more detail, with the model then tested in the final part of the chapter through a comparative analysis of contemporary rural politics in Britain, Australia and France.

2.1 Introduction

The end of the twentieth century and early years of the twenty-first century have seen an unexpected return of rural issues to the centre stage of politics in the global north. After decades in which rural areas were popularly perceived to be docile and largely apolitical – at least in comparison with the confrontational and politically charged social environment of Western cities – and in which rural issues such as agriculture were considered to be marginal to mainstream political debate, the years since the early 1980s have witnessed a proliferation of rural protests, new rural movements, and new political parties drawing on rural discontent across countries in Europe, North America, Australia and New Zealand. Yet, although this rural reawakening has occurred against a shared background of neoliberal policy reforms, globalisation and extensive social change in the countryside (Woods, 2007, 2011), the resulting actions and groupings

D. Strijker, G. Voerman and I.J. Terluin (eds.) **Rural protest groups and populist political parties**
DOI 10.3920/978-90-8686-807-0_2, © Wageningen Academic Publishers 2015

have lacked the ideological and organisational coherence of urban social movements or of the rural movement that has emerged concurrently from the global south, framed around issues of land rights and trade justice (Borras *et al.*, 2008; Woods 2003, 2008), with the possible exception of France (this volume, Chapter 5).

Not only have rural activists in the global north largely failed to develop and mobilise transnational networks of solidarity and coordinated action, instead remaining within national political parameters, they have also varied in their character, focus and methods. Pressures on agricultural incomes and proposed agricultural policy reforms have been a widespread trigger for protests, but the focus for these has ranged from national governments to supermarkets and transnational corporations to international trade agreements. Protests have also been mobilised in opposition to perceived threats to the 'rural way of life', notably from restrictions on hunting, emanating either from national legislation (as in Britain, the Netherlands and New Zealand), or from the implementation of European directives (as in France and Belgium). In Britain, the Countryside Alliance has attempted to integrate diverse rural concerns as an umbrella movement, but in many countries protest groups focused on different rural issues continue to be fragmented or only informally linked. Mass public rallies and marches have captured headlines in Britain and France, and 'tractorcade' demonstrations in Ireland and Germany have echoed earlier demonstrations in the USA in the 1970s. In France, and to a lesser extent Britain and several other European countries, mainstream demonstrations have been supplemented by radical direct action by farmers or pro-hunting activists, including blockades and vandalism. Nevertheless, in Australia attempts to mobilise farmers' protests in opposition to radical agricultural policy reforms have failed, with the last sizeable farmers' demonstrations occurring at the start of the neoliberal reforms in the mid-1980s. While the capture of discontented rural votes has contributed to sporadic electoral success for 'rural' political parties such as the Chasse, Pêche, Nature et Tradition (CPNT) party in France, Independent candidates in Australia (this volume, Chapter 7) and broader populist parties such as Law and Justice in Poland (Chapter 9), in other countries, including Britain, rural discontent has had limited impact on electoral politics.

This chapter accordingly seeks to understand and explain these divergent trajectories of rural protest in different countries of the global north. It proposes that the determination of 'trigger issues', the organisational forms adopted by protest movements, and the forms and tactics of mobilisation chosen, all reflect a combination of factors, including historical policy regimes and patterns of rural representation, the dynamics of urban-rural relations and influence, and domestic political systems and the opportunities that they provide for different forms of intervention. This model is developed further and the historical context discussed in the next section, and the model is then illustrated in the final part of the chapter through a comparative analysis of contemporary rural politics in Britain, France and Australia, linking with the substantive contributions on these three countries in this book by Matthew Reed, Ivan Bruneau and Darren Halpin (Chapters 4, 5 and 7 respectively).

In addition to the chapters by Reed, Bruneau and Halpin, this discussion also draws directly on a series of research projects conducted in the period since 2001 and funded by the UK Economic and Social Research Council (ESRC) and the Aberystwyth University Research Fund. These have variously involved: interviews with rural leaders, activists and commentators in Britain, Australia and the USA; a questionnaire survey of Countryside Alliance members in Britain; ethnographic observation at rural protests in Britain; and library and archival research in Britain, Australia, France, New Zealand and the USA, as well as the analysis of election data from Britain, Australia and France.[1]

2.2 Placing rural protests in historical context

The popular representation of contemporary rural protests as the unexpected and atypical out-pouring of anger that sits at odds with prevailing discourses of the rural idyll (Anderson, 2006; Woods, 2010a; Woods *et al.*, 2012) disregards the long history of protests and political mobilisations that were commonplace in the countryside up until the early twentieth century. Not only had the countryside been a recurrent site of rebellion and resistance against state authority over centuries, including the thirteenth-century Peasants' Revolt in England and the Jacqueries in fourteenth-century France, but the sometimes violent politics of rural society had intensified with industrialisation and modernisation. Conflicts around land rights, workers' rights, mechanisation, taxation, prices and trade policies were frequent throughout the eighteenth and nineteenth centuries, with rural protesters often espousing a radical left-wing political position. It was agricultural workers who provided the first heroes of the trade union movement in Britain – in the Tolpuddle Martyrs, sentenced to be transported to Australia for their attempts at labour organisation in 1834 – and it was agricultural workers who mounted Australia's first significant industrial action, the Shearer's Strike of 1891. In North America, farmers mobilised against the nascent state in Shay's Rebellion of 1786 and the Whiskey Rebellion of 1793-1794 among other insurrections, and later formed the electoral base for the Populist movement in the late nineteenth and early twentieth centuries, including the People's Party, the Farmers Alliance and the Nonpartisan League (Shannon, 1957; Stock, 1996).

[1] This chapter draws on three funded research projects: 'Grassroots Rural Protest and Political Activity in Britain', funded by the ESRC, 2006-7 (Grant RES-000-23-1317); 'The Emergent Politics of Rurality in Britain, France and the USA', funded by the Aberystwyth University Research Fund, 2001; and 'Contesting the Global Countryside: Policy responses and political reactions to globalization in rural economies and societies', funded by the Aberystwyth University Research Fund, 2004. I am indebted to my co-researchers on the ESRC project, Jon Anderson, Steven Guilbert and Suzie Watkin, particularly for data collection and analysis; and grateful to colleagues at the University of Queensland, especially Lynda Cheshire and Geoff Lawrence, for guidance on rural politics in Australia and sharing research materials and data.

While this radical heritage has been largely forgotten in popular culture, it has been recalled by a number of contemporary rural movements, in part as an effort to establish their patriotic credentials and position themselves as part of a long tradition. The Countryside Alliance in Britain, for example, consciously evoked the Peasants' Revolt, Levellers, Diggers and Chartists in publicity for their mass rallies (Woods, 2005). However, such claims to continuity are disingenuous to the say the least, given that the petering out of rural protests in the early twentieth century can be linked to the assertion of authority by the class of land-owning capitalist farmers that have formed the core group in contemporary rural protests. Indeed, the nineteenth-century rural protests had been expressions of a struggle over the shape of the modern countryside, pitching industrial, capitalist agriculture against tenant farmers, subsistence farmers and farm workers. As capital and technological advances favoured the former group, the capacity for radical rural protest was gradually drained by the mechanisation of agricultural labour and the out-migration of the rural working class, while land reforms helped to consolidate the position of larger, capitalist farmers. It was this cohort of larger, capitalist farmers who assumed leadership of rural communities in North America, Australia and New Zealand, and stepped into the rural power vacuum left by the decline of the aristocracy in Europe (Newby *et al.*, 1978; Woods, 2005).

Equally importantly, the rural political settlement was informed by national economic considerations. The geopolitical instabilities of the early twentieth century raised concerns about food security that prompted greater state intervention in the management of rural economies (Friedmann and McMichael, 1989), framed by a discourse that constructed the rural as a space of production, supplying the nation, and positioned farmers as key agents in managing the land and producing food. As such, the interests of farmers merged with the national interest, underpinning an 'agricultural exceptionalism' in which agriculture came to be treated differently to any other industry (Skogstad, 1998). Within this settlement, economic stability was guaranteed by subsidies, price controls and tariffs on imports, a package that was adopted across the global north and has been described as the 'agricultural welfare state' (Sheingate, 2001). At the same time, political stability was achieved by granting privileged access for rural interest groups to policymaking, as part of a small, stable 'policy community' that shared a discursive consensus on agricultural policy and the management of the countryside and excluded dissenting views (Browne, 2001; Sheingate, 2001; Smith, 1993; Winter, 1996). With favourable policies and direct representation in policymaking, the need for protests and demonstrations to further rural interests was limited, and efforts focused instead on marshalling the rural population behind the formal representative bodies.

Significantly, though, the precise means of rural representation differed between countries. In Britain, the key organisation was the National Farmers Union, which counted nine in ten farmers as members during the 1950s (Self and Storing, 1962) and enjoyed a close relationship with the Conservative Party, which in turn built an electoral near-monopoly in rural areas, assisted by a discourse that presented the countryside as 'apolitical' and discouraged socialism and trade unionism (Woods, 2005). This arrangement was paralleled in Japan, where rural interests were incorporated by the mainstream conservative party, the Liberal Democratic Party, supported by close links to the Central Union of Agricultural Cooperatives (*Nokyo*) (Sheingate, 2001). Similarly, in the Netherlands, the three agricultural unions each had close relations with conservative political parties, with both unions and parties organised along religious lines (Chapter 3). In contrast, rural interests in several northern and eastern European countries (including Denmark, Hungary, Finland, Norway, Poland and Sweden), as well as Australia, were primarily represented through expressly rural or agrarian parties that negotiated deals in coalition politics to exert control over agricultural politics. The Country Party in Australia, for example, held the Ministry for Primary Industries in coalition governments from 1958 to 1972 (Connors, 1996). In France and the USA, however, the mechanisms were less clear-cut. Although the major French farmers' union, the Fédération nationale des syndicats d'exploitants agricoles (FNSEA) became increasingly aligned to Gaullist parties, it was never able to deliver an electoral monopoly for the right in rural areas, and was repeatedly challenged by left-wing rival unions and internal factions (Naylor, 1994; Sheingate, 2001, this volume, Chapter 5). Similarly, rural electoral support in the USA was divided between Democrats and Republicans, and farmer representation divided between the Democrat-leaning National Farmers' Union and Republican-leaning Farm Bureau. Partisan politics at times frustrated rural policymaking in the USA, but in the absence of a non-agricultural rural lobby, the institutional power of the US Department of Agriculture, together with interest group lobbying of individual members of Congress representing rural states, effectively secured a political consensus behind agricultural exceptionalism (Browne, 2001; Sheingate, 2001; Woods, 2010b).

The differences in the form of rural representation between countries were to become significant as pressures on the rural political settlement intensified in the last decades of the twentieth century. In part, these pressures were arguably a product of the settlement and its success in protecting and promoting the agricultural industry. I have elsewhere called this the 'productivist paradox': productivist policies introduced by agricultural policy communities succeeded in modernising agriculture and protecting farm incomes, but they also reduced the farm workforce, stretched supply chains and disconnected farms from rural communities, thus creating the conditions in which agricultural exceptionalism could be challenged (Woods, 2011). Out-migration from rural areas reduced the electoral significance of rural regions in national elections and undermined the voter base for rural political parties. In regions where counter-urbanisation reversed the migration trend in the 1970s and 1980s, the importance of the agricultural population in rural communities was diluted by the in-migration of largely middle-class settlers with no connection to farming. Many of the in-migrants were attracted by ideas of the rural idyll and understood the countryside in ways that often clashed with modern agricultural practices. Consequently, local-scale conflicts proliferated around access to land for recreation, land management, the erection of new buildings, pollution and other issues, which led to newcomers challenging established elites for positions of local leadership (e.g. Munters, 1989; Woods, 1997, 1998), and developed over time into wider pluri-local movements (Mormont, 1987, 1990; Woods, 2005).

On the national scale, productivist policies had secured food supply, but at a cost. Food surpluses had replaced food shortages, but interventionist policies meant that surplus product was being bought by the state to prevent market forces driving down prices. By 1984, about 70% of the European Community budget was being spent on supporting farming in this way through the Common Agricultural Policy (Winter, 1996). In the context of a wider economic recession and cut-backs in other areas of public spending, the special treatment afforded to agriculture looked increasingly indefensible, not only to politicians on the left, but also to a new breed of neoliberal conservative politicians. Furthermore, growing public awareness of environmental, animal welfare and consumer issues raised questions about the impact of industrialised farming, and made the exclusion of these lobby groups from rural policymaking difficult to maintain.

The first cracks appeared in the USA towards the end of the 1970s. Rising interest rates, combined with depressed food prices, costlier inputs and drought placed pressures on farms that had borrowed heavily to invest in modernisation, especially small farmers (Dyer, 1998; Stock, 1996). However, small farmers facing bankruptcy received little support from the major farm unions or the US Department of Agriculture, whose policies were shaped by the lobbying of agribusiness. Frustrated by the apparent failure of the established mechanisms of rural representation, disenchanted farmers founded the American Agriculture Movement (AAM), which, like later rural protest groups, adopted a rhizomic structure that enabled it to spread rapidly (cf. Woods *et al.*, 2013). Activists in the AAM drew on the tradition of American rural radicalism and mounted protests, most notably two 'tractorcades' in 1978 and 1979 involving over 3,000 farmers who demonstrated in Washington D.C. However, the AAM was not able to counter the lobbying power of agribusiness, and thus had little impact in changing agricultural policy (Stock, 1996).

The next pressure points followed in Australia and New Zealand, triggered by the loss of traditional export markets following Britain's accession to the European Community in 1973. In both countries, radical neoliberal agricultural reforms were introduced in the early 1980s, including the dismantling of state subsidies, as part of packages to reorient their economies away from agricultural exports and to restructure agriculture to compete more efficiently in global markets. Farmers initially mobilised against the reforms, with 40,000 farmers demonstrating in Canberra in July 1985 (Connors, 1996; Lawrence, 1987), and 10,000 in Wellington in 1986 (Howie and Ramsay,

1998). However, the effectiveness and longevity of the protests was compromised by two factors. Firstly, political connections through the rural National Party in Australia (as the Country Party had been renamed), and the mainstream conservative National Party in New Zealand were inadequate to influence the Labour governments that introduced the reforms. Secondly, the farming lobbies were not united: as in the USA, large farmers and agribusiness increasingly had divergent interests to small farmers, and both the National Farmers' Federation (NFF) in Australia and Federated Farmers in New Zealand contained advocates for the neoliberal reforms (Bremner and Brooking, 1993; Campbell and Wards, 1992; Connors, 1996).

Meanwhile, in Europe the rural political settlement still held sufficiently to restrain the pace of agricultural reforms. Modest reforms were made to the Common Agricultural Policy, such as the introduction of milk quotas in 1984, and were met with protests by farmers, notably in France (Naylor, 1994), but radical reforms were held back by the coherence of the farm unions, and by continuing political influence – including in the mainstream conservative parties of Britain, Belgium, the Netherlands and Germany (which countered neoliberal agitation) – especially the importance of the countryside as an electoral battleground in France (Greer, 2005; Woods, 2005). However, the settlement was strained by farmers' dissatisfaction at the handling of food safety scares and by the growing momentum of environmentalism. In Britain, it was the proposed introduction of legislation to ban hunting with hounds in a Labour-dominated parliament after the 1997 election that provided the eventual spark for the rural protest movement, as discussed below. Elsewhere in Europe, implementation of the EU Habitats Directive and designation of Natura 2000 sites with restrictions on the hunting of birds and wildfowl prompted mobilisations including Platform Buitengebied in Belgium, and the CPNT party in France, also discussed below.

2.3 Towards a comparative model of rural protest

The brief historical sketch above has shown that while rural policies in developed countries have followed a broadly similar trajectory over the last century, and have been largely subject to the same overarching trends and pressures – including food security concerns in the mid-twentieth century, and the later rise of the neoliberal deregulation and free trade agenda, as well as the consequences of agricultural modernisation, rural population change and the growth of environmentalism – the form and expression of rural politics has varied. A number of inter-linked factors can be identified as informing these differences, as summarised in Figure 2.1.

First, the form of the twentieth-century rural political settlement continues to exert a legacy over contemporary political mobilisation. As discussed briefly above, different mechanisms were adopted to represent rural interests within the political

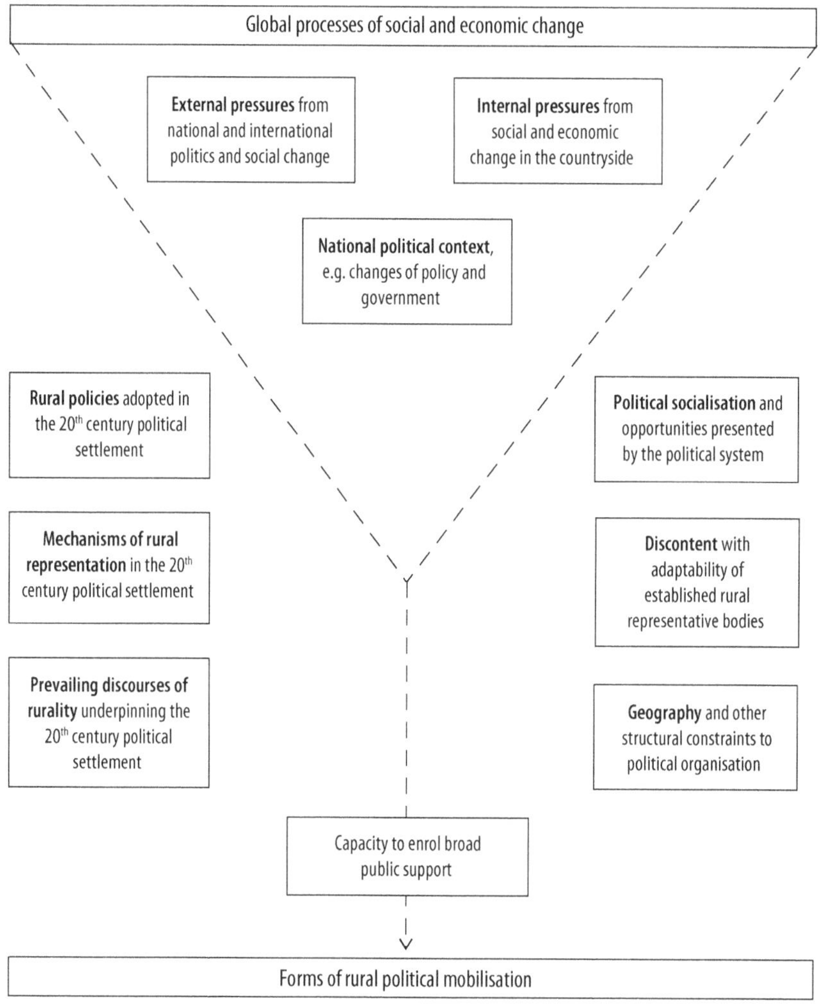

Figure 2.1. Model of factors in contemporary rural mobilisation.

settlement, and these have an ongoing influence both in terms of their continuing legitimacy and effectiveness, and in terms of the expectations about different forms of political representation held by rural activists. The precise policies adopted by national governments during this period also varied, which in turn has meant that they have come under pressure at different times for different reasons. Additionally, differences also existed between countries in the prevailing discourses of rurality that underpinned the political settlement, and consequently in the ways in which these discourses have been challenged by alternative representations, and in their potency for mobilising rural activism.

Second, although pressures on existing rural policies have been framed by shared global trends, the way in which these have been articulated within particular countries has also varied. Internal pressures on established policies and power structures within rural areas have differed between regions with significant in-migration and those characterised by depopulation – with implications for the ignition of local conflicts that may have the potential to upscale, and for the capacity of rural protest movements to mobilise across the rural population. Equally, external pressures emanating from national politics (or even from international actors and forces) have been configured differently in different countries, and are closely tied to national political contexts, including the complexion of the parties in government and their ideological agenda, with impacts on the timing and focus of rural protests.

Third, a number of further factors also have a bearing on the capacity of rural actors to mobilise and the organisational form that such mobilisations take. These include the adaptability of the established rural political organisations, and the extent to which they are perceived to have become 'out of touch' or ineffective; the opportunities for intervention provided by the form of the national political system, including the electoral system, and linked to this the political socialisation of individuals as to what constitutes appropriate political activity; the geography of the country, including the rural-urban balance of the population and the significance of distance and settlement patterns in constraining or facilitating protest organisation and mobilisation; and the capacity of the movement to engage with the wider public in enrolling support for its cause.

Collectively, these factors have shaped the timing and focus of contemporary rural protest movements in different countries, the form of political engagement that they have adopted, and their success in mobilising activists and influencing policy outcomes. This can be demonstrated by comparing the application of the model to three countries which have experienced different manifestations of rural politics over the last quarter-century: Britain, Australia and France. Each of these countries is discussed in more detail in other chapters in this book; in the brief sketches below, I focus on working through the model presented in Figure 2.1 to identify the critical factors shaping the forms of rural political mobilisation witnessed in each country.

2.4 Britain: an unexpected rural uprising?

Rural politics erupted into national consciousness in Britain at the end of the 1990s with a series of large-scale demonstrations that shocked politicians and grabbed media attention with their unexpectedness and contrast to the rather staid, quiet and conservative politics of the British countryside, at least as perceived in the popular imagination (Woods, 2005, 2010a; Woods *et al.*, 2012). The protests included three mass rallies in London organised by the Countryside Alliance, primarily to oppose legislation on hunting – the Countryside Rally (1997), the Countryside March (1998) and Liberty and Livelihood March (2002) – the last of which involved some 450,000 participants, as well as smaller protests and demonstrations around the country by hunting supporters and farmers, including radical direct action such as blockades, most notably the blockade of fuel depots by farmers in September 2000 (Table 2.1; see also Chapter 4).

These protests caught the government and media off-guard because they marked a dramatic departure from the forms of rural representation that had been integral to the rural political settlement through most of the twentieth century. As Reed argues in Chapter 4, the ebb and flow of rural protests in Britain can be understood in terms of shifting dynamics of interactions between social movements and parliamentary politics. Although Reed describes instances of rural protests earlier in the twentieth century, these were commonly small scale and short-lived and were secondary to the main emphasis on parliamentary politics in representing rural interests. The primary

Table 2.1. Key rural protests and related events in Britain, 1997-2005 (data collated from media reports).

Date	Protests and events
March 1997	Countryside Alliance formed by three pro-hunting groups
2 May 1997	Labour government elected
27 July 1997	120,000 pro-hunting supporters join Countryside Rally in London
Nov. 1997-Jan. 1998	Blockades by farmers at ports in England, Scotland and Wales
1 March 1998	250,000 demonstrators join Countryside March through London
14 June 1998	10,000 farmers demonstrate at EU Summit in Cardiff
Sept.-Oct. 1998	Farmers protest over lamb prices, including blockade of Severn Bridge
September 1998	Farmers demonstrate outside Labour Party conference
Autumn 1999	Regional Countryside Alliance rallies attract 10,000-20,000 participants
February 2000	Farmers protest over milk prices, including Carlisle to London March
May 2000	Farmers for Action formed
7-14 September 2000	Blockades of fuel depots by farmers and hauliers
Feb.-Aug. 2001	Protests against government handling of Foot & Mouth Disease
16 December 2001	Countryside Alliance 'March on the Mound' in Edinburgh
May 2002	Breakaway 'Real Countryside Alliance' launched
22 September 2002	408,000 demonstrators join Liberty & Livelihood March in London
14 February 2003	Real CA activists hang 'Love Hunting' banner from Angel of the North
16 September 2004	Hunting protestors invade House of Commons chamber
18 November 2004	Hunting Act passed, banning hunting with hounds in England and Wales
18 February 2005	Ban on hunting comes into force, mass turnout of hunts

mechanisms for rural representation in twentieth-century Britain were the National Farmers' Union (NFU) and other more specialised interest groups such as the pro-hunting British Field Sports Society (BFSS), which worked closely with government in tight-knit policy communities, but significantly were also closely aligned with the Conservative Party, which provided parliamentary representation for rural interests (Figure 2.2). These relationships underpinned a rural policy consensus around agricultural exceptionalism and productivism, assisted by prevailing discourses of the 'apolitical countryside' (Woods, 2005) – which discouraged militancy – and the centrality of the rural to British national identity – which allowed rural interests to be conflated with national interests.

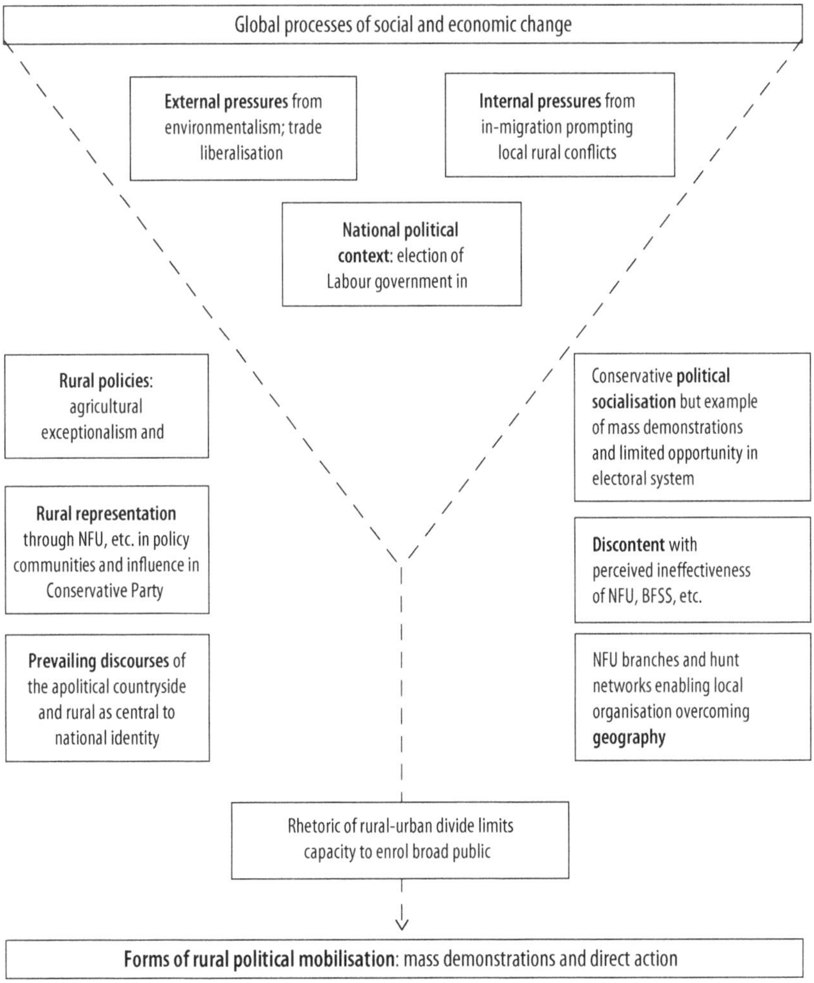

Figure 2.2. Factors in contemporary rural mobilisation in Britain.

The challenge to this rural political settlement arose against the background of global processes of social and economic change, but was encountered largely in terms of domestic processes and politics. Within the countryside, internal pressures on the established rural power structure and its agricultural exceptionalism came from in-migration and gentrification, sparking a plethora of local conflicts, some of which contributed to a weakening of Conservative electoral support in some rural constituencies (Woods, 1997, 2003, 2005). At the same time, external pressures came from the growth of new social movements concerned with environmental issues and animal welfare, as Reed describes. These processes eroded the rural political settlement, but the tipping point came with the election of a Labour government in 1997, not because of its policies per se, but because of the problem it created for a system of rural representation that had become too tied to the Conservative Party.

The hardening of rural militancy in Britain after 1997, therefore, is best explained as an expression of discontent by farmers and hunting enthusiasts with their traditional representative organisations. In the hunting community, criticism had been mounting of the BFSS's emphasis on parliamentary lobbying and its reliance on Conservative Minister Presidents (MPs) and peers voting out anti-hunting legislation. The election of a parliament with an anti-hunting majority in 1997 derailed this strategy, leading grassroots activists to force a change of tack towards public demonstrations and protests under the new banner of the Countryside Alliance' (with some moving further into direct action with militant splinter groups such as the Countryside Action Network and the 'Real CA'). Similarly, there had been growing discontent among farmers over the effectiveness of the NFU during the BSE crisis of 1996 – the handling of which was seen by some farmers as an act of betrayal by the Conservative government – and when agriculture was plunged into crisis by falling farm-gate prices in the winter of 1997-1998, many farmers had little faith in the NFU to represent their interests to a Labour government with which they had few connections. Instead, grassroots farmers organised spontaneous blockades of ports to protest against meat imports, launching a protest movement that coalesced into the radical Farmers for Action group (Woods, 2005).

Nevertheless, the protest movement was implicitly dependent on the structures of the old representative bodies for their capacity to overcome the problem of geography in rural mobilisation. It was connections through the branch network of the NFU (and the Farmers Union of Wales [FUW]) that enabled militant farmers to organise direct action; while the Countryside Alliance incorporated the structure and resources of the BFSS and utilised the network of local hunts to mobilise participants across the country:

> The extraordinary thing about the hunting infrastructure is that you can press a button in most hunt counties and you can get x-thousand people very quickly indeed, you can get to every farmer, you can get to every shoot captain, you can get to all the members of the hunt supporters' club.

> So when you add all of those people up you are getting tens of thousands of people. It's why we always said that...you could go to Parliament Square, you pressed a button and in twenty-four hours you could get two thousand people, in forty-eight hours you could get four thousand people, and in a few months you had the organization to get nearly half a million.
>
> Countryside Alliance official (interview, 2007)

Moreover, the protests had to overcome the conservative political socialisation of rural Britain, which had dismissed the legitimacy of political demonstrations as part of an ideologically driven and troublesome urban politics. The organisers of the rural protests now looked to these same demonstrations (such as anti-nuclear marches and rallies against the 'poll tax') as inspiration, but sought to sanitise events such as the large pro-hunting rallies in London for participants that had little experience of direct political activity (our survey of Countryside Alliance members found that 90% had not taken part in a protest before 1997, but that 75% had joined the Liberty and Livelihood March in 2002). Publicity around the events accordingly emphasised their patriotic and orderly character, as well as the 'ordinariness' of the participants (Burridge, 2008), with one sympathetic newspaper commentator describing the 1998 Countryside March as 'the most respectable public protest in British history' (quoted by Woods, 2005: p. 119).

The sudden militancy of rural activists was therefore presented as an emotional response to a perceived failure of parliamentary representation which left little option but to protest to protect true national interests (Woods *et al.*, 2012). This formula also placed limits on the action. The farmers' blockade of fuel depots in September 2000, for instance, was called off at the point that it seemed that the government might be brought down, with its leader telling a newspaper that 'we weren't there to topple a government' (quoted by Woods, 2005: p. 151). As such, the protests aimed to inform public opinion and thus place pressure on parliament through voting behaviour, thus ceding primacy to the parliamentary process.

Given this qualification, it is striking and significant that rural discontent in Britain was not expressed through electoral support for a rural protest party. A 'Countryside Party' was formed in 2000 and described itself as 'the only political party run by country people for country people' (Countryside Party, 2004), but its potential was constrained by the 'first-past-the-post' electoral system in national elections. It contested the 2003 Scottish Parliament and 2004 European Parliament elections held using proportional representation systems, yet its best result was in the South West England region for the European elections, where it polled just 2.1% of the vote. Attempts to exploit rural discontent by populist right-wing parties, the UK Independence Party and the British National Party also failed to gain momentum.

Instead, the alignment of the rural lobby with the Conservative Party was reaffirmed as the Conservatives – in opposition after 1997, and with a new leader who represented one of England's most rural constituencies – came out strongly in support of the countryside protests. Indeed, although the Countryside Alliance presented itself as non-partisan, and appointed a Labour MP as its chair and a Labour peer as president, 52% of Countryside Alliance members surveyed were also members of the Conservative Party (Table 2.2). One chief executive of the Countryside Alliance was subsequently elected as a Conservative MP, and its deputy chief executive was appointed as a Conservative member of the House of Lords. The leader of the farmers' protests in North Wales was similarly elected as a Conservative member of the National Assembly for Wales. The alignment was further consolidated by Vote-OK, an initiative that targeted anti-hunting MPs by mobilising campaigning support for their opponents:

> It was very targeted. There were half a dozen seats where we did work with an existing supportive MP, who was being seriously threatened by an anti-hunting second placed candidate, and we called that bolster. So you had target seats, 120 odd of them, and about 10 bolster seats. So we were working with about 130 constituencies, and we built up a structure where we had one person, which we put in touch with the candidate, and they were then detailed to...we then allocated the surrounding hunts to help that person, to supply the needs of the campaign. And we stressed to the candidates that we weren't going to campaign on hunting. We just wanted to help them.
>
> Vote-OK coordinator (interview, 2006)

Vote-OK claims to have mobilised around 7,500 people to campaign in targeted constituencies in the 2005 general election, helping with canvassing, delivering leaflets, erecting poster boards and addressing mailings. Just under 30% of the Countryside Alliance members that we surveyed had been involved with Vote-OK, mostly by helping candidates with leaflet delivery and posters (Table 2.2). Not all candidates supported were Conservatives (some were Liberal Democrats or nationalists), but the large majority were, as were almost all of the 29 MPs that Vote-OK claimed credit for electing.

Even given these proclaimed successes, however, the electoral impact of the rural protest movement in Britain was small, as Reed notes. Certainly, expectations that the numbers for the 1997 and 1998 countryside marches would translate into a significant loss of rural constituencies by Labour in the 2001 general election were misplaced, as analysis shows Labour suffering little damage to its rural vote (Woods, 2002). This points to the failure of the rural protest movement to build the broad coalition of public support required for effective political mobilisation. The branding of the mass demonstrations

Table 2.2. Political activity of members of the Countryside Alliance.[1]

Activity	%
Participated in Liberty and Livelihood March, Sept. 2002	75%
Participated in Countryside March, March 1998	56%
Participated in Countryside Rally, July 1997	56%
Participated in direct action (e.g. blockades, pickets) after 1997	17%
Had participated in a protest prior to 1997	10%
Member of the Conservative Party	52%
Member of the Liberal Democrat Party	1%
Member of the Countryside Party	0.2%
Member of the Labour Party	0.1%
Participated in Vote-OK campaign in target constituencies, 2005 election	30%
Helped with canvassing voters	10%
Helped with delivering leaflets	21%
Helped with erecting posters	20%

[1] Author's survey of Countryside Alliance members, 2006 (n=1,243). A postal questionnaire survey was sent to 4,344 members of the Countryside Alliance resident in four case study areas: Cheshire, Exmoor and North Devon, Mid Wales and Suffolk, with 1,243 usable responses returned. The survey was undertaken with the assistance of the Countryside Alliance.

as 'countryside protests' was intended to suggest a broad campaign extending beyond hunting to encompass issues such as rural services and housing development, but it obscured the fact that most participants were drawn from a narrow farming- and hunting-centred community. At the same time, the rhetoric of the countryside mobilising against urban ignorance and political correctness that featured heavily in publicity and reportage for the protests militated against enrolling urban support for their cause. Thus, as Reed observes, the rural protest movement in Britain ultimately failed to achieve its objectives.

2.5 Australia: marginalised militancy

In comparison with the changes that prompted rural protests in Britain, the scale of transformation experienced by rural Australia since the 1980s has been deep and dramatic. The adoption of radical neoliberal (or 'economic rational') policies by both Labor and Liberal governments in response to global economic and political challenges transformed rural Australia by removing agricultural subsidies and deregulating markets, leading to the disappearance of over 106,000 farms (or 40% of Australian farms) in the three decades to 2011 (ABS, 2012) (see also Lockie, 2000). The closure of processing plants, as a further consequence of neoliberal reforms and heightened competition, and the rationalisation and privatisation of public services additionally contributed to the decline of rural communities and widening economic disparities between rural and urban regions. Together with environmental pressures such as drought, and struggles over aboriginal rights, these trends fuelled rural discontent and perceptions of a 'great divide' between urban and rural-regional Australia (Pritchard and McManus, 2000). Yet, initial protests against the reforms in the 1980s were soon stymied, and ongoing restructuring has been confronted with only limited and sporadic resistance that has failed to make an impact in mainstream politics. Once again, the reasons for this apparent acquiescence can be revealed through the model of rural mobilisation (Figure 2.3).

The structure of rural representation and policy in twentieth-century Australia is described in detail by Halpin in Chapter 7. For much of the century, the most prominent mechanism of rural representation was the Country Party, supported by a prevailing discourse of 'countrymindedness', which identified the Australian national character with the struggles of country people and positioned primary production as central to the Australian economy. It believed that because power resided in the city 'there has to be a separate political party for country people to articulate the true voice of the nation' (Aitkin, 1985: p. 35; Wear, 2009). Through coalition politics, the Country Party controlled the Department for Primary Industries through much of the mid- to late-twentieth century, implementing strongly protectionist rural policies that were described as 'one of the most remarkable demonstrations of agrarian socialism in the western world' (Connors, 1996: p. 32).

However, as Halpin (Chapter 7) demonstrates, the rural political settlement in Australia was not as coherent or as consensual as its equivalent in Britain. Different sectors of agriculture, and different regional groupings of farmers, had their own, sometimes conflicting interests, and not all considered themselves adequately represented by the Country Party. The formation of the NFF in 1979 was an attempt to create a coherent national body to represent agriculture independent of the Country Party (by this time renamed the National Country Party, and soon to be renamed the National Party) (Connors, 1996; Lawrence, 1987), but tensions remained within the affiliated groups it encompassed. As such, the internal pressures on the rural political settlement in Australia came from the divergent interests of agricultural industries, and this fragmentation in turn weakened the capacity of the rural lobby to confront external pressures resulting from the loss of British markets for agricultural exports following the UK's entry into the European Community, and from proposals by the Labour government elected in 1983 to cut agricultural subsidies as part of a wider neoliberal restructuring of the state.

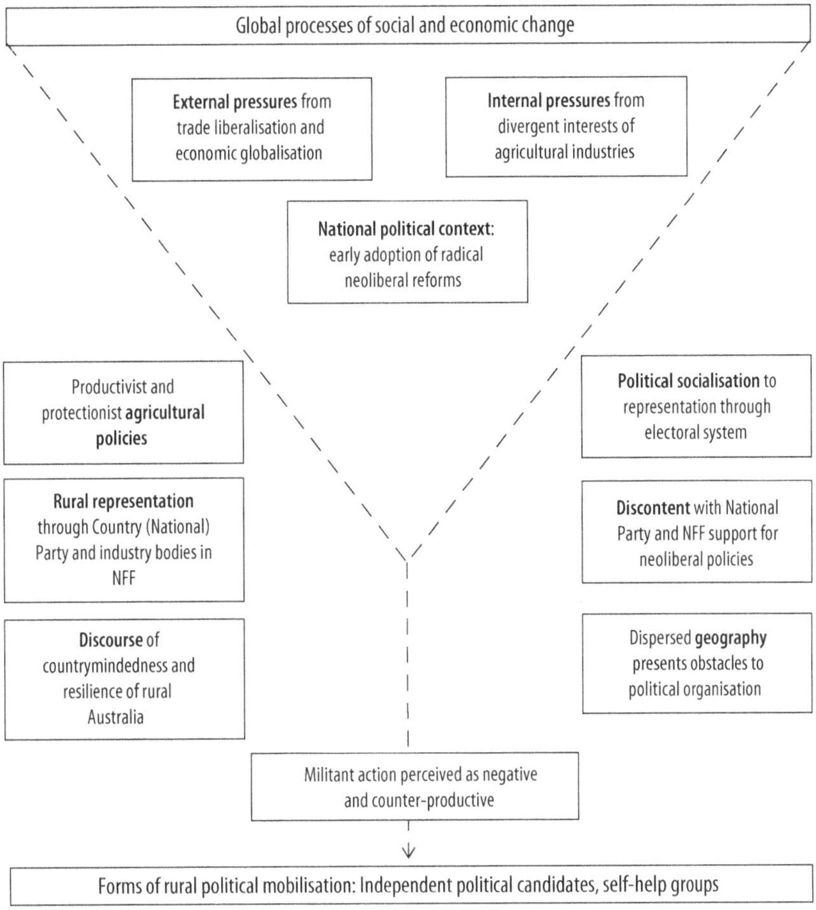

Figure 2.3. Factors in contemporary rural mobilisation in Australia.

Different rural representative groups disagreed over strategies to oppose agricultural policy reforms, creating a space for militancy which was manifested in a mass demonstration in Canberra in 1985, from which the National Party was excluded (see Chapter 7). Yet, although the NFF backed the rally, and its willingness to engage in militant direct action arguably strengthened its hand with government, it also included elements that supported neoliberal deregulation, and as Halpin argues, its desire to construct a corporatist relationship with government led it to contain farmer militancy by the end of the 1980s.

Parallels might perhaps be drawn between the NFF's tactical tolerance of farmer militancy to strengthen its strategic position in the 1980s and the role played by groups such as Farmers for Action and the Countryside Action Network as militant out-riders for the NFU and Countryside Alliance respectively in Britain (see also Woods *et al.*, 2013). However, the difference between rural protests in Australia and Britain is more than just timing. Although Halpin shows that mainstream farmer militancy in Australia had been quashed by the end of the 1980s, the pressures on rural communities did not stop. Further deregulation of domestic markets continued to impact on smaller farmers, and the privatisation of industries such as the telecommunications company Telstra affected rural communities. Moreover, neoliberalism had left rural Australia exposed to global competition as new trade deals were signed. By the turn of the century, rural Australia was perceived to be facing a new crisis (Pritchard and McManus, 2000). New radical groups such as Concerned Dairy Farmers and the Australian Milk Producer's Association emerged and engaged in protests, including public meetings and rallies, marches, picketing of supermarkets and political meetings, and stunts such as milk-dumps (Table 2.3), but these actions remained small scale, highly localised and failed to coalesce into a wider protest movement.

The non-revival of rural militancy in the 2000s in Australia can be attributed, following our model, to critical factors of resources, geography, coalition-building and political socialisation. The absence of backing from a national association such as the NFF made it difficult for protest organisers to overcome the problem of geography in mobilising activists; while the traditional organisation of agricultural representation around commodity groups meant that in some (but not all) states there were no equivalents to the local branches of the NFU in the UK that could provide a structure for local organisation and solidarity of farmers across sectors. Protests in consequence tended to be sector-specific and lacked a coherent strategy, variously targeting supermarkets, processing plants, state governments and the federal government, and presenting differing positions on, for example, the merits of free trade. An attempt at constructing a more comprehensive movement was made in 'Resurrecting Rural and Regional Australia' rallies in 2004, which were addressed by representatives from the beef, pork, sugar, dairy and egg industries, but arguably failed to connect with the concerns of small farmers, by focusing on issues of trade liberalisation and foreign investment that reflected the interests of their corporate backers.

Moreover, the lack of engagement by farmers in the protests reflected their political socialisation. The ultimate failure of the 1980s demonstrations to block agricultural reforms had reinforced conservative antipathy towards protests, and the pool of smaller farmers with a greater tendency towards militancy was gradually diminished as many gave up the struggle and left farming. Additionally, in regions such as Queensland, political socialisation still emphasised the National Party as the vehicle for rural representation. As such, electoral politics became the more significant vehicle for rural discontent from the late 1990s, much of it directed at the perceived abandonment of rural interests by the National Party:

Table 2.3. Protests by farmers in Australia, 2001-2004 (data collated from media reports).

Date	Protests and events
6 Feb. 2001	Milk tip in protest at dairy deregulation on three farms in Queensland
7 Feb. 2001	Dairy farmers carry coffin symbolising death of industry in march through Brisbane
1 Aug. 2001	400 potato farmers blockade McCain factory in Tasmania in protest at low price paid to farmers
12 April 2002	1000 banana growers march through Cairns in protest at trade deal with Philippines
13 Aug. 2002	Sugar cane growers picket cabinet meeting in Cairns
3 Sept. 2002	Sugar cane growers picket meeting of Queensland Parliament in Townsville
10 Dec. 2002	Rally by sugar cane growers in Townsville, Queensland
12 Feb. 2003	Sugar cane growers picket office of Queensland Minister in Cairns
10 March 2003	100-150 protesters blockade farm in Nyngan, New South Wales (NSW) to stop officers from the Department of Land & Water Conservation checking suspected illegal land clearing
July-Aug. 2003	Cattle Council of Australia place 50 plastic cows with slogans outside Japanese Embassy in Canberra in protest at Japan's increase in tariffs on beef imports
1 Oct. 2003	Rally in Dubbo, NSW, opposing privatisation of telecommunications company Telstra
21 Dec. 2003	500 farmers blockaded Hume Highway between Melbourne and Sydney in protest against proposed toxic waste site
27 Feb. 2004	Banana growers rally in Cairns over trade deals
4 March 2004	500 farmers protest outside Victoria Parliament in Melbourne against toxic waste site
24 March 2004	2 tonnes of hemp dumped at Federal Parliament in Canberra in protest at lack of government support for industrial hemp industry
31 March 2004	Farmers protests outside NSW Parliament in Sydney against closure of 10 rail branch lines
13 April 2004	60 tractors blockade Princes Highway in Trafalgar, NSW to protest against low milk returns
April 2004	Protests by dairy farmers in Brisbane lead to arrest of one demonstrator
1 June 2004	Apple and pear farmers protest in all state capitals against trade deal allowing imports from New Zealand
June 2004	Forum of 1200 beef producers in Roma, Queensland, passes vote of no confidence in Agriculture Minister
July 2004	Farmers protests outside MPs offices in Queensland against national competition policy
9 Aug. 2004	'Resurrecting Rural and Regional Australia' Forum in Taree, NSW
6 Sept. 2004	300 farmers attend 'Resurrecting Rural and Regional Australia' forum at Gympie, Queensland

> The Nationals are supposedly the farmers' party but they should be seen as they are, the wolf in lamb's clothing. They are solely responsible, as without their support, the Federal Government couldn't pass the enabling legislation. Farmers should wake up to the fact that they are being led to the slaughter by the Judas sheep.
>
> Letter in 'Queensland Country Life', 7 October 2004, p. 17

> Our stand is solely that the National Party has deserted us. We therefore have to ensure that as many National Party seats as possible are lost at the next election.
>
> Speaker at Resurrecting Rural and Regional Australia rally, Roma, Queensland, quoted by the 'Manning River Times', 11 August 2004

There was, however, no single clear alternative to the National Party and disaffected voters fragmented in three directions. Firstly, discontented rural voters contributed to a surge in support for the populist One Nation party in the late 1990s, drawing rural concerns to national prominence for the first time (Forrest *et al.*, 2001; Pritchard and McManus, 2000). However, the electoral base for One Nation was peri-urban more than rural (Davis and Stimson, 1998), and fed on a broader set of anxieties with its xenophobic and socially conservative platform (Badcock, 1998; Pritchard and McManus, 2000). It also failed to hold this vote, especially following the defeat of its controversial leader, Pauline Hanson. Secondly, periodic attempts were made to launch a new 'rural' party to rival the Nationals, including the Rural Australia Party and the Enterprise Party in the 1980s and the New South Wales Country Summit Alliance in the 1990s (Halpin, 2003; Lawrence, 1987). Some individuals in the 'Resurrecting Rural and Regional Australia' movement were linked to the launch of a New Country Party (NCP) in 2004, but despite contesting seven electorates in the 2004 election the NCP failed to poll above 4% of the vote in any (Table 2.4). Thirdly, and most successfully, rural opposition to agricultural and trade policies has generated support for National Party defectors standing as Independents, notably Peter Andren and Tony Windsor in New South Wales and Bob Katter in Queensland. These three Independent Members of Parliament (MPs) consolidated local support through populist policies and an appeal to a 'non-partisan' rural politics (Costar and Curtin, 2004), but their success has not translated into a wider movement, despite Katter's flirtation with the NCP and more recently Katter's Australian Party, and the capture of a safe National Party electoral seat by an Independent candidate in a 2008 by-election (Woodward and Curtin, 2009). Two Independent candidates backed by the 'Resurrecting Rural and Regional Australia' movement failed to unseat National MPs in the 2004 federal election, as did a third Independent representing forestry workers in Victoria (Table 2.5).

Table 2.4. Electoral divisions contested by the New Country Party, 2004 Australian Federal Election (data from Australian Electoral Commission, www.aec.gov.au).

Electoral division	New Country Party candidate	% of first preference vote	Position	Party winning seat
Lyne (NSW)	Robyn Murphy	3.39	4th	Nationals
Maranoa (QLD)	Rick Benham	2.09	6th	Nationals
Blair (QLD)	Alan Price	2.07	6th	Liberal
Forrest (WA)	Ken Vagg	1.34	8th	Liberal
Pearce (WA)	Jeanette Radisich	1.21	7th	Liberal
O'Connor (WA)	Jan Hough	0.88	8th	Liberal
Dawson (QLD)	Debbie Lowis	0.79	7th	Nationals

Table 2.5. First preference votes polled by notable rural Independent candidates in Australia federal elections, 2001-2010 (data from Australian Electoral Commission).

Electoral division	Candidate	2001	2004	2007	2008 (by election)	2010
Calare (NSW)	Peter Andren	51.4%[1]	50.2%[1]			
	Gavin Priestley			23.1%		
Kennedy (QLD)	Bob Katter	47.1%[1]	40.1%[1]	39.5%[1]		46.7%[1]
New England (NSW)	Tony Windsor	45.0%[1]	57.3%[1]	61.9%[1]		61.9%[1]
Lyne (NSW)	Rob Oakeshott				63.8%[1]	47.1%[1]
Dawson (QLD)	Margaret Menzel[2]		6.9%			
Gippsland (VIC)	Peter Kelly		3.7%			
Wide Bay (QLD)	Lars Hedberg[2]		12.2%			

[1] Elected.

[2] Candidate endorsed by Resurrecting Rural and Regional Australia campaign.

Attempts to mobilise rural protests and protest votes in Australia have therefore been limited by structural factors and political socialisation, but they have also failed to connect with dominant discourses of rurality to enrol broader public support. In particular, the very notion of rural protests conflicts with prevailing conceptions of rural Australians as stoic, adaptable and innovative individuals. Many farmers and other rural residents disliked the negative tone of the movement and perceived protests as counterproductive. Instead they responded to rural restructuring with individual entrepreneurship, community self-help and promotional networks such as the Regional Women's Alliance (Cheshire, 2006; Halpin, 2003).

2.6 France: fragmented rural radicalism

In contrast to both Britain and Australia, rural protests have been a more regular occurrence in France over the last century, reflecting in part the more balanced electoral politics of the French countryside. Both right-wing and left-wing parties have drawn support from rural areas, and the status of rural France as an electoral battleground has kept agricultural and rural issues prominent in French politics and militated against movements towards substantial policy reforms. However, partisan and regional cleavages have also fragmented the rural lobby. These factors have produced a distinctive model of rural mobilisation, as depicted in Figure 2.4, and as discussed in more detail by Bruneau in Chapter 5.

The mechanisms for rural representation during the post-war period were more unstable in France than they were in Britain or many other European countries. Partisan competition initially frustrated efforts to establish a single farmers' union, but during the 1940s and 1950s the FNSEA gradually assumed this role by uniting farmers in opposition to price restraints on agricultural goods (Naylor, 1994). Over the ensuing decades, the FNSEA occupied a position akin to that of the NFU in the UK, working closely with the Ministry of Agriculture in a tight-knit policy community that supported both productivist and protectionist rural policies (Sheingate, 2001), reflecting the cultural significance afforded to farming in the prevailing discourse of *la France profond*.

However, the primacy of the FNSEA was not unchallenged. During the 1950s, 1960s and 1970s the FNSEA became increasingly closely aligned with right-wing Gaullist parties, and actively encouraged farmers to vote for Gaullist candidates. However, the union continued to include socialists and communists who controlled some local and regional-level associations, and its drift to the right was also contested by radicals within the young farmers' association, the CNJA, as well as by breakaway regional groups, as Bruneau describes (see also Naylor, 1994; Sheingate, 2001). Protests and demonstrations were used to aid factional agendas, and became accepted as part of the repertoire of rural politics in France.

Accordingly, as in Australia, internal pressures on the post-war rural political settlement in France came from the divergent interests of different groups of farmers, as well as from competing ideological and party-political positions. These tensions in turn produced disunity in the response of farmers to external pressures, particularly the threat posed to smaller farms by purported reforms to the European Community's Common Agricultural Policy (CAP). Furthermore, the cleavages were deliberately

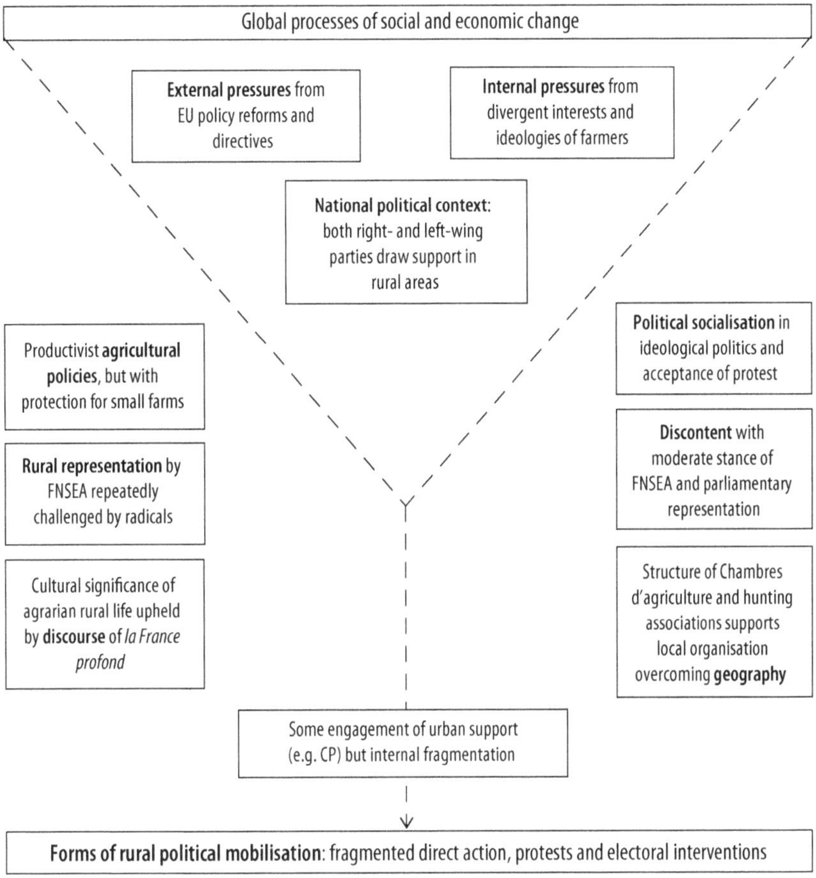

Figure 2.4. Factors in contemporary rural mobilisation in France.

exploited by Socialist Prime Minister Edith Cresson in the early 1980s, who sought to break the monopoly of the FNSEA by widening access to policy meetings to include other, smaller unions (Naylor, 1994).

While the influence of the FNSEA on government fluctuated with the changing political context during the 1980s and 1990s, its weakened position and moderate line generated grassroots discontent that spawned rival unions, including the Co-ordination Rurale on the right, and the Confédération Paysanne (CP) on the left, as discussed in more detail by Bruneau. Importantly, despite Cresson's efforts to expand the policy community, the new unions had limited access to government, and thus tended towards direct action to get their message across, including 'closing roads, motorways, or railway lines with burning tyres or piles of surplus agricultural produce, (and) blocking off towns and cities with slow moving convoys tractors and farm machinery' (Naylor, 1994:

p. 268). This mobilisation of rural protest activity was hence fuelled by discontent with established mechanisms of rural representation, but it was also facilitated by the strong local structure of departmental Chambres d'agriculture, which enabled obstacles of geography to be overcome and scales to be jumped, and by a political socialisation of French farmers that viewed rural politics in ideological terms and which accepted protests and demonstrations as legitimate tactics.

At the same time, the ideological character of French rural politics prevented the various actions coalescing into a single protest movement. Ideological differences were even evident within unions such as the CP, as Bruneau details, with debates around strategy and the positioning of rural concerns within wider ideological struggles. As Bruneau shows, under the leadership of José Bové and allies, the CP adopted a stance that linked farm incomes to the counter-globalisation movement and garnered supporter from *neo-ruraux* in-migrants and urban residents, but at the expense of building coalitions with other rural campaign groups.

The fragmentation of the French rural protest movement is further illustrated by the different strategy adopted by supporters of hunting in response to pressures on France to implement two EU directives: the Birds Directive, introducing a closed season for hunting, and the Habitats Directive, requiring the designation of Natura 2000 protected sites in which hunting would be forbidden. Opposition to Natura 2000 was initially mobilised through political lobbying by the 'Group 9' alliance of agricultural, forestry, game and fish-breeding organisations, with some success (Alphandéry and Fortier, 2001), but the election of a Socialist-Green coalition in 1997 reaffirmed government determination to implement the directives, and led to a fracturing of the alliance when hunting supporters adopted more confrontational tactics. These included protests, notably a demonstration by 150,000 hunting supporters in Paris in February 1998, but also electoral mobilisation.

The CPNT party had been founded in 1989, but gained support as debates around the two directives achieved public prominence, and it unexpectedly polled 6.8% of the vote in the 1999 European elections, ahead of the Front National, electing six Members of the European Parliament (MEPs). This was followed by party leader Jean Saint-Josse polling 4.2% in the first round of the 2002 presidential election. With more than a million votes secured on both occasions, these results were enough for the CPNT to be taken seriously as an electoral force, not least because it had taken votes from both left and right. Analysis of the 2002 presidential election, for example, found that 4% of voters who identified themselves with the Socialist Party, 5% who identified with the main conservative Rassemblement pour la République (RPR) party, and 7% of voters identifying with the Greens had voted for the CPNT in the first round (Table 2.6).

Table 2.6. Level of support for Chasse, Pêche, Nature et Tradition (CPNT) in first round of French Presidential Election 2002, by voter characteristics (data from Ipsos polling, www.ipsos.fr).

All voters	4.3%
Farmers	3%
Small business owners	8%
Rural residents	9%
Voters identifying with Socialist Party	4%
Voters identifying with RPR	5%
Voters identifying with Greens	7%
Voters identifying with Front National	2%
Voters identifying with political right	5%
Voters identifying with political left	4%
Voters not identifying with any party	9%

Notably, the CPNT presented itself not just as pro-hunting party, but as a rural party, responding to the perceived failure of mainstream parties to adequately represent rural interests, particularly against urban-based environmentalism. Yet, its emphasis on hunting did not necessarily reflect the concerns of farmers, and in the 2002 presidential elections its level of support among farmers was significantly lower than in the electorate as a whole. Moreover, the CPNT's anti-urban rhetoric contrasted sharply with the approach of José Bové, whose appeal to urban dwellers and *neo-ruraux* in-migrants, with a defence of small farms that evoked the embedded discourse of *la France profond*, led to him being represented as a 'bridge' between rural and urban France. Bové's election to the European Parliament as a Green MEP in 2009, and his presidential campaign as an alter-globalisation candidate in 2007, further reinforced the contrast between his politics and those of the CPNT.

As such, although rural activists in France achieved notoriety through direct action, their failure to form a coherent movement compromised their potential to enrol wider public support and to achieve their objectives.

2.7 Conclusion

The apparent upsurge in the number of rural protests and political movements across developed countries at the turn of the century has not been a coincidence. They reflect the impact of large-scale transnational processes – including economic and social globalisation, the ascendency of neoliberalism, and the spread of the global environmental movement – in disrupting the settled political and policy structures that had dominated rural affairs since the end of the Second World War. However, the way in which this disruption has occurred and the responses developed have differed between countries, as demonstrated in the case studies above. These variations show that while the influence of global processes is pervasive, national factors can be decisive in shaping outcomes. In particular, the national political context, including electoral geography, political culture and the complexion of the national government, is important in determining the issues that trigger rural protests and their timing, as well as the form that the protests take. Rural activists are conditioned by the legacy of the post-war settlement, by their political socialisation and by the example of other protest movements, and these factors, combined with geography and the opportunities afforded by the political system, shape the strategies that they adopt and the success of attempted mobilisations. However, more protests and larger protests do not necessarily translate into successful outcomes. For all the noise and media coverage of the British countryside protests, or the militancy of the French rural protests, ultimately neither has had any more effect on policy than the more muted and limited rural mobilisation in Australia. To really change policy, rural activists need to build coalitions with urban voters, something that those involved in all the cases studied (with the possible exception of José Bové) conspicuously failed to do.

References

Aitkin, D., 1985. Countrymindedness – the spread of an idea. Australian Cultural History 4: 34-41.

Alphandéry, P. and Fortier, A., 2001. Can a territorial policy be based on science alone? The system for creating the Natura 2000 network in France. Sociologia Ruralis 41: 311-328.

Anderson, A., 2006. Spinning the rural agenda: the Countryside Alliance, fox-hunting and social policy. Social Policy and Administration 40: 722-738.

Australian Bureau of Statistics (ABS), 2012. Australian social trends: Australian farming and farmers. Release 4102.0. Available at: http://tinyurl.com/lkq4lyb.

Badcock, B., 1998. Forgotten places, excluded citizens and the rise of One Nation. Urban Policy and Research 16: 241-246.

Borras, S.M., Edelman, M. and Kay, C. (eds.), 2008. Transnational agrarian movements confronting globalization. Wiley-Blackwell, Oxford, UK.

Bremner, R. and Brooking, T., 1993. Federated farmers and the state. In: Roper, B. and Rudd, C. (eds.) State and economy in New Zealand. Oxford University Press, Auckland, New Zealand, pp. 108-127.

Browne, W.P., 2001. The failure of national rural policy. Georgetown University Press, Washington, DC, USA.

Burridge, J., 2008. 'Hunting is not just for blood-thirsty toffs': the countryside alliance and the visual rhetoric of a poster campaign. Text and Talk 28: 31-53.

Campbell, H. and Wards, S., 1992. Farm politics in New Zealand: pushing on to 'drier' ground? Rural Society 2: 12-15.

Cheshire, L., 2006. Governing rural development. Ashgate Publishing, Farnhem, UK.

Connors, T., 1996. To speak with one voice: the quest by Australian farmers for federal unity. National Farmers' Federation, Canberra, Australia.

Costar, B. and Curtin, J., 2004. Rebels with a cause: independents in Australian politics. UNSW Press, Sydney.

Countryside Party, 2004. European parliament elections 2004 manifesto. Countryside Party, Perth, UK.

Davis, R. and Stimson, R., 1998. Disillusionment and disenchantment at the fringe: explaining the geography of the One Nation party vote at the Queensland election. People and Place 6: 69-82.

Dyer, J., 1998. Harvest of rage: why Oklahoma City is only the beginning. Westview Press, Boulder, CO, USA.

Forrest, J., Alston, M., Medlin, C. and Amri, S., 2001. Voter behaviour in rural areas: a study of the farrar electoral division in Southern New South Wales at the 1998 federal election. Australian Geographical Studies 39: 167-182.

Friedmann, H. and McMichael, P., 1989. Agriculture and the state system: the rise and fall of national agricultures, 1870 to the present. Sociologia Ruralis 29: 93-117.

Greer, A., 2005. Agricultural policy in Europe. Manchester University Press, Manchester, UK.

Halpin, D., 2003. The collective political actions of the Australian farming and rural communities: putting farm interest groups in context. Rural Society 13: 138-156.

Howie, C. and Ramsay, C., 1998. The reforms: riding the rollercoaster. Otago Daily Times, Dunedin, New Zealand.

Lawrence, G., 1987. Capitalism and the countryside. Pluto Press Australia, Sydney, Australia.

Lockie, S., 2000. Crisis and conflict: shifting discourses of rural and regional Australia. In: Pritchard, B. and McManus, P. (eds.) Land of discontent. UNSW Press, Sydney, Australia, pp. 14-32.

Mormont, M., 1987. The emergence of rural struggles and their ideological effects. International Journal of Urban and Regional Research 7: 559-578.

Mormont, M., 1990. 'What is rural?' or 'how to be rural': towards a sociology of the rural. In: Marsden, T., Lowe, P. and Whatmore, S. (eds.) Rural restructuring. David Fulton, London, UK, pp. 21-44.

Munters, Q.J., 1989. Farmers and local leadership in Dutch agricultural communities, 1917-1986. Netherlands' Journal of Social Sciences 25: 114-128.

Naylor, E., 1994. Unionism, peasant protest and the reform of French agriculture. Journal of Rural Studies 10: 263-274.

Newby, H., Bell, C., Rose, D. and Saunders, P., 1978. Property, paternalism and power. Hutchinson, London, UK.

Pritchard, B. and McManus, P. (eds.), 2000. Land of discontent: the dynamics of change in rural and regional Australia. UNSW Press, Sydney, Australia.

Self, P. and Storing, H., 1962. The state and the farmer. George Allen and Unwin, London, UK, 251 pp.

Shannon, F.A., 1957. American farmers' movements. D. Van Nostrand Company, Inc., Princeton, NJ, USA.

Sheingate, A.D., 2001. The rise of the agricultural welfare state. Princeton University Press, Princeton, NJ, USA.

Skogstad, G., 1998. Ideas, paradigms and institutions: agrarian exceptionalism in the European Community and the United States. Governance 11: 463-490.

Smith, M.J., 1993. Pressure, power and policy. Harvester Wheatsheaf, Hemel Hempstead, UK.

Stock, C.M., 1996. Rural radicals: righteous rage in the American grain. Cornell University Press, Ithaca, NY, USA, 219 pp.

Wear, R., 2009. Countrymindedness and the nationals. In: Botterill, L.C. and Cockfield, G. (eds.) The national party. Allen & Unwin, Crows Nest, Australia, pp. 81-99.

Winter, M., 1996. Rural politics. Policies for agriculture, forestry and the environment. Routledge, London, UK.

Woods, M., 1997. Discourses of power and rurality: local politics in Somerset in the 20th century. Political Geography 16: 453-478.

Woods, M., 1998. Researching rural conflicts: hunting, local politics and actor-networks. Journal of Rural Studies, 14: 321-340.

Woods, M., 2002. 'Was there a rural rebellion? Labour and the countryside vote in the 2001 general election'. In: Bennie, L., Rallings, C., Tonge, J. and Webb, P. (eds.) British elections and parties review: volume 12 – the 2001 general election. Frank Cass, London, UK, pp. 206-228.

Woods, M., 2003. Deconstructing rural protest: the emergence of a new social movement. Journal of Rural Studies 19: 309-325.

Woods, M., 2005. Contesting rurality: politics in the British countryside. Ashgate Publishing, Farnhem, UK, 209 pp.

Woods, M., 2007. Engaging the global countryside: globalization, hybridity and the reconstitution of rural place. Progress in Human Geography 31: 485-507.

Woods, M., 2008. Social movements and rural politics. Journal of Rural Studies 24: 129-137.

Woods, M., 2010a. Reporting an unsettled countryside: the news media and rural protests in Britain. Culture Unbound 2: 215-239.

Woods, M., 2010b. 'Representing rural America: the reconstruction of a political space'. In: Winchell, D., Ramsay, D., Koster, R. and Robinson, G. (eds.) Sustainable rural community change: geographical perspectives from North America, the British Isles and Australia. Rural Development Institute, Brandon, Canada, pp. 298-312.

Woods, M., 2011. Rural. Routledge, London, UK.

Woods, M., Anderson, J., Guilbert, S. and Watkin, S., 2012. 'The country(side) is angry': emotion and explanation in protest mobilization. Social and Cultural Geography 13: 567-587.

Woods, M., Anderson, J., Guilbert, S. and Watkin, S., 2013. Rhizomic radicalism and arborescent advocacy: a Deleuzo-Guattarian reading of rural protest. Environment and Planning D: Society and Space 31: 434-450.

Woodward, D. and Curtin, J., 2009. Beyond country to national. In: Botterill, L.C. and Cockfield, G. (eds.) The national party. Allen & Unwin, Crows Nest, Australia, pp. 27-41.

3 Rural protest groups in the Netherlands

D. Strijker[1*] *and I.J. Terluin*[2*]

[1]*Faculty Spatial Sciences, Groningen University, Groningen, the Netherlands;* [2]*LEI Wageningen UR, The Hague, the Netherlands; d.strijker@rug.nl; ida.terluin@wur.nl*

Abstract

This chapter examines whether a shift in the orientation of farmers' groups in the Netherlands has occurred, from sector-specific interests to a focus on the defence of rural identity. Attention is also paid to the question of whether rural protest groups have evolved into political parties. The chapter distinguishes four types of rural protest group found in the Netherlands in the postwar period. The first are the anti-modernist farmers, known as the Free Farmers, who were dissatisfied with continuing government regulation of the agricultural sector and who strongly opposed agricultural modernisation policy. The second group were action-oriented farmers, who challenged the existing farmers' unions, arguing that the official organisations were too soft and too much involved with government. They aimed at strict production controls and higher agricultural prices. The third group, progressive farmers' coalitions, had deep concerns about the environmental consequences of agricultural practices and the impact of the CAP on farmers in less developed parts of the world. The fourth group consists of non-farmers' coalitions, including both rural and urban residents who wished to conserve those aspects of the countryside which represented their specific interests. Protests by these coalitions concerned issues such as biodiversity, national nature conservation areas, high-speed railways, wind farms and mega farms for intensive husbandry. The chapter concludes, firstly, that the clear shift in the orientation of rural protest groups from sector-specific interests to a focus on the defence of rural identity which took place in the UK, France and the USA is only partially reflected in the Netherlands. To a certain extent this can be explained by the fact that the Dutch rural space has always been close to urban centres; the rural has never been isolated. Secondly, the chapter also finds that the Free Farmers are the only group among the protest groups discussed that evolved into a political party: the Farmers' Party. The main reason for this was that their position was not represented by any of the existing political parties in parliament. Other protest groups, in particular the direct action-oriented farmers and the progressive farmers' coalitions, appeared to be more successful in achieving their aims through existing political parties.

D. Strijker, G. Voerman and I.J. Terluin (eds.) **Rural protest groups and populist political parties**
DOI 10.3920/978-90-8686-807-0_3, © Wageningen Academic Publishers 2015

3.1 Introduction

In his pioneering article about rural protest, Michael Woods (2003) states that a clear shift from rural political action targeted at particular sector-specific interests (farming, forestry, etc.) to a focus on the defence of rural identity can be perceived in the UK, France and the USA in the post-war period. This shift could partly be explained by a wider societal move from deliberation to action. In the Netherlands, for instance, from the 1950s on, the predominantly passive political stance of the Dutch changed gradually towards a more activist one, reflected in the use of less conventional methods and operating outside established channels, for example through strikes, mass demonstrations and the occupation of universities (Lijphart, 1975). According to Woods, in the early part of the post-war period, farmers' organizations still tended to support paternalistic structures. Representation worked through formal, sectoral organizations, generally with close ties to mainstream political parties (such as the Conservative Party in the UK and the Christian Democratic parties in Germany, Austria and Belgium). At a later stage, groups of discontented farmers came more to the foreground. These groups, often relatively small and without ties to the key players in the political field, tended to use direct action, such as blockades or the destruction of imported produce, as a weapon.

In this contribution, we will investigate whether this shift in the orientation of farmers' groups also occurred in the Netherlands. In addition, we will also pay attention to one of the main questions in this book, namely whether rural protest groups evolved into political parties. The Netherlands is an interesting case, as the intensity and the degree of specialization of agricultural production is among the highest in the world, resulting in modern agriculture being prominently visible in the landscape. The Netherlands is also one of the most densely populated countries in the world. Combined with a settlement structure of many scattered medium-sized towns and villages, the Dutch rural and urban are always only a short distance from each other.

Green Front

The representation of rural interests in the first few decades after the Second World War in the Netherlands can be denoted as sectoral and deliberate, similar to Woods' findings for the UK, France and the USA (Woods, 2003). In the early years after the Second World War, agriculture was the cornerstone of the general economic recovery in the Netherlands. The Minister of Agriculture Sicco Mansholt – who would later, as member of the European Commission, become the main driving force behind the Common Agricultural Policy (CAP) – actively controlled the supply and prices of agricultural products. At that time, the agricultural sector employed about one fifth of the Dutch labour force (Huizinga and Strijker, 1986).

Until the mid-1970s, Dutch society was 'pillarized': social organization was shaped along religious and ideological lines (Blom, 2000). As a consequence, agricultural representation in the Netherlands was also organized that way. This meant that there were several farmers' unions: the protestant NCBTB, the catholic KNBTB and the liberal KNLC. In addition, the three national agricultural labour unions were built up along comparable lines (Frouws, 1994). Both the farmers' and the labour unions had close relations with their respective political parties: the protestant ARP and CHU, the Catholic KVP and liberal VVD. Until the late 1960s, the farmers' unions represented about two thirds of farmers, and to a large extent they also represented the interests of the related rural population. There was close cooperation between the agricultural sector representatives, agricultural spokesmen of political parties, and the Minister of Agriculture. This cooperation resulted in a strong agricultural lobby within Dutch politics, which is reflected in its nickname, the Green Front (*Groene Front*). This strong lobby was not only a force at the national level, but also at the regional and local levels (the provinces and municipalities) (Munters, 1989).

Agricultural Board enhanced farmers' interests within the Green Front

The position of the Green Front was enhanced by a specific arrangement in Dutch agriculture: the Agricultural Board (*Landbouwschap*), in which the farmers' unions and the agricultural labour unions were represented. The Agricultural Board – founded in 1954 – was a quasi-public body with a legal right to issue regulations for the sector. The Board was the product of the broader neo-corporatist approach advocated by the first post-war governments, based on the idea that the pre-war political currents (socialist, liberal, Christian-democratic) had lost their ideological relevance. Consensual policies fitted that approach (Kickert, 2003). However, it was soon discovered that this ideal world did not exist in reality. Nevertheless, the Agricultural Board survived for quite a period. The Board showed many similarities to the French *Chambres d'Agriculture*, and the German and Austrian *Landwirtschaftskammer* (see Chapters 4 and 5), although its members were not elected but appointed by the participating organizations (farmers unions and labour unions). In the Agricultural Board, the three farmers' organizations tried to speak with one voice as much as possible. A process of coordination was developed to achieve mutual consensus, in which the specific interests of subsectors (dairy farming, horticulture, etc.) and regional interests were taken into account. In doing so, the Agricultural Board established a strong position from which to represent general agricultural interests in relation to other societal structures and policies.

The Green Front represented the interests of farmers. It definitely had a conservative image, but at the same time it was also an advocate of moderate modernization of the agricultural sector (Frouws, 1994). Dutch structural agricultural policy was directed at farm modernization, mechanization, farm enlargement, farm termination, research, education and vocational training (see e.g. Van der Ploeg, 2003). In this process the speed of modernization sought by the Minister of Agriculture tended to be higher than that favoured by the Agricultural Board.

Four types of rural protest groups

The insider position the Green Front achieved since the 1950s has been challenged by various protest groups, which can be classified into four types (Table 3.1):
1. Anti-modernist (Luddite) farmers (small, traditional farmers).
2. Direct activist farmers (discontented and impatient members of farmers' unions).
3. Progressive farmers' coalitions (cooperatives of progressive farmers and concerned citizens focusing on environmental and Third World issues).
4. Non-farmers' coalitions (defence of the rural identity by rural and urban residents).

These four different types of protest groups will be discussed in Sections 3.2-3.5.

3.2 Anti-modernist farmers

In the first years after the Second World War, anti-modernist farmer sentiments were channelled by the *Landelijke Vereniging voor Bedrijfsvrijheid in de Landbouw* (BVL). The BVL was established in 1946 by three farmers from the Veluwe, a region with poor, sandy soils in the central Netherlands. They were dissatisfied with the continuing government regulation of the agricultural sector. The members called themselves 'Free Farmers' for two reasons. First, they rejected government intervention in the farming business and the agricultural sector and second, they were not associated with one of the three farmers' unions in the Netherlands. As a consequence, they had no formal or informal say in the Agricultural Board. Nevertheless, they had to pay the Board's quasi-taxes.

The Free Farmers aimed at a 'free economy' in which the supply of agricultural products would originate only from small mixed farms and in which the market mechanism would be left undisturbed by government intervention. This ideal, however, was threatened by all kinds of governmental regulations and taxes and by the Dutch national government's farm modernization policy. This policy encouraged small farmers to leave the agricultural sector, so that the remaining farmers could enlarge their businesses. The members of the BVL came from different branches of agriculture, and from all over the Netherlands, but the focal point of the BVL was among the small, marginalized farmers on the sandy soils of the provinces of Drenthe, Gelderland and Overijssel in the eastern Netherlands (Terluin, 2012).

The Free Farmers represented the rural as a place where small farmers are the backbone of the community, undisturbed by government intervention. The protests of the Free

Table 3.1. Different types of rural protest groups in the Netherlands and their links with political parties.

Type of protest group	1. Anti-modernist farmers	2. Direct activist farmers	3. Progressive farmers' coalitions	4. Non-farmers' coalitions
Rural protest groups	BVL of the Free Farmers (1946-2005)	Dutch Arable Union (1993-) Dutch Pig Producers Union (1994-) Dutch Dairymen Board (1996-) Glass Force (*Glaskracht*) (2005-)	Farmers' Group (*Boerengroep*) (1971-) Veenkoloniën Working Group (1981-1993) A Better Dairy Policy Working Group (WBZ) (1974-1998) Regional groups for protection of the traditional landscape (1970s-) Agricultural Nature Cooperatives (1990s-)	Specific action groups for the conservation of local landscapes Action groups against bio-industry NIMBY protest groups against windfarms/ high-speed railways/ motorways, etc. (1990s-)
Type of protest	Demonstrations Refusal to pay the quasi-taxes of the Agricultural Board	Mass demonstrations Strikes Negotiations Establishment of sectoral organizations	Building coalitions with non-farm interest groups Meetings and debates	Local NIMBY protests Regional and national advocacy platforms
Evolved into a political party	Yes: Farmers' Party	No	No	No
Interests represented by existing political parties	No	Yes: Christian democrats (CDA), right-wing liberals (VVD)	Yes: Green progressive parties (PPR, EVP, *GroenLinks*)	Yes: local parties
Representing broader rural concerns	A bit: traditional values/rural idyll	Hardly	A bit: concern about environment and landscape changes	Yes: core business

farmers were especially focused on the quasi-taxes levied by the Agricultural Board and the government's modernization policy. Their protests included the refusal to pay the quasi-taxes, demonstrations during public auctions of the property of farmers who had not paid their quasi-taxes, activism to enable the establishment of campsites on farms and the direct sale of farm produce. These multifunctional activities were not allowed on farms in the Netherlands at the time. The group was not successful in abolishing government involvement in the agricultural sector; however, the BVL did contribute to the emancipation of small farmers, to the awareness among the general public of the repression of small farmers by the government, and to the creation of multifunctional agriculture (Terluin, 2012).

In contrast to the three farmers' unions, the Free Farmers had no relationship with specific political parties. In order to have their ideas heard in Parliament, they established their own Farmers' Party in 1958 (see Chapter 13). Contrary to its name, it was presented as a party for all members of society concerned about the loss of traditional rural values, and not just for farmers alone. According to its principles, the Farmers' Party represented itself as a Christian, democratic right-wing protest party. The party was represented in the Dutch parliament from 1963 to 1981. It was disbanded afterwards, but the BVL continued its activities until 2005, although with a very low profile in its later years. Today, in 2015, the BVL is a 'sleeping organization'. The reason for the termination of its activities was that there were increasingly fewer Free Farmers, both due to natural deaths and the winding up of farm businesses.

3.3 Action-oriented farmers

In addition to the Free Farmers, who were strongly opposed to the direction of Dutch agricultural modernization policy, ambivalent feelings about agricultural policy were found among larger parts of the Dutch farming community. These sentiments rather originated from the fact that the price protection, subsidies and regulations for modernization provided by the government were judged as insufficient by farmers. In the late 1960s and subsequent years, the first Dutch agri-environmental laws were passed, limiting the freedom of certain farmers – in particular the intensive livestock farmers – even further. As one of the partners in the Green Front, the Agricultural Board faced increasing difficulties with protests against policy intervention in farming, which was in fact part of its policy. The opposition to the established agricultural organizations became clearly visible in 1974, when the farmers' unions proclaimed a mass demonstration against the policy of the national centre-left government and against the price policy in the CAP. However, a huge demonstration in a football stadium (Galgenwaard) in Utrecht against policymakers in The Hague and Brussels, turned into a mass protest against the leaders of the three farmers' unions. Soon after this demonstration, the leaders of the unions were forced to resign. The mass protest was wholly focused on the farmers' discontent, there were no broader rural issues on the agenda and non-farmer rural interest groups were not involved. The rebellion of the rank-and-file against their own leaders was not without result: in the ensuing years, the farmers' unions tried to follow a more independent course from the Ministry of Agriculture, but without much success (Kraijenbrink, 2005).

Dissatisfaction from members resulted in the end of the Agricultural Board

The late 1980s saw another crisis when, after a long period of price cuts, many farmers sought to take action. Local committees emerged and organized blockades of power stations and sit-ins, and in 1990 a mass rally of mainly arable farmers was held in The Hague. This rally was not organized by the farmers' unions, but by local committees, with thousands of tractors on the main highways. The government threatened to deploy the army if farmers blocked the national airport, Schiphol. These mass protests lasted several weeks (Hees, 2000).

This mass protest reflected once again the erosion of the authority of the Agricultural Board and its most prominent members, the three farmers' organizations. The declining number of farmers meant fewer members for the farmers' unions and fewer people to bear the costs of the Agricultural Board and the farmers' unions. In addition, both dissatisfaction from specific agricultural branches – who felt their interests were not sufficiently recognized in the general agricultural policy of the Board – and dissatisfaction in several regions, further weakened the Agricultural Board's position. In the 1990s, several agricultural branches established separate single-issue unions with partial, specific interests. An important background for this tendency was the increasing specialization of farms and regions in the Netherlands (Van Leeuwen *et al.*, 2010). The three farmers' unions, who despite their cooperation in the Agricultural Board still had their own extensive organizational structures, decided in 1994 – mainly for financial reasons – to merge into one agricultural and horticultural farmers' union, the LTO. This amalgamation was facilitated by the depillarization of Dutch society, which had begun in the 1960s (Bryant, 1981). In 1995, a seemingly minor conflict between the farmers' unions and the labour unions over wages for agricultural workers developed into a more fundamental debate about the functioning of the Agricultural Board. This resulted in the collapse of the model of agricultural representation as reflected in the Green Front and in the dismantling of the Agricultural Board, effectively from 1996, formally in 2000.

Start of sector-specific unions

The agricultural action groups which emerged in several sectors in the 1990s are denoted in this chapter as direct activist farmers. They challenged the existing farmers' unions, and LTO after amalgamation, arguing that the official organizations were too soft and too close to the government. The first direct activist farmers' group was the Dutch Arable Union (*Nederlandse Akkerbouw Vakbond*, NAV) in 1993, partly the successor of the Veenkoloniën Working Group, which was dissolved in the same year (Section 3.4). Compared to this group, the Dutch Arable Union had and continues to have a much more narrow objective of production control and higher arable prices: idealistic objectives for the environment and Developing World issues are largely absent (www.nav.nl). Today, the themes on their website omit 'global' issues. The basic ideas underpinning it were and continue in 2015 to be: (1) that the Union should be directed at the defence of the immediate interests of its members and not for the agricultural sector as a whole; and (2) that it behaves as a labour union, with negotiations, action and strikes, and without undertaking responsibility from semi-public bodies.

More or less comparable groups to the Dutch Arable Union were created in the pig, dairy and horticultural sectors. The pig producers established the Dutch Pig Producers Union (*Nederlandse Varkenshouders Vakbond*, NVV) in 1994; the dairy sector founded the Dutch Dairymen Board (*Nederlandse Melkveehouders Vakbond*, NMV) in 1996; and in the horticultural sector the Glass Force (*Glaskracht*) union was created in 2005. The Dutch Dairymen Board joined with similar farmers' groups in other EU member states to found the European Milk Board in 2004. In contrast to the three farmers' unions, which maintained close relations with relevant political parties, such formal ties were lacking in the four new direct activist farmers' groups. An exception was the former chairman of the NVV, Wien van den Brink, who was a member of parliament for some time for the populist LPF party. Despite the absence of these direct formal links, the aims of these groups overlap to a certain extent with the right-wing parties in Parliament, as their demands tend to be directed at self-interest, instead of the interests of the broader agricultural sector or society at large.

3.4 Progressive farmers' coalitions

In the late 1960s, progressive farmers' coalitions emerged as another challenge for the traditional Green Front. Like the Free Farmers, these protest groups were concerned about the virtues of agricultural modernization. However, the progressive farmers' coalitions operated at the other end of the political spectrum. In contrast to the Free Farmers and its political wing in the Farmers' Party, which defended traditional agricultural and rural values, the often more highly educated farmers belonging to the progressive farmers' coalitions had deep concerns about the environmental consequences of their own practices. They also had misgivings about the impacts of their current agricultural practices and the general agricultural policy – at that time already the CAP – for farmers in less developed parts of the world. Their concerns about environmental issues were inspired by Rachel Carson's book *Silent Spring*, published in 1962, which documented the detrimental effects of the indiscriminate use of pesticides on the environment, and the report by the Club of Rome (Meadows and Meadows, 1972), which predicted that economic growth could not continue indefinitely because of the limited availability of natural resources, particularly oil. The impact of Western European agriculture and the CAP for developing countries became more visible, including to many farmers, after the United Nations Conference on Trade and Development of 1972 in Santiago de Chile. The common characteristic of all the groups and organizations labelled here as progressive farmers' coalitions is that they distanced themselves from the traditional (splendid) isolation of the agricultural sector (Woods, 2003) and built coalitions with non-farm interest groups such as local residents, consumers, environmentalists, Third World activists, intellectuals, urban pressure groups and protectionists (Frouws and Leroy, 2003). Below we will elaborate on the protests of some of these progressive farmers' coalitions.

Farmers' Group

The Farmers' Group (*Boerengroep*) was established in 1971 as a grass-roots research group by radical students from Wageningen Agricultural University (www.boerengroep.nl/HTML/Tekst/Boerengroep/Tekstboerengroepgeschiedenis.html). The group was connected to local action committees of farmers who sought fairer incomes for farmers within and outside the European Communities (EC) and more generally for a better world. The role of the Farmers' Group mainly consisted of supporting local groups by providing them with scientific input. The aim of some of the supported local action committees was in line with the ideals of the previously mentioned Free Farmers: for example, local action groups opposed land consolidation projects. These projects were perceived by these groups as a means of undesirable agricultural rationalization, as they harmed the interests of small farmers and destroyed the characteristic small-scale agricultural landscape. Although not all local action groups were successful in preventing land consolidation projects due to a complicated and therefore contested voting procedure, some groups were. The best-known example of successful act of resistance was that of farmers in Tubbergen in the eastern Netherlands in 1971 (TubbergenNieuws.nl, 2014). Due to mass riots on the day of voting on the land consolidation project, at which the deployment of 200 riot police led to violent scenes, fewer than 1% of the nearly 3,000 voters cast their votes. More than half of these few voters were in favour, which meant under the then prevailing Land Consolidation Act that the land consolidation project was approved. Despite this legal approval, the authorities decided to refrain from the execution of the land consolidation project in Tubbergen.

Veenkoloniën Working Group

The abovementioned Veenkoloniën Working Group, founded in 1981, organized farmers in the Veenkoloniën region in the north-eastern Netherlands. This region is known for the production of starch potatoes for the cooperative starch industry

(Strijker, 2008). The Veenkoloniën Working Group had a number of objectives: a better position for local farmers in the starch cooperative, introduction of new crops to make the region less dependent on the monoculture of starch potatoes, introduction of environmentally friendly agricultural production techniques in the region, reasonable prices for agricultural products through production control, and more attention in the CAP for farmers in developing countries. This last aspect was closely linked to the protection of EC farmers' income, as strict production quotas would prevent subsidized exports and would result in higher incomes 'there and here'. Politically, the members of the group were generally progressives or from the left: they tended to favour the PPR party (a left-wing offshoot of the Catholic KVP party) and Green Left (with which the PPR merged in 1990, along with a few other small left-wing parties). Although the group was sometimes involved in direct action for higher prices and better incomes, their most important instrument was meetings and debates. Representatives of governments, institutions and businesses were often invited and critically interviewed about their role in agricultural and development policies, and the consequences for farmers in Europe and Third World countries. The focal point in these debates was the policy of their own potato starch cooperative, AVEBE. Since farmers knew that the meetings of the Working Group were often lively, critical and influential, the group was able to attract relatively large audiences. The group lost its subsidy from the Ministry of Development Cooperation in 1993, as the Ministry changed its policy and terminated subsidies to domestic groups. As a consequence, the Working Group had to cease its activities.

A Better Dairy Policy Working Group

In the same period a comparable group emerged in the dairy sector: the A Better Dairy Policy Working Group (*Werkgroep Beter Zuivelbeleid*). This group operated at the national level and advocated strict production control for its own and Third World interests. They were active from 1974 to 1998 (Internationaal Instituut voor Sociale Geschiedenis, 2014).

Protection of the traditional landscape

In the late 1970s, a number of regional groups emerged, aiming at the protection of the traditional landscape, sometimes supported by radical scientists. These groups were active in areas including the wetlands north of Amsterdam (Waterland) and in the 'Green Heart' area (between Amsterdam, Rotterdam and Utrecht). The increasing scientific interest and support for farmers' action was also visible in the foundation of the Centre for Agriculture and the Environment (*Centrum voor Landbouw en Milieu*, CLM) in 1981, which to this day focuses on applied research at the cutting edge of agriculture and the environment (www.clm.nl/over-clm).

Agricultural nature associations

In the 1990s the first agricultural nature associations at the local level were founded by farmers participating in nature and landscape conservation schemes. Some of these associations also allow membership to ordinary citizens. In 2013 there were about 160 of them in the Netherlands (SCAN, 2013), with over 10,000 farmers (Prins and Smit, 2011). In order to mitigate the external effects of intensive agricultural production on the landscape and the environment, the associations aim to apply agricultural practices which contribute to the conservation of nature, landscape, soil quality, water and air (Slangen *et al.*, 2008). By participating in agri-environmental schemes, such practices can be implemented in an economically beneficial manner. The association often organizes the administrative and legal requirements for the schemes. Association activities can also include the organization of study clubs on agri-environmental friendly practices, and the provision of information on multifunctional farming and sustainable agriculture (Prins and Smit, 2011).

3.5 Non-farmers' coalitions

As the progressive farmers' coalitions sometimes mobilize both farmers and non-farmers around agricultural issues, they form a kind of bridge to the non-farmers' coalitions, directed at rural issues. In the slipstream of the Club of Rome in the 1970s, and fed by the rising prosperity in Dutch society, rural and urban residents started to organize themselves to conserve those aspects of the countryside which represented their specific interests. These groups, which are denoted here as non-farmers' coalitions, embody the general shift in rural action away from agriculture towards the defence of rural identity, as outlined by Woods (2003) (Section 3.1). These groups might include farmers and might include consideration of agricultural issues, but they are not the essential characteristics of the non-farmers' coalitions. The causes these coalitions advocate are national and international issues including biodiversity, national nature areas and opposing high-speed railways, and regional/local issues such as opposing wind farms and mega-farms for intensive husbandry.

The local rural issues can often be recognised as simple not-in-my-backyard-ism (NIMBYism) (Devine-Wright, 2009). In the political arena, these interests are primarily represented by local parties, and by local branches and the representatives of protest parties at the national level (in the past by the Communist Party, more recently by parties such as the socialist SP and the populist PVV). However, a number of groups, especially the ones operating at the regional level, are primarily concerned about the changing rural landscape, ecology and the environment. This group includes the Association for the Preservation of the IJsselmeer (*Vereniging tot behoud van het IJsselmeer*) (www.ijsselmeervereniging.nl/geschiedenis.htm), the Association for the Wadden Sea (*Waddenvereniging*) (www.waddenvereniging.nl) and the Dune Conservation Foundation (*Stichting Duinbehoud*) (duinbehoud.nl). These groups have often existed for many years and are recognized by government as interlocutors with specialized knowledge.

Protests against wind farms

One issue which often brings people together in their defence of rural identity is the development of wind farms. In the Netherlands there are numerous examples of such protests, from the southwest of the country in Zeeland (www.windmolensnee.nl) to the northeast (www.tegenwind.nl). These protests can be characterized as NIMBY protests. The local groups are sometimes loosely united in regional or national platforms, but without much impact on national policies. One reason is that local contexts such as landscape, population and political conditions differ strikingly. Another factor is that decisions on wind farms are partly taken at the local and regional political level. However, since a national plan was launched in 2013 for a number of large-scale wind projects distributed across the entire country, local protest groups have tended to combine their plans to target these 'large-scale' projects. This shift to the national level has not calmed the local and regional protests, rather inflaming them as the legitimacy of the national government in this field has been brought into question as well.

Uneasy relationship between non-farmers' coalitions and national political parties

A peculiar aspect in the relationship between the non-farmers' coalitions and the political parties is that the nature of most of the action groups tends to be somewhat to the left of the political spectrum (Social Democrats, D66, Green Left). On the one hand, the aims of the NIMBY protest groups could overlap with those of the political parties, such as in their opposition to developments regarded as detrimental to the landscape or to other aspects of the rural identity. On the other hand, the aims of the NIMBY protest groups can conflict with the same political parties' goals in the achievement of a sustainable world (e.g. through the use of renewable energy sources such as wind and solar energy, sustainable water management projects and public transport projects such as high-speed railways and inter-urban tramways). As a result of these partly overlapping and opposing aims, the relationship between NIMBY protest groups and political parties tends to be rather uneasy, leaving an opening for populist parties to represent these interests at the national level.

3.6 Concluding remarks

From the analysis in this chapter it appears that the clear shift from agrarian political action focusing on particular sector-specific interests to a focus on the defence of rural identity, which occurred in the UK, France and the USA (Woods, 2003), is only partially reflected in the Netherlands. To a certain extent this can be explained by the fact that the Dutch rural has always been close to urban centres: the rural was and is not isolated at all. At least since the 1960s, the Dutch rural has become physically and culturally part of the wider urban system. Although protest groups classified as progressive farmers' coalitions and non-farmers' coalitions defending the traditional rural landscape and culture entered the protest scene, these groups have barely a link to agrarian protest groups classified as anti-modernist farmers and direct activist farmers. A possible explanation for the somewhat inward-looking position of the activist farmers groups is the high degree of specialization in agriculture in the Netherlands. The interests of the owners of large, intensive and specialized farms, and those of the people and organizations aiming to protect the traditional countryside or global interests, conflict. In addition, within the traditional farmers' organization, the LTO – which emerged from the three farmers' unions cooperating in the Agricultural Board – a difficult and painful process of restructuring occurred. As a result of this process, the LTO also tends to be inward-looking.

We also conclude that the Free Farmers is the only group among those protest groups discussed which evolved into a political party: the Farmers' Party. The main reason for this was that their arguments were not put forward by any of the existing political parties in Parliament. Other protest groups, in particular the direct activist farmers and the progressive farmers' coalitions, were more successful in achieving their aims through the existing political parties. However, although the NIMBY protest groups have definite access to local political parties, they have serious problems in gaining access to political parties at the national level, due partly to their opposing goals, especially in the field of renewable energy and transport.

Our analysis also revealed that both right-wing and left-wing groups entered the rural protest scene. Whereas the anti-modernist farmers and the direct activist famers could be placed to the right of the political spectrum, the progressive farmers' coalitions tended to occupy the left. This is mainly due to the emphasis of the progressive farmers' coalitions on the negative effects of modern agriculture on the environment and their stance on Third World issues.

Finally, a shift from deliberation towards action did not really occur in the Netherlands. The deliberation model in agriculture, as reflected in the Green Front, was in the early decades after Second World War accompanied by protests and direct actions by the Free Farmers. From the 1970s, actions were organized by the direct activist farmers' groups, the progressive farmers' coalitions and the non-farmers' coalitions. After the dismantling of the Agricultural Board in the 1990s, the deliberative role was continued by the LTO. Although a clear shift from deliberation towards action is absent, we can conclude that the early twenty-first century seems to be marked by a greater emphasis on action than on deliberation compared to the 1950s.

References

Blom, J.C.H., 2000. Pillarisation in perspective. West European Politics 23: 153-164.

Bryant, C.G.A., 1981. Depillarisation in the Netherlands. The British Journal of Sociology 32: 56-74

Carson R.L., 1962. Silent spring. Riverside Press, Cambridge, MA, USA.

Devine-Wright, P., 2009. Rethinking NIMBYism: the role of place attachment and place identity in explaining place-protective action. Journal of Community and Applied Social Psychology 19: 426-441.

Frouws, J. and Leroy, P., 2003. Boeren, burgers en buitenlui. Over nieuwe coalities en sturingsvorming in het landelijk gebied. Tijdschrift voor Sociaal-wetenschappelijk onderzoek van de Landbouw 18: 90-102.

Frouws, J., 1994. Mest en macht. Een politiek-sociologische studie inzake de mestproblematiek in Nederland vanaf 1970. Phd thesis, Landbouwuniversiteit Wageningen, Wageningen, the Netherlands.

Hees, E., 2000. Trekkers naast de trap: een zoektocht naar de dynamiek in de relatie tussen boer en overheid. PhD-thesis, Wageningen Universiteit, Wageningen, the Netherlands, 181 pp.

Huizinga, W. and Strijker, D., 1986. Two lectures on the historical development of Dutch agriculture, 1600-1985. Landbouw-Economisch Instituut, Stafafdeling, Mededeling no. 351. LEI, The Hague, the Netherlands, 31 pp. Available at: http://edepot.wur.nl/264446.

International Institute of Social History (IISH), 2014. Archief Werkgroep Beter Zuivelbeleid. Available at: http://tinyurl.com/nls4kmw.

Kickert, W.J.M., 2003. Beneath consensual corporatism: traditions of governance in the Netherlands. Public Administration 81: 119-140.

Krajenbrink, E.J., 2005. Het Landbouwschap; 'zelfgedragen verantwoordelijkheid' in de land- en tuinbouw 1945-2001. Nederlands Agronomisch Historisch Instituut, Groningen, the Netherlands, 464 pp.

Lijphart, A., 1975. The politics of accommodation: pluralism and democracy in the Netherlands. University of California Press, Berkeley, CA, USA.

Meadows, D.H. and Meadows, D.L. (eds.), 1972. The limits to growth: a report for the Club of Rome's project on the predicament of mankind. Earth Island Limited, London, UK.

Munters, Q.J., 1989. Farmers and local leadership in Dutch agricultural communities, 1917-1986. Netherlands' Journal of Social Sciences 25: 114-128.

Prins, H. and Smit, A.B., 2011. Vermaatschappelijking van het GLB; Kunnen Agrarische Natuurverenigingen inspiratie opleveren? LEI raport Onderzoeksveld Sector & Ondernemerschap 2011-023. LEI, The Hague, the Netherlands, 66 pp.

Stichting Collectieven Agrarisch Natuurbeheer (SCAN), 2013. Plan van aanpak professionalisering collectieven. Het voorbereiden van gebiedscollectieven op hun nieuwe rol van eindbegunstigde in het Vernieuwd Stelsel Agrarisch Natuurbeheer onder het GLB vanaf 2016. Available at: http://tinyurl.com/q38zdjv.

Slangen, L.H.G., Loucks, L.A. and Slangen, A.H.L., 2008. Institutional economics and economic organisation theory. An integrated approach. Wageningen Academic Publishers, Wageningen, the Netherlands, 430 pp.

Strijker, D., 2008. Globalization of a potato starch co-operative: the case of AVEBE. In: Pellenbarg, P. and Wever, E. (eds.) International business geography. Routledge, Abingdon, UK, pp. 169-185.

Terluin, I.J., 2012. 100 jaar boer Koekoek. Rijksuniversiteit Groningen, Faculteit der Ruimtelijke Wetenschappen, Groningen, the Netherlands, 160 pp.

TubbergenNieuws.nl, 2014. Veertig jaar na dato: boerenopstand in Tubbergen vanwege 'gedwongen' ruilverkaveling. Available at: http://tinyurl.com/lll5qg8.

Van der Ploeg, J.D., 2003. The virtual farmer: past, present, and future of the Dutch peasantry. Van Gorcum, Assen, the Netherlands.

Van Leeuwen, E.S., Strijker, D. and Terluin, I.J., 2010. Regional concentration and specialisation in agricultural activities in EU-9 regions (1950-2000). European Spatial Research and Policy 17: 23-39.

Woods, M., 2003. Deconstructing rural protest: the emergence of a new social movement. Journal of Rural Studies 19: 300-325.

4 Actors in the distance: rural protests in the UK and the parliamentary parties

M. Reed

The Countryside and Community Research Institute, University of Gloucestershire, Gloucester, United Kingdom; mreed@glos.ac.uk

Abstract

With various forms of protest becoming part of the lexicon of British political activity, this chapter considers rural protests in relation to parliamentary politics with the aim of tracing the interlinkages, if any, between the activities of political parties and social protest movements. The chapter focuses on the period between 1991-2011, with reference to the historical development of political groups and protest movements in British rural areas, arguing that after a period in which rural areas were the site of a range of protest movements a transition has now taken place to a new form of protest activity. During the period 1991-2003, mass mobilisation tactics were used to represent rural issues or issues in rural areas, seeing a spike of activity during this period that was distinctive in its form, scale and the range of actors. These have now given way to protests that involve smaller numbers of people but are amplified by their connection to other more formal forms of politics. This shift signals not only a change in tactics and organisation by social movements engaged in politics but also changes to the British polity, as the locus of decision-making shifts and the remit of the state changes. By considering the scope of British rural protests, from the Tithe Tax and Kinder Scout Trespasses through to contemporary protests against renewable energy schemes, fox hunting and genetically modified food, the chapter takes a broad perspective on the significance of protest in, and about, rural areas.

4.1 Introduction

Protest has become a prominent feature of British political and social life since the state bail-out of the collapsing banks and the descent into austerity economics. Students have fought running battles with the police over university fees, the summer of 2011 saw rioting by the disaffected greedy, the forecourt of St Paul's Cathedral was occupied

D. Strijker, G. Voerman and I.J. Terluin (eds.) **Rural protest groups and populist political parties**
DOI 10.3920/978-90-8686-807-0_4, © Wageningen Academic Publishers 2015

by anti-banking protestors, and flash mobs protesting against corporate tax evasion stalked company HQs (Mason, 2012; The Guardian *et al.*, 2011). Urban protest, in various forms, has returned to Britain just as parliamentary politics is being convulsed by scandal and corruption allegations. As the political situation has become uniquely demanding, the professional politicians who dominate British parliamentary politics look increasingly inept and remote.

It would be a mistake to think that because the national media are not reporting rural protests that all is quiet in rural Britain. In the place of the mass mobilisations that came to characterise an earlier period of protest, at present there are separate processes of small, localised, attritional conflicts and internet-mediated reflections on the sinews of power in a globalised world. Rather than understand these conflicts and epistemic experiments through the lens of individual psychologies or amorphous concepts such as 'nativism', I suggest that these be understood as part of the lifecycle of social movements, in tension and dialogue with the established systems of parliamentary democracy. The contemporary forms of rural protest do not yet offer either a route to mass mobilisation or a link to parliamentary parties but they do suggest the contours of new forms of politics.

This paper looks at the interactions between rural protest, social movements and the parliamentary political parties, particularly in the period 1991-2011, through reference to the mid-twentieth century. It begins with a consideration of social movement theory and the evolution of the British political party system. It then considers rural protest between the two World Wars as an example that points towards later mobilisations. The mass social mobilisations of the period between approximately 1991 to approximately 2003 represented the high point of rural protest movements in the late twentieth century in the UK. These mobilisations can be seen in the rise of the protests against expanding the road network, the counter-movement against the banning of hunting with hounds and then opposition to the introduction of genetically modified (GM) crops. After this period there were no mass mobilisations but rather a series of connected contests about the future of particular rural localities, through either opposition to the siting of wind farms or the preparation for post-carbon communities. This latter development, although in some instances diametrically politically opposed, is analytically similar to the previous period, as both rest on a renewed discussion of rural life, at each time not initially linked to collective protest. If these movements were to discard their populism or become part of a wider populist movement they may yet become influential political actors.

This social movement activity at first appears to be separate from the formal politics of parties and elections in the UK, but as this paper makes clear, it has often been intimately intertwined with electoral politics and the programmes of governments. Certainly during the 1990s the aims of social movements were represented in government policy, but more recently that linkage has become less clear. In part this reflects the shifting forms of governance in the UK, with devolution to multiple levels of government, but also the transition of the major political parties towards a consensus position that excludes many other voices. Rather than being perpetual outsiders, social movements weave in and out of policy networks, exchanging ideas and influence with electoral forms of politics.

4.2 Social movements and political parties

Dynamics of British rurality

Michael Woods has suggested that, with respect to rural politics, the twentieth century can be split into three periods. In the pre-1914 period, control was devolved to an aristocratic elite, which in the post-1918 period was superseded by the British state and the centralisation of power (Woods, 2005; see also Chapter 2). This period has been characterised in the past as 'corporatism', particularly the 1945-1979 period, whereby a closed policy community, represented institutionally by bodies such as the National Farmers Union (NFU) and the Conservative Party, monopolised the representation of the rural population in Parliament (Winter, 1996). In the last quarter of the twentieth century, policy became increasingly de-nationalised as supra-national bodies such as the EU and World Trade Organisation grew in influence. Simultaneously, some powers of the state also devolved and private corporations became increasingly influential. To this picture I would suggest some refinements, partly to contextualise, but also of theoretical value. Before the current round of neo-liberal globalisation, the British state had an imperial structure and domestic agricultural policy was viewed within that context. Governmentally, the period between 1945-1975 was unusual, as the British state controlled agriculture primarily with regard to national interests, and not with regard to the imperial flows of the pre-war period or the neo-liberal apparatus of 1979 onwards (Marsden *et al.*, 1999). The evacuation of the state in favour of the corporations has been largely confined to agriculture, as the state remains firmly in control of the process of land-use planning. Culturally, as Woods argues, the most powerful discourse at play in the British countryside is that of 'rurality', which remains fiercely contested (see Chapter 2). The state has relinquished control of large areas of activity, for example food supply and regulation, but it firmly retains it in other areas, such as spatial planning, which shapes contemporary patterns of protest (Curry, 2012).

Social movements

Recent global uprisings and revolutions have demonstrated once again the power of social movements to effect social change, just as the emulation of the tactics adopted by these movements caught the British police off-guard when adopted by criminals in the UK's cities in the summer of 2011. An analytical understanding of social movements and the forms of protest associated with them allows these forms to be differentiated from others, such as rioting and criminality, although at times the boundaries are fuzzy. Della Porta and Diani (2006) have a fourfold definition of a social movement. First, the 'informal interaction networks' of movements are not hierarchically organised and are without the formal structures typically associated with political parties or pressure groups. Second, 'shared beliefs and solidarity' mean that those in the movement share a common set of ideas, sometimes only in a particular area, and will support other believers. Third, 'collective action focusing on conflicts' means that social movements are generally involved in contesting a social stake that is considered to be important materially and/or symbolically. Other groups are perceived as seeking to control, or as already controlling that stake, and those in the movement work together to wrest control of that stake, leading to conflict. Four, the 'use of protest': for many years, the use of unconventional protests, beyond the norms of formal political participation, defined scholarly accounts of social movements (Della Porta and Diani, 2006). Activities that signal dissent, build alternatives and change culture may not be equivalent to the public displays typically identified with movements, but they are central not only to sustaining a movement but also the social changes it wishes to see realised (Crossley, 2002; Micheletti and Stolle, 2007; Rochon, 1998).

What Tarrow and Tilly (2012) described as the 'repertoires of collective action' change over time and provide those engaged in protest with a set of practices that people, broadly speaking, know how to take part in, with known aims and outcomes. Street protests demonstrate the number of people who feel strongly about the issue, and when marshalled and peaceful they also convey the disciplined determination and organisation of the protestors. For example, 'boycotts' strike at the economics of the target by withdrawing from the exchange in question. Protests are also culturally specific, in the UK dressing as a rubber chicken, phoning the police and then damaging property in pursuit of a cause will be treated leniently by the courts, whereas in the US the perpetrator would be shackled and goaled. Yet, the public bearing of firearms in the UK will be met with severe sanctions, while in certain contexts in the US this would be unremarkable.

There is a contest with the police and security forces in protests, as each tries to out-manoeuvre the other. During the present cycle of protests and revolutions, the use of social media to coordinate, record and disseminate protests has seen security forces across the world wrong-footed. As the journalist Paul Mason (2012: p. 66) noted, 'the protestors seem more in tune with modernity than the methods of their rulers', observing how mobile telephony is the 'sliced white bread' of the present uprisings. Yet, many of the forms of protest discussed in this paper are not about trying to change the governing regime but an attempt to engage in a dialogue with an elected government;

the protests need to communicate a more nuanced message than 'just leave'. This takes the discussion back to the protestor dressed as rubber chicken, playing on the British sense of the ridiculous, demonstrating sincerity and also respect for the law by breaking it deliberately – the protest is wrapped in semiotics and thrown into the mass media.

As Wright (2007) noted in his study of the US militia movement, unknowingly echoing Melucci (1996), social movement studies have not studied 'right-wing' or 'reactionary' movements sufficiently. Wright draws attention to the importance of 'threat' in mobilising right-wing movements, rather than political opportunity, and to the processes through which threat is attributed to certain actions – particularly those of the government. He draws convincingly on the threat/opportunity spiral of McAdam *et al.* (2001), demonstrating that the actions of the state or its actors can lead to an escalation, as right-wing movements feel increasingly threatened. This is particularly important in relation to rural protest in the UK, which tends to occur in response to a threat, rather than acting to capitalise on a political opportunity.

Populism and the post-political

Populism – while the term 'populist' is often used loosely in journalistic reporting, it is a slippery concept, and its meaning in political philosophy is tightly contested. The contest is between those who argue that populism is an empty 'mode of articulation' (Laclau, 2005) that is available to a range of ideologies or political content – a historically novel argument – and those influenced by Zizek (2006: p. 557), who argued that populism 'harbors in the last instance a long-term protofascist tendency' and should therefore be avoided by the political Left because it will ultimately un-do their aims (see Chapter 8 for a more extensive discussion of populism).

The British political party system became increasingly complex during the twentieth century, from the zenith of a two-party system in the early 1950s, when the Conservative Party and the Labour Party divided up the vote between them. By the 1990s the electoral map was far more complex, with the Conservative and Labour Parties continuing to dominate Westminster politics but with an increased presence from the centrist Liberal Democrats, the nationalist parties – Plaid Cymru, the Scottish National Party – and the Northern Irish parties, the Ulster Democrats, the Ulster Loyalists, the Social Democratic and Labour Party (SDLP) and Sinn Féin. Outside Westminster, particularly in elections for the European Parliament, based on larger constituencies and latterly different voting systems, other parties have flourished, with Green Members of Parliament (MEPs), as well as members of the UK Independence Party (UKIP) being elected. Also, if votes are aggregated, the far-right British National Party (BNP) has attracted significant support but not representation. Thus, although the Conservative and Labour Parties remain dominant at Westminster, it cannot be described as a purely two-party system (Deacon, 2010).

The Labour Party was born out of the Labour movement at the turn of the twentieth century, and for many years was closely associated with working-class politics and the Trade Union movement. Before 1945, this was matched by a rural working-class movement that was vibrant and focused on radical change in the countryside, which was arguably undermined by the policies of the first majority Labour government (1945-1951) (Wright, 1996). The immediate post-war settlement saw an increase in the mechanisation of British agriculture, with farmers becoming increasingly owner-operators and with farm sizes steadily increasing. Farm workers declined in number and with them 'the old proletarian culture of the countryside' (Howkins, 2003: p. 173). The fact that the British countryside had a working-class history was eventually eclipsed by more powerful narratives of rurality.

As Woods has argued (Chapter 2), for much of the post-war period, arguably until 1997, the Conservative Party was the rural party, enmeshed with networks of power and influence, such as the National Farmers Union and landowners. The decline and to some extent revival of the Conservative Party provides the backdrop to many mobilisations and counter-mobilisations over the last twenty years. Their comprehensive defeat in 1997 was not only political but also organisational: the aging membership of the Party was concentrated in safe parliamentary seats rather than in the marginal seats that the Party needed to win. In the years after 1997 the Party collapsed in on itself as it re-debated UK membership of the EU, the role of the market and social policy, as well as how to respond to the new politics of devolution. After decades of the Conservative Party being the route to rural, if not national, influence, it was clearly marginal and unlikely to be in power for some years. Matters framed as 'rural' could no longer be channelled through the Conservative Party and needed other routes to gain influence. The failure to win the election of 2010, making a coalition with the Liberal Democrats necessary, demonstrates that organisationally and politically it has not recovered or re-invented its grass roots (Evans, 2011; Pattie and Johnston, 2009).

4.3 Rural protests between the wars

Although rural protest in the UK during the twentieth century was diverse, a constant theme was the interplay between protests that were generated by substantially rural issues and protests that were used to illustrate national topics: what Woods describes as 'the politics of the rural'. For example, the Kinder Scout trespasses in 1932 concerned access of urban workers to recreation in rural areas (Table 4.1). At that time, the British Workers' Sports Federation (BWSF) was largely comprised of members of the Communist Party of Great Britain (CPGB). The jailing of the leaders of the mass trespasses highlighted questions of inequitable patterns of landownership and access to land that have remained the locus of radical protest and agitation (Monbiot, 1998; Tichelar, 2002). From the same period, the Tithe Tax protests were generated within

Table 4.1. Key events in the development of rural protest groups in the UK, 1930-2008.

Year	Event
1932	Kinder Scout trespass – a wilful trespass by walkers protesting at restrictions on access to countryside
1930s	Tithe Tax protests, small farmers protesting at paying archaic land taxes
1934	British Union of Fascist 'Blackshirts' take part in Tithe Tax protests
1945-1979	Corporatist Countryside – a closed policy community of the Conservative Party, the National Farmers Union (NFU) and the Ministry of Agriculture
1980s	Rise of anti-hunting protests
1992	Twyford Downs anti-road building protest, first of a wave of roads protests
1995	Brent Spar oil platform protest
1997	The Conservative Party beaten in rural areas in Parliamentary elections
	The Countryside Alliance (CA) founded
	Prince Charles warns of the dangers of genetically modified (GM) crops
1998	The first arrests at anti-GM crop protests that continue for six years
	The first countryside march in London organised by the CA
2000	'Fuel Tax' protests blockade oil refineries by Farmers For Action
2002	The CA organises the 'Livelihood and Liberty' March in London
2003	The Land Reform Act (Scotland) giving crofters and communities the right to buy their land from landlords
2004	'The Hunting Act' bans hunting with hounds in the UK
2008	The Climate Change Act – cross-party agreement on the role of renewable technologies, including wind turbines

rural areas and were led largely by farmers, who were angry that an archaic tax paid to the Church of England remained in place even during the agricultural recession. These protests offered political opportunities to those seeking publicity and what was perceived to be a new constituency, with the press barons Lords Beaverbrook and Rothermere attempting to form the Agricultural Party, which met with little success. Oswald Mosley sent his newly formed 'Blackshirts' from the British Union of Fascists (BUF) to try to take advantage of the same situation (Cullen, 1994; Twinch, 2001) and the populism of the fascists had a consistent audience in rural areas during the 1930s, with members of the far-right being influential in post-war agricultural groups. Former members held senior posts in post-war agricultural bodies: Jorian Jenks worked in the Soil Association and Robert Saunders became a Vice-President of the National Union of Farmers (Dorril, 2007; Reed, 2010). At that time various political actors recognised rural Britain as potentially fertile ground for their radical or populist messages (Matless, 1998; Wright, 1996). These interventions represented a pattern that was followed through the twentieth century, with the rural being a zone of protest for urban protestors, and rural issues being a field of opportunity for other political actors.

4.4 Mass mobilisation (1991-2003)

Animal welfare, 1997 and the counter-movement

The early 1990s saw rural areas become a new zone of protest as a number of flows of politics and social change came together. Although not widely discussed, and surprisingly absent from the academic literature, one constant theme of protest during this period that included rural areas concerned animal welfare and rights (Munro, 2005). This movement covered a wide spectrum of activities and beliefs, from those looking for legal sanctions against activities such as hunting with hounds, intensive farming methods and animal experimentation, through to those prepared to use violence, intimidation and the destruction of property to those ends. The rural protests in support of hunting foxes and deer with hounds captured both academic and media interest, because it symbolised issues of class and culture to many. However, the hunts were largely a counter-movement to reassert the existing social order that was threatened by the success of the animal-rights movement. For many in these opposing movements all the issues were fraught with symbolism: animal welfare was about class issues, or hunting was about the neglect of rural communities.

In parallel to the protests in favour of and contesting hunting (see below), protests and various forms of direct action were taken against different forms of farming. Again, a spectrum of protests took place, from the bombing of an abattoir and the destruction of trucks used to move meat and vandalism against butcher shops (Henshaw, 1989), through to the overwhelming peaceful majority who held peaceful protests and petitioned elected members for changes to the law while adopting a vegetarian or vegan diet. Those who produced animals for use in scientific experimentation were particular targets for sustained and vociferous protests. During this protest period, one of the most high-profile campaigns was against the exportation of live animals to continental Europe, principally calves for veal production. One protestor was killed in an accident at one such protest, at a port from which the animals were being exported, which brought the topic even wider publicity and legislation to end the practice.

The animal welfare lobby had managed to successfully influence the New Labour administration, with the payment of a donation of £1 million in 1997 held by many to be especially significant (Fisher, 1997). The proposal to ban hunting with hounds and farmed animal welfare standards attracted the headlines, but the period also saw the suppression of militant and violent animal-rights groups. Throughout the 1980s and 1990s, members of the Animal Liberation Front (ALF) had been given stiff prison sentences when caught, but this was focused on those responsible for bombings. In the post-2003 period the police started to arrest and convict those responsible for intimidation, harassment and organising others to carry out these acts (Walby and Monaghan, 2011). Although welcomed loudly by opponents of the animal-rights movement, there is reason to believe more moderate groups within the animal rights movement also welcomed it, as Henshaw (1989: p. 105) noted in the 1980s:

> Just as the revolutionary left would reserve its deepest vitriol for the 'traitors' within the Labour Party, so the Liberationists always made time to fight their own private war against the respectable welfarists they held in similar contempt.

The Countryside Alliance

Those who hunted either foxes or deer saw the election of a Labour government with a large majority as a direct threat to their activities. As Toke (2010) noted, the Labour Party had been engaged with questions of animal welfare from the 1920s but had not had the parliamentary focus to act on that interest. Formed out of three existing groups, the British Field Sports Society, the Countryside Business Group and the Countryside Movement in 1997, the Countryside Alliance (CA) began to literally and metaphorically rally those who supported hunting against the proposed ban. This was manifested in major marches in London in 1998 and 2002, which mobilised hundreds of thousands of people to support hunting as part of a raft of broader concerns about the neglect of rural life. Hunting with hounds was severely restricted by the Hunting Act 2004, which became effective in 2005 after legal challenges by the CA. Nominally, the CA was a non-partisan body, chaired by a Labour MP, and was interested in articulating a wider range of issues afflicting rural life. In doing so it attracted others who were less focused on hunting, but more on issues pressing on agriculture, with the group Farmers For Action instrumental in the blockading of fuel distribution depots in 2000 as a protest against rises to excise duty on fuel (Doherty *et al.*, 2002).

This conflict between movements for and against hunting with hounds occurred in the context of a radically realigned political geography between the major political parties. New Labour won a crushing general election victory in 1997, gaining 418 seats out of the 658 in Parliament. It had more MPs in rural and semi-rural constituencies (180) than the Conservative Party had MPs (165) (Ward, 2002). After 18 years in office the Conservative Party had lost to New Labour, which had deliberately built a 'big tent' to include as many different groups aspiring to change as possible in order to secure electoral victory. In relation to the lasting power of the New Labour project in UK politics it has to be noted that in 2010, after 13 years in office, including the invasion of Iraq, Tony Blair stepping down, the economic crisis and the Murdoch press turning against it, the Labour Party still won 258 Parliamentary seats. Part of New Labour strategy was to break the Conservative Party, and it nearly succeeded, with the unwitting assistance of internal Conservative Party factionalism. It certainly saw the end of the Conservative Party as the dominant or 'natural' political organisation of rural England, and the rise of the Countryside Alliance was in part a reaction to the vacuum which was created. It meant that for the period from 1997-2010, New Labour's policy agenda, or more significantly, institutional reform programme, began to reshape the rural UK (Ward, 2008).

Because the CA were successful in the first few years in framing their dispute in terms of the broader concerns of rurality, their opponents found it hard to characterise and criticise it. The populist discourse of the people was so wide as to allow many different groups to find accommodation with it, particularly when it championed the neglect of rural issues. Farm incomes were plummeting just as the rest of the economy was entering a boom, leaving many rural areas in acute economic distress (Reed *et al.*, 2002). The economic critique was also widely shared, and many radical environmentalists would echo the criticisms levelled at supermarkets (Simms, 2006) by the Farmers For Action (FFA). Although allied with the mobilisation of the CA and not directly part of its campaign, the blockade of oil refineries by farmers and hauliers as a protest against the burdens placed on the rural economy, also marked how far many in the CA would be prepared to go, and the limit of how far the state would permit them (Bush and Simi, 2001; Doherty *et al.*, 2002).

The electoral politics of hunting was such that it was only salient to a very small group, with most other groups being moved by other concerns. The banning of hunting transformed the tactics of the Countryside Alliance, as it began to lobby in a more conventional manner and to target those who opposed hunting in the general election of 2005. Toke (2010: p. 209) argued that this was to the benefit of the Conservative Party, as voting on hunting was highly partisan: 'the practical effect is to help Tory candidates rather than candidates from other parties'. He contended that the purpose of the CA was to bind its supporters to the Conservative Party, which was the only realistic way of reversing the hunting ban, and that framing hunting within wider rural issues was a way of making it more palatable to rural people not sympathetic to hunting. Toke mistakenly argued that rural people are 'naturally conservative' but their support needs to be translated from a small 'c' of disposition to the large 'C' of party support, which was the role of the CA. In the 2010 election, the Conservative Party pledged to hold a free vote to allow hunting to be legalised, a position that even according to the CA's own polling was not popular, but it provided the Conservative Party with a committed group to campaign for them in marginal electoral seats.

Nevertheless, the mobilisation of the CA was not very effective, with hunting being banned, and with the shift to supporting electoral candidates and conventional lobbying being thwarted by the formation of a coalition government in 2010. The CA faced the problems shared by all counter-movements – trying to reverse attitudinal and legislative changes, without finding a repertoire of protest that they could sustain or a purchase on the polity. Ultimately, they brought their campaign back into the fold of the Conservative Party, where it has been given symbolic but not legislative importance.

Roads and seeds

The second strand of protest active in this period was that of radical environmentalism, which manifested itself in protests, firstly against the extension of road building and later against the introduction of GM crops. Although different concerns – one the conservation of cultural landscapes and the other security of the food system – they were both attempts to influence the broader technological development of British society: diverting it away from an ever-greater dependence on road transport and opposing a new tranche of agricultural technologies. Both were partially successful insofar as the New Labour government elected in 1997 did not build the network of roads that had been proposed in the early 1990s, and struggled to rein back the growth of road traffic (Bryant, 1995; Doherty, 1999). The anti-GM movement succeeded in removing GM products from the shelves of UK supermarkets, but was less successful in its opposition to the legislation enabling those crops to be planted, or the appearance of GM crops in the extended food chain (Reed, 2006; Toke, 2004).

The roads protests occurred in rural sites designated to be the routes of new roads, and took new forms – particularly tunnelling and tree sitting, used to delay construction. These tactics saw the protestors make themselves deliberately vulnerable to injury, forcing contractors and the police to go to great lengths to stop the protest. Such dramatic and committed tactics to 'bear witness' were also exclusively the preserve of the young and those with young families (Doherty, 1999). They were also part of a larger lexicon of protest that emphasised playfulness, even childishness, in an attempt to diffuse and avoid violent confrontation. Protestors erected camps on the sites, built tree houses for accommodation, playing with tropes of playfulness in the countryside. This repertoire of protest was brought into the GM protests that followed, as one of the many roads that protestors took in seeking new ways of living according to their environmental ethics (Seel *et al.*, 2000).

In the opposition to GM crops the repertoire of protest tactics was combined with more conventional and collective ones, such as boycott, lobbying, petitioning and public debates. The targets of the protest were also more diffuse; rather than government policy, the protests were aimed at the companies that formed and governed the food supply system. Europe-wide boycotts against Shell over the Brent Spar oil platform demonstrated the efficacy of these tactics – the company had conceded when the UK government had not (Dickson and McCulloch, 1996). The same pattern was repeated:

multiple retailers removed GM products from their shelves and supply chains, while the government resisted setting out a process to scientifically review the crops (Reed, 2002). The trials of GM crops became the focus of sustained protests and sabotage long after GM products were no longer on sale. To a very large extent the publication of the findings from the trials in peer-reviewed journals was only of academic interest (BBC Online, 2004). This cycle of protests taught other groups the political power of direct action tactics, but also that electorally these issues were of low voter saliency. As a result, the political parties missed the broader signals of social change which these issues represented.

4.5 Into the post-political (2003-2011)

The period after 2003 was one in which the mass mobilisations of the preceding ten years started to fade from the politics of the rural, as protests focused on issues such as the intervention in Iraq. In part this can be explained by the changes taking place in the governance of the rural areas in the UK, the New Labour programme of institutional reform and devolution began to see significant change in how rural issues interacted with government. Scottish and Welsh devolution, with a Parliament and Assembly respectively taking up their responsibilities in 1999, saw rural affairs in those nations increasingly managed away from London. At the same time, the initial enthusiasm of New Labour to make rural areas economically competitive fell away to be replaced by a greater focus on the agricultural sector, particularly after the Foot and Mouth Disease (FMD) outbreak of 2001 (Shucksmith, 2008). Ward, who worked inside the Prime Minister's office at that time, sees 2000 as the highpoint of interest in rural policy, which became an interest in agricultural policy after the FMD outbreak (Ward, 2008). In England, which lacks a national assembly, responsibility for rural affairs was devolved to Regional Development Agencies, ensuring that, as Goodwin (2008: p. 57) argued, the rural became 'embedded in the regional'. It became increasingly difficult to discuss 'the rural' in the UK, as it had become more like a series of ruralities governed by a range of bodies (Goodwin, 2008: p. 57): 'this does mean that we have an institutional structure that promotes very different understandings of the rural and rural policy'.

The rural protests bring into question the political alignment of those who live in rural areas, which is a complex geography of convictions, cultures and opportunities represented by a complex electoral array. Analyses of the regional electoral geography generally find that region is not important in English general elections, apart from the distinct political sub-culture found in southwest England (Johnston, 2005). Since 1992, southwest England has become the heartland of the Liberal Democrats, now the junior partner in the Coalition government. Although the literature has been dominated by a discussion of Labour Party support in rural areas, the Liberal Democrats are at least as significant in terms of the number of MPs returned to Parliament (Table 4.2). This

Table 4.2. Number of Members of Parliament returned for rural constituencies by major party (Johnston, 2005).

Party	Election year			
	1992	1997	2001	2005
Conservative	50	40	43	43
Labour	19	28	27	23
Liberal Democrat	29	27	26	28

meant that it became increasingly difficult to focus on one centre of policy to which protests could be directed, but also that the borders between the state and civic society became increasingly porous.

The second shift that moved discussion of rurality in the UK towards the 'post-political' was the agreement among the major political parties about the future of energy policy, which began to re-cast rural areas as providers of energy in various forms. All of the major parties have backed the measures taken under the Kyoto agreement, with EU agreements to dramatically reduce the UK's CO_2 emissions being enshrined in the UK Climate Change Act of 2008. Polling shows that global warming and energy security have been a consistent concern of the British public, particularly before the global recession (Toke, 2011). The mechanism chosen by the UK government has been to oblige the major energy companies to obtain a proportion of the energy they sell from renewable sources. This has been backed by interventions to develop new technologies or refine existing ones, in order to provide low carbon sources of energy. The chosen vehicles for these developments are the large corporations, often ex-nationalised bodies that are seen as being able to deliver both the technology and low prices. This is broadly a technocratic process where the market price is seen as a proxy for the public interest, even though that market is highly controlled. Competition is seen as a vital enabling tool in making the 'right' choices, as well as reducing prices, generally seen as a proxy for customer interests (Mitchell and Woodman, 2010). As discussed below, the major route taken has been to install wind turbines, with development focused in coastal waters due to concerted opposition through the land planning systems, rather than the cheaper option of being sited in rural areas. As Mitchell and colleagues have observed in a series of papers on the development of these policies, they have emphasised market mechanisms and assessments that ignore many aspects of the social impact of energy generation (Mitchell and Connor, 2004; Mitchell and Woodman, 2010).

Into this complex web of institutions, with questions of rurality becoming intermingled with those of energy generation and climate change, two threads of protest have come to the fore, with the societal impact of energy at their core. The move to renewable energy described above is a victory of the wider environmental movement, as their concerns have become part of mainstream politics and technocratic management is increasingly common around issues of climate change (Swyngedouw, 2010). Social movements are often deeply involved in not only contesting but also generating new technologies, with the organic food and farming movement being a leading example of this process (Hess, 2004; Reed, 2010). Toke (2011) argued persuasively that the same can be said of renewable technologies in the 1970s and 1980s before they were taken up as part of what he characterises as the ecological modernisation of the state. In this context, the movement resisting wind farms, and to a lesser degree other forms of renewable energy, is a counter-movement contesting these technologies and its underpinnings. The Transition Town (TT) movement, or increasingly, the Transition movement (see below), is focused on extending these technologies in preparation for an energy descent that it argues is both inevitable because of the exhaustion of fossil fuels and necessary because of climate change.

The Transition Town movement

The arguments about Peak Oil have sparked the mobilisation of a new movement looking to address a transition to the end of a carbon-based economy (Heinberg, 2007). Initially called Transition Towns, but increasingly the Transition movement, it is based on the premise that peak oil has either occurred or is imminent, so immediate measures must be taken to prepare for the end of an oil-based society. In this context, the aim of the TT process is to engage local communities in responding to the end of an oil-based society with creativity and optimism, through looking to make their area more self-sufficient and self-reliant. As part of this process the movement calls for each community to create an Energy Descent Action Plan (EDAP) to put into place the measures needed to ensure that it can make this transition. Founded in the Devon town of Totnes in 2005, the movement reflects many aspects of southwest England's political sub-culture, although it has spread widely in the UK and beyond (McKay, 2000).

Comparable developments are found in many other suburban contexts in Western Europe. By creating a series of groupings, the community creates its own EDAP; through mapping their resources and envisioning their community in 15-20 years, they create milestones; then by engaging with the local government's plans for the area the community can finalise the EDAP and seek to act on it. The informal leader of the movement, Rob Hopkins[1], is clear that this is a deliberate process of creating cultural change and reversing the common prejudice against the rural. Hopkins uses models drawn from the psychology of addiction as a guide, apparently viewing oil usage as a literal dependency. The movement argues that the process of change needs to be an

[1] www.transitionnetwork.org/blogs/rob-hopkins.

inclusive one, not based on creating conflict or opposing factions but an all-embracing process of community development.

Many elements suggested as part of the EDAP are familiar to most environmental or Green programmes of the past 30 years: a call for a re-localised economy based on local currencies and time-sharing, high-density, eco-efficient housing that can make use of systems of mass, public transport and food grown through a Cuban-style integration of housing with horticulture, and, more broadly, low-carbon organic farming (Holden, 2007). In some of their arguments they make explicit reference to the environmental thinking of the early 1970s, which focused on an imminent resource 'crunch' following the Club of Rome's modelling. In doing so they draw in part on the networks and history of local action, of nuclear free and GM free zones, of the Steiner schools, local food and currency experiments, but these are reframed within the urgency of a peak oil context. The movement is self-conscious and reflective about creating cultural change, but it frequently uses populist elements in its arguments, not least attempts to create consensus about the EDAP rather than recognising the importance of differences. It tends to see itself as a post-political or civic actor rather than its programme being explicitly political.

Anti-wind farm activists

The anti-wind farm, or rather the anti-renewable power, movement conforms to most of the criteria of a 'movement', although it rarely comes together in displays of unity, instead remaining a generally localised clustering of campaigns. The coordinating website, Country Guardian, claims links to 285 active campaigns across the UK, with a presence in the national press and the increasing support of free-market orientated lobby groups. These groups, and their effective use of the planning system, have had a number of effects on the deployment of these technologies, as well as the political system (Mitchell and Connor, 2004; Stevenson, 2009). All of the major political parties support the move to non-carbon forms of energy generation, so broadly speaking the development of renewable energy policy has enjoyed a broad continuity. Because of the difficulty of gaining permission for terrestrial wind farms, the UK government has chosen to develop wind capacity off-shore. This process is streamlined, as the government is effectively able to give itself permission to build. It has also led to the proposal of significant changes to the planning system, which will centralise control over planning under the rhetoric of devolving decision-making to the community (Mitchell and Woodman, 2010).

Discursively, the arguments of those opposed to terrestrial wind farms are more complex than the cliché of self-interested opposition suggests (Van der Horst, 2007). Their attack, exemplified here by that of Etherington (2009), seeks to reject wind turbines from a number of directions simultaneously: they do not generate the energy claimed, are not economically viable, are not safe and damage the landscape. Significantly, Etherington (2009: p. 96) does not argue that the landscape is natural but rather that it is a precious and fragile cultural artefact that needs to be preserved:

> This is akin to demolishing the great cathedrals for road stone or shredding the contents of the National Gallery to make wall insulation. What hatred of national identity could envisage such a body blow?

In addition, Etherington (2009: p. 100) argues that the semi-wild areas where turbines have tended to be sited are 'havens of peace for the mending of broken souls'. The rural is a place of leisure, repose and repair in this romantic vision, menaced by the industrial, hectic ('twitching crucifixions') and exploitative economy. In his efforts to prove that wind farms are unnecessary, Etherington's logic leads him to question the science behind global warming. In doing so he finally chooses nuclear power (Etherington, 2009: p. 190): 'nuclear satisfies every power demand of a power system and it also 'tackles climate change', if saving CO_2 emission is truly necessary'.

The TT movement and the anti-wind farm movement are poised between the populist and the genuinely political in their discourse. Those opposed to wind farms struggle to attain more than the status of a network, as they are confined to campaigning location by location without the ability to organise into overarching bodies. It may be that the planning system has served to prevent the necessity of this by providing a form of national coordination, as the network has not had to challenge the policy by pitting one part of government policy – land-use planning – against another – energy policy. In its appeal to romantic notions of the rural it has attempted to mobilise the British/Welsh/Scottish/English nation as 'the people', against the 'others' of rapacious corporations and duped government. Their critique is not of modern industrial life or capitalism, but rather the specific technologies being adopted – to this degree they are localised populist protests. That the major political parties largely ignore these protests in the rush to engage technologies that will reduce the UK's CO_2 emissions and national energy independence, demonstrates in part the new status of the rural in the politics of the UK.

The transition movement echoes the local focus of the anti-wind farm movement, and is firmly rooted in the small and market towns of rural Britain. It represents continuity in the discourse with the environmental movement, of imminent ecological collapse and the necessity of action through local, participatory small-scale actions. In its insistence on consensus and participation it is post-political; it suggests that the answers are so obvious that it would not be possible to dissent – the issue is the management of a transition. There is no questioning that the local is the best geographical level for this to be achieved or that the resources necessary, such as renewable energy systems, are beyond the means of small communities. In this it points to what Giddens (2009) has noted: that there is no logical connection between the political means and the ends in Green thinking. The anti-wind farm activists are using localised protests and forms of participation to thwart the technologies that the Transition movement deems to be vital, and it is only the national state, using its powers to circumvent consultation, that is achieving the necessary infrastructure to combat climate change.

Scottish Land Reform – the one that got away

The Land Reform movement in Scotland is something of a coda to the examples of the interaction of rural protest, policy and social change. Land reform has been a long-standing goal of the British Left, from the Kinder Scout trespasses in the 1930s to discussion of land nationalisation in the 1940s. It appeared again in the early 1990s under the banner of the network 'The Land is Ours' (TLIO), when an area in London was occupied, on which an eco-village was constructed as a demonstration and exemplar (Monbiot, 1998). The 2003 Land Reform (Scotland) gave communities the right to buy feudal estates, and crofters were given the right to buy their land with community support. Although confined to the former sporting estates and Isles of Scotland, this offered an important example of what community land ownership means for rural areas and that land redistribution can be achieved in an advanced democracy. By taking land out of the global market place and returning it to the community, an important step has been taken in configuring rural resources outside the circuits of global capitalism. Although currently geographically and politically marginal there is no reason why the transfer of land into community ownership should remain isolated to the north of Scotland.

4.6 Conclusion

Twenty years ago I was introduced to an elderly man in my home town whose answer to the matter of local dispute at the time was that some 'riks needed burnin', referring to the traditional tactic of striking farm workers – burning the boss's store of hay. In his account of the 'Freedom Riders' of the early 1960s and their pivotal struggle to desegregate public transport in the southern states of the USA, Arsenault (2011) argued that the Riders were more successful in teaching others the political value of direct action than they were in teaching non-violence: 'Both inside and outside the movement, the primary legacy was the efficacy of direct action, not the moral rectitude of nonviolence'. Many of the uprisings in and around the Arab Spring would appear to confirm that analysis, that many groups have learnt the power of direct action and that many polities are unable to accommodate this. I would suggest that the evidence of this paper points to a decline in the use and utility of direct action in the UK, and a turn to longer term and more diffuse politics of contesting the grounds of understanding. Rural areas are no longer an advantageous space in which to launch or sustain a protest. While this may appear as a retreat from radicalism, as argued above it also has the potential to open up new areas of politics.

The anti-wind farm network and the Transition movement have very different visions of rural life and the future of rural communities. The Transition movement has a vision in which rural communities are the centre of social innovation that will be emulated and adopted by urban ones, where the locality will provided the resources to sustain those communities. In contrast, the anti-wind farm movement proposes a model of rural life based on a more romantic vision of leisure, retreat and conservation. Jointly they represent the conflict between the environmental movement and the conservation movement: their potential zone of conflict is rural Britain. Although not aggressively populist, at present both shelter behind elements of populist argument, insofar as they view their arguments as being unproblematic: who could reject the unanimous community/national patrimony? In this they are repeating a claim often made in the past that rural areas are apolitical or that there is a form of social unity in rural areas that makes politics unnecessary. In so doing they appear to be in danger of emulating the populist politics that dominates much of the discourse about environmental change; in other words, that it does not require social change, only technocratic and scientific managerialism. Their insistence on local-level action in part evades this; however, to become more radical they need to be prepared to contest the future in a debate of ideas rather than consensus.

In a provocative paper, Swyngedouw (2010) argued that most policies targeted at climate change lack a genuine sense of urgency, and the idea of apocalypse is used as part of a populist politics to foreclose debate about genuine alternatives. However, the anti-wind farm and transition movements both have a genuine sense of urgency that is lacking in the consensus position of the political parties. If the two movements were to enter into sustained conflict, either through abandoning the populist elements of their arguments or changes in government policy, they could become the seedbed for a new rural politics. This is especially the case because both of these movements are not only contesting rurality but, more generally, the intersection of rurality and technology, which means that they will have a potentially wider reach than the politics of the rural might otherwise suggest.

References

Arsenault, R., 2011. Freedom riders: 1961 and the struggle for racial justice. Oxford University Press, Oxford, UK, 320 pp.

BBC Online, 2004. Predictable divide on crop decision – the UK is finally open to the cultivation of genetically modified crops. BBC Online March 9, 2004. Available at: http://tinyurl.com/ptesyot.

Bryant, B., 1995. Twyford down: roads, campaigning and environmental law. Spon Press, London, UK, 352 pp.

Bush, E. and Simi, P., 2001. European farmers and their protests. In: Imig, D. and Tarrow, S. (eds.) Contentious Europeans. Protest and politics in an emerging polity. Rowman and Littlefied, London, UK, pp. 97-125.

Crossley, N., 2002. Making sense of social movements. Open University Press, Buckingham, UK., 192 pp.

Cullen, S., 1994. The development of the ideas and policy of the British Union of fascists 1932-1940. Journal of Contemporary History 22: 115-136.

Curry, N., 2012. Sustainable rural development in England: policy problems and equity consequences. Local Economy 27: 95-102.

Deacon, R., 2010. The British party system: an introduction. Parliamentary Affairs 63: 577-581.

Della Porta, D. and Diani, M., 2006. Social movements: an introduction. Blackwell, London, UK.

Dickson, L. and McCulloch, A., 1996. Shell, the Brent Spar and Greenpeace: a doomed tryst? Environmental Politics 24: 122-129.

Doherty, B., 1999. Paving the way: the rise of direct action against road-building and the changing character of British environmentalism. Political Studies 47: 275-292.

Doherty, B., Paterson, M., Plows, A. and Wall, D., 2002. Explaining the fuel protests. Working paper school of politics, international relations and the environment. Keele University, Keele, UK.

Dorril, S., 2007. Black shirt: Sir Oswald Mosley and British fascism. Penguin, London, UK.

Etherington, J., 2009. The wind farm scam: an ecologists evaluation. Stacey International, London, UK.

Evans, E., 2011. Two heads are better than one? Assessing the implications of the conservative–liberal democrat coalition for UK politics. Political Science 63: 45-60.

Fisher, J., 1997. Donations to political parties. Parliamentary affairs 50: 235-245.

Giddens, A., 2009. Politics of climate change. Polity Press, London, UK.

Goodwin, M., 2008. Rural governance, devolution and policy delivery, In: Woods, M. (ed.) New Labour's countryside. Rural policy in Britain since 1997. Policy Press, Bristol, UK, pp. 45-59.

Heinberg, R., 2007. The essential re-localisation of food production. In: Hopkins, R. and Holden, P. (eds.) One planet agriculture. The case for action. The Soil Association, Bristol, UK, pp. 13-17.

Henshaw, D., 1989. Animal warfare: the story of the Animal Liberation Front. Collins, London, UK.

Hess, D.J., 2004. Organic food and agriculture in the US: object conflicts in a health-environmental social movement. Science as Culture 13: 493-513.

Holden, P., 2007. One planet agriculture. The Soil Association, Bristol, UK.

Howkins, A., 2003. The death of rural England, a social history of the countryside since 1900. Routledge, London, UK, 260 pp.

Johnston, R., 2005. The election results in the UK regions. Parliamentary Affairs 58: 786-801.

Laclau, E., 2005. Populism: What's in a name? In: Panizza, F. (ed.) Populism and the mirror of democracy. Verso, London, UK.

Marsden, T., Flynn, A. and Harrison, M., 1999. Consuming interests: the social provision of food. UCL Press, London, UK.

Mason, P., 2012. Why it's kicking off everywhere. The new global revolutions. Verso, London, UK, 224 pp.

Matless, D., 1998. Landscape and englishness. Reaktion Books, London, UK.

McAdam, D., Tarrow, S. and Tilly, C., 2001. Dynamics of contention. Cambridge University Press, London, UK.

McKay, G., 2000. Glastonbury: a very English fair. Victor Gollanz, London, UK.

Melucci, A., 1996. Challenging codes: collective action in the information age. Cambridge University Press, Cambridge, UK, 441 pp.

Micheletti, M. and Stolle, D., 2007. Mobilizing consumers to take responsibility for global social justice. The Annals of the American Academic of Political and Social Science: 157-173.

Mitchell, C. and Connor, P., 2004. Renewable energy policy in the UK 1990-2003. Energy Policy 32: 1935-1947.

Mitchell, C. and Woodman, B., 2010. Towards trust in regulation – moving to a public value regulation. Energy Policy 38: 2644-2651.

Monbiot, G., 1998. Reclaim the fields and the country lanes! The land is ours campaign. In: McKay, G. (ed.) DIY culture: party and protest in nineties Britain. Verso, London, UK, pp. 174-186.

Munro, L., 2005. Strategies, action repertoires and DIY activism in the animal rights movement. Social Movement Studies 4: 75-94.

Pattie, C. and Johnston, R.O.N., 2009. The conservatives' grassroots 'revival'. The Political Quarterly 80: 193-203.

Reed, M., 2002. Rebels from the crown down: the organic movement's revolt against agricultural biotechnology. Science as Culture 11: 481-504.

Reed, M., 2006. Turf wars: the attempt of the organic movement to gain a veto in British agriculture. In: Holt, G. and Reed, M. (eds.) Sociological perspectives of organic agriculture. CAB, Wallingford, UK, pp. 37-56.

Reed, M., 2010. Rebels for the soil – the rise of the global organic movement. Earthscan, London, UK, 178 pp.

Reed, M., Lobley, M., Winter, D.M. and Chandler, J., 2002. Family farmers on the edge: adaptability and change in farm households. Countryside Agency, Exeter, UK.

Rochon, T.R., 1998. Culture moves: ideas, activism, and changing values. Princeton University Press, Chichester, UK, 282 pp.

Seel, B., Paterson, M. and Doherty, B., 2000. Direct action in British environmentalism. Routledge, London, UK.

Shucksmith, M., 2008. New labour's countryside in international perspective. In: Woods, M., (ed.) New labour's countryside. Rural policy in Britain since 1997. Policy Press, Bristol, UK, pp. 59-78.

Simms, A., 2006. Tescopoly. How one shop came out on top and why it matters. Constable & Robinson Ltd., London, UK.

Stevenson, R., 2009. Discourse, power, and energy conflicts: understanding Welsh renewable energy planning policy. Environment and Planning C: Government and Policy 27: 512-526.

Swyngedouw, E., 2010. Apocalypse forever? Theory, Culture & Society 27: 213-232.

Tarrow, S. and Tilly, C., 2012. Contentious politics. Oxford University Press, New York, NY, USA.

The Guardian, Roberts, D., Lewis, P. and Newburn, T., 2011. Reading the riots: investigating England's summer of disorder. The Guardian Shorts Books, London, UK, 86 pp.

Tichelar, M., 2002. The Labour Party and land reform in the inter-war period. Rural History 12: 85-101.

Toke, D., 2004. The politics of GM food. A comparitive study of the UK, USA and the EU. Routledge, London, UK.

Toke, D., 2010. Foxhunting and the conservatives. The Political Quarterly 81: 205-213.

Toke, D., 2011. Ecological modernisation, social movements and renewable energy. Environmental Politics 20: 60-77.

Twinch, C., 2001. Tithe war 1918-1939. Media Associates, London, UK.

Van der Horst, D., 2007. NIMBY or not? Exploring the relevance of location and the politics of voiced opinions in renewable energy siting controversies. Energy Policy 35: 2705-2714.

Walby, K. and Monaghan, J., 2011. Private eyes and public order: policing and surveillance in the suppression of animal rights activists in Canada. Social Movement Studies 10: 21-37.

Ward, N., 2002. Representing rurality? New labour and the electoral geography of rural Britain. Area 34: 171-181.

Ward, N., 2008. Rethinking rural policy under New Labour. In: Woods, M. (ed.) New Labour's countryside. Rural Policy in Britain since 1997. Policy Press, Bristol, UK, pp. 29-45.

Winter, D.M., 1996. Rural politics: policies for agriculture, forestry and the environment. Routledge, London, UK.

Woods, M., 2005. Contesting rurality: politics in the British countryside. Ashgate Publishing, Farnhem, UK.

Wright, P., 1996. The village that died for England – the strange case of Tyneham. Cape, London, UK.

Wright, S.A., 2007. Partriots, politics and the Ohlahoma city bombing. Cambridge University Press, Cambridge, UK, 256 pp.

Zizek, S., 2006. Against the populist temptation. Critical inquiry 32: 551-574.

5 The Confédération Paysanne and the political field: a conflicted history

I. Bruneau

Department of Political Science, Triangle (Research centre), University of Lyon 2, Lyon, France; ivan.bruneau@univ-lyon2.fr

Abstract

This chapter[1] concerns the Confédération Paysanne (CP), a French agricultural union that has protested against the dominant position held by the two majority unions, the FNSEA and the CNJA, since its founding in 1987. The CP became known to the general public due to its prominent role in the anti-globalisation movement in the early 2000s. Although the union's engagement in this nebulous protest movement may be explained by the activists' 'leftist' political sensibilities and their opposition to international agricultural policies, one can only fully understand the implications of this engagement by taking into account the history of the union's relationships with leading political parties, whether to the right or left. This contribution thus retraces the steps constituting the organisation's history, having successfully established itself at a lasting distance from the political field. Nevertheless, the anti-globalisation period (1999-2005) also led to the emergence of internal conflict, in part due to differences over the question of the appropriateness of relationships with the political field. The second part of the chapter thus turns to this specific context, aiming to understand how this issue might correlate with other distinctions connected to activist practices or professional identities. These internal differences are first made coherent in department-level organisations, as illustrated by fieldwork from the department of Aveyron, which also reveals that these localised interactions play a decisive role in national-level discussions.

5.1 Introduction

In France, as in other European countries (Reed, 2008), rural protests are not only farmers' business. As a social group, farmers are actually a relatively small minority in

[1] This chapter was translated by Juliette Rogers, who I warmly thank for her availability and commitment to the project.

D. Strijker, G. Voerman and I.J. Terluin (eds.) **Rural protest groups and populist political parties**
DOI 10.3920/978-90-8686-807-0_5, © Wageningen Academic Publishers 2015

rural spaces today, where labourers far outnumber farmers and represent one third of the working population (Mischi, 2013b). This heavy labour presence is nothing new (Renahy, 2005), and it is still regularly accompanied by union mobilisations and protests aiming to defend working-class leisure activities such as hunting (Mischi, 2013a). Despite all of this, rural spaces remain marked by struggles led by various organisations claiming to represent the interests of 'the agricultural profession', or at least some of its social strata.

This article concerns one such organisation, the Confédération Paysanne (CP), a minority union that challenges the dominance of the Fédération nationale des syndicats d'exploitants agricoles (National Federation of Farmers' Unions; FNSEA) and the Centre national des jeunes agriculteurs (National Young Farmers' Centre; CNJA). Since the period following the Second World War, the FNSEA and CNJA coalition have consistently won over 50% of the vote in elections for the 'Chambers of Agriculture', which are public agricultural advising organisations found in every department and region, led by elected bodies representing the range of people working in the local agricultural sector and influencing local agricultural policy and implementation. Elections to renew the 21 elected seats of each department are held every six years, and they also serve to assess the unions' relative weight nationwide. By holding on to the majority of votes at the national level and by controlling nearly all Chambers of Agriculture through their own department-level organisations, the FNSEA and CNJA have a legitimacy that allows them to play a determining role wherever agricultural policy is produced, be it alongside European authorities and the French government or in institutionalised regional and departmental-level consultations. This representative power has weakened since the early 1980s, due in part to the growing visibility of the CP's opposition to the system, embodied in its activists.

The CP, formed in 1987 and composed mainly of farmers from farming families and a minority of so-called *néo-ruraux* ('new rurals'), has garnered around 20% of the vote since the late 1980s. Although French farmers are predominantly positioned to the right on the political spectrum (Gombin and Mayance, 2010), activists in the CP distinguish themselves by an attachment to the left or extreme left. This singularity is also apparent in the CP's unique repertoire of demands, the main lines of which are presented as alternatives to national and international agricultural policies.

Firstly, to stem the decline in the number of farms and to protect farmers in the most vulnerable situations, the CP favours a policy of controlling production (the only policy they think might regulate markets and stop brutal market fluctuations) and calls for a cap on public aid, its fair distribution, and the equitable distribution of production rights, while denouncing the ineffectiveness of policies for new farmers that encourage larger farms. Secondly, since the late 1980s, and much more explicitly than the FNSEA or CNJA, the CP has highlighted the impact of intensive production methods on the environment and consumer health, and has based its 'peasant agriculture' agenda on promoting practices that involve breaking with modernisation models of the 1970s. Thirdly, this leftist opposition to majority unionism is characterised by consistent support for farmers in countries in the global South, which might in part explain the decisive role it played in the formation of the anti-globalisation movement in France in the late 1990s. Rural protest groups concerned with support for farmers in the global South also emerged in other countries, such as the Working Group Veenkoloniën in the Netherlands (Chapter 3; Strijker and Terluin, 2015). More precisely, as I have demonstrated elsewhere (Bruneau, 2004), this anti-globalisation engagement is in line with an evolution in how 'the international' was framed following the 1992 MacSharry reform of the Common Agricultural Policy (CAP). Quickly seen as the first act of 'the integration of the CAP into GATT' (General Agreement on Tariffs and Trade) by CP activists, this reform modified perceptions of the agricultural policy decision-making space, although interpretations of its consequences were not uniform. 'The international' thus became an issue and a resource in internal struggles, and a group of activists succeeded in imposing its radical critique of the CAP and the World Trade Organization (WTO) on the union.

However, an analysis of the internal uses of 'the international' is inadequate to take full measure the ensemble of conditions that allowed the union to inscribe itself in the nebulous protest scene of the early 2000s. Indeed, the history of the CP is also one of a gradually established but profound defiance toward the dominant organisations in the political field (Bourdieu, 2000). This is the history I aim to tell in this chapter, based on a lengthy period of research among activists in the CP, primarily conducted between 2000 and 2004 and with supplementary fieldwork in 2010 and 2011. This research not only addressed the CP's relationships with political parties, but also more generally studied the conditions of CP engagement (Bruneau, 2006), primarily in two administrative departments: the Orne (in the region of Basse-Normandie, northwestern France) and the Aveyron (Midi-Pyrénées, central southern France). The ethnographic fieldwork was supplemented by research in the organisation's archives at the national level. Both kinds of data will be used in this chapter.

The history of the peasant left[2] predates the CP's official founding in 1987 (Lagrave, 1990). Preceding left-leaning peasant movements had been squeezed out of access to representation on agricultural matters thanks to an alliance between right-leaning governments and the majority unions. Relations with political parties and activists' attitudes towards political power were thus strongly determined by a collective struggle

[2] In using the word 'peasant' (*paysan*), I do not intend to socially qualify these activists or to distinguish them from *agriculteurs*, another word for 'farmer', with its own sociopolitical connotations. I am merely taking up the term used by the actors themselves, who have designated themselves as peasants since the early 1970s, not out of a refusal of innovative production methods, but in opposition to a discourse pushing farmers to become 'entrepreneurs', and to draw attention to the economic and social consequences of 'productivist' policies implemented since the 1960s.

that aimed to modify the conditions of access to decision-making authorities and to obtain official recognition of their representative legitimacy. This 'national' history, the key moments of which I will review in the first part of this chapter, is fundamental to an understanding of the CP's earlier positions, since it consists of a series of events experienced by the CP's founders, who had led activist groups throughout the 2000s in various departments. The second part will be devoted to the CP's anti-globalisation period (1999-2005), in the course of which the union benefited from unprecedented media attention. However, this particular context also led to the emergence of internal conflict, partly spurred on by differences in activists' relationships with the political field. In shifting the analysis to the department-level organisations, however, we will see that the various positions held with respect to political parties cannot be separated from other distinctions running through the activist group.

5.2 Facing majority unionism and political authorities: the history of a double confrontation (1970-2000)

In March 2011, *Campagnes Solidaires*, the CP's national magazine, devoted a special issue to what had become a topic of discussion and concern in the CP: in 2009 and 2010, four former members of the union's national leadership earlier in the decade were elected as either European Deputy or Regional Councillor from slates put forward by the Europe Ecologie/Les Verts party. Writing in the issue, the four elected officials presented their political engagement as being in line with their previous struggles in the CP. By their reckoning, it allowed them to bring the same demands to another level. At the same time, the title of the special issue ('From the CP to politics: they went over to the other side ...') and the introduction to the issue (written by an activist) implied a different interpretation, concluding as follows (Curade, 2011):

> Even today, the CP must remain an autonomous counter-power to political parties, which it should try to inspire without being eaten. The example of many leaders from the majority agricultural union [FNSEA], forgetful of peasants and spineless accomplices of the most reactionary right, can only incite us to great vigilance.

Another article, even more critical, forcefully reaffirmed the principle of the separation of political parties and unions: 'the party logic is the logic of power, the logic of paying its staff, and it's not the logic of activist union members, [which is a logic] of the construction of relations of force' (Duntze, 2011).

This interpretation, insistent on the necessity of union autonomy and the CP's independence from political parties, was shared by the great majority of activists, and was the result of several contentious experiences with political parties that had held dominant positions in the political field. These experiences can be divided into three periods. The first, from the early 1970s through the early 1980s, is typified by the simultaneous emergence of this opposition to the FNSEA and the CNJA 'from the left' and by the closing of access to the realm of professional representation. The left's arrival in power in 1981 came with a renaissance in farmers' union pluralism, but the 1980s were also rife with a series of disappointments that transformed initial expectations into a lasting distrust of left-leaning political parties that had acceded to power in the government. The organisational context of the 1990s would thus be marked by the search for 'the right distance' from the political field, stemming from a dual need to accede to the authoritative bodies in which agricultural policy was defined and to become part of protest movements beyond the boundaries of the agricultural space.

Confédération Paysanne opposition in the face of the political legitimation of a monopoly of representation

The CP's pre-1987 origins cannot be divorced from the formation of a 'new peasant left' in 1968 (Martin, 2005; Bruneau, 2008). Initially, these diverse currents were predominantly composed of 'modernised' livestock farmers in western France who opposed the national-level leadership of the dominant CNJA and FNSEA. The most emblematic movement of this post-1968 radicalism was the Association nationale des Paysans-Travailleurs (Peasants-Workers, known as the PT), whose Marxist-inspired analyses aimed to reveal the existence of conflicts of interest in the agricultural world. This analysis of relations of domination throughout the agricultural sector was heavily inspired by Bernard Lambert's book *Les paysans dans la lutte des classes* (1970, Peasants in the class struggle). Lambert was a farmer from the western department of Loire-Atlantique and a former member of the CNJA national leadership in the late 1950s, who became a member of the Parti Socialiste Unifié (Unified Socialist Party; PSU). According to Lambert, 'exploited peasants' include 'impoverished or poor peasants', those who have not undertaken 'the investments allowing them to industrialise production', and 'proletarianised peasants [who] are engaged in intensive modernisation, are heavily in debt, and find themselves in fact dominated by the industries and cooperatives that supply them and provide the market for what they produce' (Lambert, 1970: pp. 78-79).

Over the course of the 1970s, this denunciation of disparities in conditions and revenues came to be accompanied by a new array of actions. The designated adversaries were not distant or abstract: they were to be sought first of all among those who profited from the 'surplus value' of agricultural labour, especially industry and dairy cooperatives. For example, in spring 1972, the PT played a major role in triggering a well-known 'milk strike' in Brittany. Similarly, activists imputed that the bank Crédit Agricole (specialised in the sector) was responsible for the excessive debt many farmers faced after making significant investments. But activists in this new peasant left mainly gained visibility from their actions related to land access. Without ever explicitly condemning private ownership of agricultural land, they equated access to land with a 'right to work'. This stance had considerable resonance in regions where demand for land was higher than the number of hectares freed through retirements or by farmers leaving the sector. PT activists were often at the head of mobilisations supporting tenant farmers in conflict with their landlords, and insisted that land access should never be an obstacle to entering into the profession. These 'land affairs', dramatised in lengthy legal battles, were occasions for large gatherings on the farmland concerned. PT activists thus succeeded in raising the awareness of farmers who, *a priori,* were otherwise unlikely to have been very open to the subversive nature of the discourse, especially their justifications for supporting striking industrial workers, illustrative of a 'new union practice' that tended to challenge the boundaries between social groups.

Nevertheless, during the 1970s, these political transgressions only rarely went as far as the creation of an independent union. The PT and other currents of this post-1968 peasant left generally favoured the creation of less formal associations while striving to 'change' the FNSEA and CNJA 'from the inside'. This paradoxical situation can be explained mainly by the fact that these activists were not unaffected by the discourse that blamed them for breaking 'peasant unity'. This catchphrase, incessantly reiterated by the two dominant unions, is certainly in continuity with a long history in politics and unionism (Cornu and Mayaud, 2007), but over the course of the 1970s the effectiveness of this construction, associating the agricultural world with a community of specific interests, was reinforced by the existence of a co-managerial relationship between right-leaning governments and the FNSEA. This alliance was initially made possible through the creation of a new political game following Charles de Gaulle's return to power in 1958. In parallel to their efforts to relegate parliamentarians to the background, Gaullist executives relied on representatives of professional organisations in a variety of sectors, including agriculture, medicine, research, and high government administration, all of which had previously had little role in the construction of public policy. It thus assured itself of the support of groups that would contribute to the institutionalisation of the Fifth Republic while also participating in shifts in power relations within the professional spaces concerned (Gaïti, 1998).

In the agricultural sector, this development first manifested itself in the early 1960s, with the promotion of the CNJA's position for policy ends (Muller, 1984) and the adoption of laws aiming to free up land occupied by the smallest farms and elderly farmers and favouring the modernisation and expansion of middle-sized farms. However, the institutionalisation of consultation between the political authorities and the main unions, and thus the stabilisation of a 'neo-corporatist' system (Keeler, 1987), did not truly take place until the late 1960s and the beginning of the 1970s. From 1969 to 1981, the Minister of Agriculture received FNSEA and CNJA leadership on the first Tuesday of every month, and in 1971 these monthly meetings were supplemented by annual conferences whose decisions committed the Prime Minister and the government to action. 'The golden age of co-management' (Coulomb, 1990: 163), the years 1971-1976, were characterised by the reinforcement of the partnership between the executive branch and FNSEA leadership (where some positions were held by former CNJA leaders) and by the strong convergence of the positions defended by Michel Debatisse, FNSEA president, and Jacques Chirac, first as Minister of Agriculture (July 1972 to May 1974), then as Prime Minister (May 1974 to August 1976).

This 'neo-corporatist' regulation not only excluded other organisations from any access to the sites of consultation and decision-making, it also helped to consolidate the FNSEA's and CNJA's work, stigmatising their union adversaries, who they systematically accused of weakening 'peasant unity' for strictly political reasons. The Paysans-Travailleurs' criticisms, for example, were called expressions of 'ideological drunkenness', foreign 'to the profession's concrete realities' (Fédération Nationale des Syndicats d'Exploitants Agricoles, 1970). Successive governments also participated in this disqualification of opposition by denying, for example, any power of representation to organisations in competition with the FNSEA-CNJA duo. Thus, when MODEF (Mouvement de coordination et de défense des exploitations agricoles familiales; Movement for the Coordination and Defence of Family Farms) received nearly one third of the vote in the 1970 and 1974 Chamber of Agriculture elections, the political authorities refused to meet with its leaders, most of whom were activists or sympathisers with the Parti Communiste (Communist Party; PC) or the Parti Socialiste (Socialist Party, PS). By legitimating the representative monopoly demanded by the FNSEA, the political right contributed to the limiting of aspirations for any official recognition of union differences. The introduction of political 'alternance' in 1981, for the first time since the 1958 establishment of the Fifth Republic, would partly modify the space of representation in the agricultural sector.

The left in power and the gradual closing of a political opportunity

In the 1970s, the activists from this 'new peasant left' were variably positioned along the political spectrum. Some placed their union engagement in an explicitly revolutionary perspective and became close to groups on the extreme left, such as Maoists and Trotskyists, while others supported the quest for political power through the PS and its allies, brought together in what would be called the Union de la gauche (Union of the Left). These differences created tensions within the Paysans-Travailleurs that would ultimately lead the association to split in 1977. However, the 1981 election of François Mitterrand as President of the Republic and the promise that farm union pluralism would be acknowledged obliged them to put political differences on the back burner. A few weeks after the Socialist government was formed, decade-long activists from the Paysans-Travailleurs movement regrouped and were joined by other dissident movements to create the Confédération nationale des syndicats de travailleurs-paysans (National Confederation of Workers-Peasants Unions; CNSTP). In spring 1982, the heads of a few departmental FNSEA organisations, who had long contested the positions held by the national leadership, broke off to form the Fédération nationale des syndicats paysans (National Federation of Peasant Unions; FNSP). At first, both organisations' expectations were more or less met. The Minister of Agriculture, Edith Cresson, received representatives of minority unions, accorded them public funding, and introduced proportional representation for the January 1983 Chamber of Agriculture elections.

However, the 'Cresson era', which activists recall fondly, would not last long. In March 1982, heeding the call of the FNSEA and the CNJA, '100,000 farmers' demonstrated against the government's policy in Paris (Champagne, 1984). Worse yet, in January 1983, 'Unité et force paysanne' (Peasant Unity and Strength) slates formed by the FNSEA-CNJA coalition would largely carry the elections for the Chambers of Agriculture, with nearly 70% of the vote. The CNSTP and FNSP obtained 7 and 6% of the votes respectively at the national level. The Socialist government drew a lesson from these elections: the majority farmers' organisations could not be ignored. In March 1983, Edith Cresson was replaced by Michel Rocard. In November of the same year, he limited the expression of union pluralism through a ministerial circular, which determined that only organisations that had obtained over 15% of the vote in the Chamber elections could be considered 'representative'. Compared to rules in place for the salaried sector, this conception of representativity seemed quite restrictive, and concretely this 15% threshold would block the FNSP and CNSTP from having seats in decision-making bodies at both the national and departmental levels.

For activists, the disappointment was even greater because Michel Rocard was the leader of the PSU throughout the 1970s, and many had thought the PSU held views consistent with their own union's orientations. The sentiment of treason was accentuated by some profound disagreements over the Socialist government's agricultural policies, in particular the establishment of dairy quotas in 1984, which they believed deprived small producers of the possibility of increasing their revenue. The CNSTP and FNSP together proposed another mechanism, the 'quantum', which would introduce a differentiated price system (inversely proportional to the volumes produced) guaranteeing an income level that was acceptable to all producers. At the national CNSTP convention in March 1985, an activist pointed to lessons learned from the past four years of Socialist government (CNSTP, 1985):

> Ministers and the president have betrayed the cause of those they represented ... And yet they are our elected officials, a vote isn't a blank check, it's a mandate ... We are the ones they should be accountable to.

The marginalisation of minority unions returned following the 1986 legislative elections, which were won by parties of the right, the Rassemblement Pour la République (Rally for the Republic; RPR) and the Union pour la Démocratie Française (Union for French Democracy; UDF). Jacques Chirac became Prime Minister and entrusted the Ministry of Agriculture to François Guillaume, who had been president of the FNSEA since 1979. This nomination broke with the usual separation of the political and union spaces, and seemed to be the ultimate step in the history of the co-management of French agricultural policy work by right-leaning parties and the main farmers' unions. François Guillaume reversed all of the arrangements Edith Cresson had made between 1981 and 1983, refusing to receive minority organisations, considerably reducing their funding, and bringing majority representation back to elections for professional bodies. This reinforcement of the representation monopoly accelerated the processes bringing the CNSTP and FNSP together, and in spring 1987 they merged to form the CP.

Finding the 'right' distance from political parties

In the late 1980s, 'structuring' the union was the national CP leaders' priority. 'Organising yourself is a form of action', read letters sent to department teams. The 1989 Chamber of Agriculture elections supported this organisational trend. Present in 74 of 95 departments, the CP obtained 18% of the vote and established itself as the second agricultural union behind the FNSEA-CNJA pair. Six years later, the 1995 elections confirmed the peasant left's progress. The union obtained over 30% of the votes in over 20 departments, and with 21% of the votes nationally, it stepped well ahead of the two other minority organisations, Coordination Rurale (Purseigle, 2010) and MODEF. This success in the farm union space took place under conditions very different from those of the early 1980s, however. In January 1995, a few months before the presidential election, the union's national spokesman Gabriel Dewalle gave an interview to the newspaper *Libération*. In reply to the question, 'Who are you rooting for during the next presidential campaign?', he said (Dewalle, 1995):

> For no-one. Less than ever before ... The attitude held by the Left, in power since the early 1980s, has made the debate in our organisation even simpler, in that it has never backed any of the options we propose. The Mitterrand years can perhaps be credited with having given greater independence to our movement.

Study of the national organisation's archives confirms the stabilisation of a line of action that involved working on defining the union's sphere of interest and neutralising divergent opinions from political parties on the left. Internally, for example, activists were regularly reminded that a mandate in the union was incompatible with having political responsibilities, whether within a party organisation or as an elected official. Above all, the fight for access to consultative bodies offered the possibility of establishing a single action strategy for different governments, left or right. With François Mitterrand's re-election in 1988, the CP pressured the government and Socialist deputies to modify the 1983 'Rocard decree'. In March 1990, following fierce discussions, the Minister of Agriculture finally drafted a decree according national representation to unions that had obtained over 15% of the vote in at least 25 departments. The CP met this criterion and could thus claim public financing and hold positions in organisations from which they had previously been excluded, such as the offices for production management. Once this status of a representative organisation was attained, activists had to ensure political leaders respected it. To give but one of many examples, when the right returned to power in 1993, its new Minister of Agriculture (also the mayor of a city in the Aveyron) refused to meet activists from the same department, so they occupied the town hall for several days.

For all that, the union's attitude towards dominant positions in the political field cannot be reduced to a single confrontation. To impose the alternative they claimed to represent, activists tried to disseminate their demands by soliciting, or accepting the solicitations of, parliamentarians, locally elected officials, and political party staff, even when, as they maintained in meetings, 'we can't expect anything to come of it'. The CP likewise participated in the elaboration of agricultural policy when invited. This ambivalence appeared very clearly following the left's victory in 1997 legislative elections.

In the early years of the 'plural left' government (referring to a period of cooperation between the PS, PC and Green parties), the CP's national leaders were frequent visitors to the Ministry of Agriculture. Louis Le Pensec, the Minister of Agriculture in Lionel Jospin's government, consulted with them most notably when developing Contrats territoriaux d'exploitation (Territorial Farm Contracts; CTE). This was a strong signal to the peasant left. With these contracts, farmers agreed to improve the quality of their products, preserve the environment, or increase the number of people working on their farm. However, the minister left the determination of CTE attribution criteria up to department-level agricultural commissions (CDOA), which were dominated by departmental groups of the FNSEA. For the CP, the 'spirit' of the CTEs was thus corrupted by the CDOA's free hand over their distribution, since CTE were often used as little more than a bonus subsidy. Generally speaking, the reformist pragmatism adopted by the CP in 1997 did not survive the test of consultation with the 'plural left' government. The fight against genetically modified organisms (GMOs), mainly led by and incarnated in the CP in the late 1990s and early 2000s (Bernard de Raymond, 2010), even led to head-on confrontation between the union and the government.

The Confédération Paysanne becomes involved in non-agricultural protests

Consequently, whatever the state of relations with political power, the CP suffered from a lack of electoral legitimacy in the agricultural world, all the more so because the system of representation of agricultural interests in France allowed little room for union opposition. From that time on, activists were perfectly aware that their power to influence agricultural policy was dependent on their capacity to raise awareness among social groups outside the farming sector. The incentives to participate in non-agricultural mobilisations were even accepted more readily internally because they were in line with a non-corporatist tradition 'invented' by Bernard Lambert and the Paysans-Travailleurs in the 1970s that had been maintained and amplified in departmental groups ever since. In 1991, the union instigated the founding of the Peasant-Environmentalist-Consumer Alliance, bringing together about twenty consumer and environmental protection associations. In addition, the 1992 MacSharry reform of the CAP reinforced this engagement in a broader space of contestation.

Indeed, with the introduction of free-trade rules – which were non-specific to the agricultural sector – into GATT agreements, the CP moved closer to other organisations in a shared rejection of 'neo-liberalism'. In December 1993, the CP demonstrated in Geneva as part of a collective called 'Act for an alternative to GATT', and over this period it generally deepened its relationship with several organisations active in the mid-1990s protest scene, including the labour union SUD-PTT and associations defending the unemployed (Agir contre le chômage) and the homeless (Droit au logement). The union's engagement in a vast movement opposing globalisation from 1999 onward was indeed consistent with a collective history characterised by a deep, shared mistrust of the sites of practise of political power. This anti-globalisation period was also marked by the emergence of internal divisions, allowing us to see how the question of relationships with political parties may, in a given context, be articulated with other logics of differentiation.

5.3 The anti-globalisation era and various uses of 'union independence' (1999-2005)

At the beginning of 1999, in response to the European Union's decision to ban the importation of 'hormone beef', the American government imposed a heavy tax on the importation of several European products. Roquefort, a sheep's-milk cheese produced predominantly in the Aveyron department, was on this list. After having exhausted legal options, activists in the Aveyron CP decided to take action. On 12 August 1999, around 100 activists descended on the site of a McDonalds restaurant in Millau (a small city in the south of the department), scheduled to open a few weeks later. They 'dismantled' the building under construction. This action in the heat of summer was widely covered by the media, both because of its form and target and because it was drawn out by the 'preventative' detention of four activists and subsequent arrest a few days later of another activist from the Aveyron, José Bové, considered to be 'the instigator', who was held until 7 September.

In the weeks following the action, the union benefited from a great number of overt expressions of support from a range of local and national organisations, for the most part on the left to extreme left of the political spectrum, who denounced the 'union repression' victimising the CP. Above all, Bové's and the CP's departmental and national leaders' justifications for their actions situated the union's demands within a surging protest movement that would come to known as the 'anti-globalisation' movement. The 'dismantling' of McDonalds was presented as a legitimate response to an arbitrary decision. The WTO, which did not oppose the American government's decision, was accused of being responsible for a situation that was a perfect example of the inequalities produced by international trade rules. José Bové's speeches to protesters gathered in Seattle for the WTO counter-summit in November 1999 were widely covered in the media. In June 2000, 100,000 people assembled in Millau in support of ten activists on trial for the McDonalds 'dismantling'. In August 2003, 300,000 people participated in a major gathering on the Larzac plateau near Millau, held a few days after José Bové's release from prison. Bové had been incarcerated between 22 June and 2 August 2003, after being convicted of the destruction of genetically modified rice fields in 1999. He had also been incarcerated the previous summer, from 19 June to 1 August, for the damage to the Millau McDonalds.

During this period, the CP would establish itself as a major actor in French mobilisations against 'neo-liberal globalisation'. The minority union had never been so politically solicited in all its 22 years; the peasant left had never had such public visibility. While this mobilisation brought the organisation new resources, it also modified the balance of internal positions. By favouring interventions that were most closely connected to criticism of the WTO, multinational companies, and GMOs, the media buried other demands that were more closely tied to the union's agricultural programme, such as the distribution of production rights, the dispensing of European aid, or the struggle against expanding farm size. In other words, most of the work undertaken in most of the departmental structures was overshadowed as those involved in the anti-globalisation dynamic were put in the spotlight, leading to an over-representation of the *néo-ruraux* and their concerns, although they were a small minority in the union.

These circumstances tended to exacerbate prior divisions and fostered the emergence of a deep schism between different conceptions of union representation. The disagreements were not really about the content of the discourse disseminated by the media, as there was consensus on the denunciation of the CAP and the WTO, but were instead predominantly based on conflicting interpretations of the consequences of the visibility of their anti-globalisation engagement. Opponents of the national board's actions were convinced that the multiplicity of images regularly associating the CP with organisations holding marginal positions in the political field could only dissuade farmers from supporting the union. To the contrary, members of the national board, backing this engagement in a wide protest movement, saw it as a confirmation of the union's 'independence' from the dominant political parties. These differences cannot be explained solely by differences between political sensibilities, nor do they automatically reflect differences between biographical trajectories. They can only be fully understood in relation to the spaces of activist positioning as they were experienced in each department. Indeed, as I will illustrate using Aveyron as an example, the logic of differentiation that structures activists' relationships with the political field first and foremost makes sense in these localised configurations. But first we should review developments in the national context, since specific events conditioned the processes defining the fault lines within the departmental organisations.

Two very distinct stages: symbolic rewards and updating differentiations

Generally speaking, after the 1999 'dismantling' of the Millau McDonalds, CP activists across France rejoiced at their union's sudden appearance on the public scene. They did not oppose the journalistic framing that reduced the CP to José Bové's positions, although previously this peasant left, strongly marked by the post-1968 origins of its movement, had been careful to control the delegation of power with respect to representing the group. In fact, they credited José Bové with having finally put an end to their isolation because, as activists often said, 'we've been saying what he says for a long time, but nobody heard us'. This visibility was all the more appreciated because it was unexpected and because the disseminated images showed and gave voice to farmers 'on the left' who were concerned about North-South imbalances, perfectly aware of environmental risks, and attentive to the experiences of other social groups. This picture coincides with the distinctive identities built and cultivated in the activist in-group, especially manifest in interviews as frequently expressed claims to 'openness', contrasted with the rest of the 'peasant world, closed in on itself'. This media attention was a source of symbolic reward (Gaxie, 1977) for union engagement because it legitimated activists' social position in the agricultural space. Indeed, members of this peasant left, whether *néo-ruraux* or the children of farmers, shared the experience of having established themselves in 'the' agricultural profession following specific social trajectories. This structured a collectively held belief in the importance of maintaining a critical distance from the agricultural world (Bruneau, 2010). Thus, to the extent that it raised interest and goodwill in non-agricultural milieus, the public recognition of the CP at first lent strength to their inclination for engagement. This time of 'euphoric' CP membership was damaged following the January 2001 Chamber of Agriculture elections when a nationwide, sometimes conflictual, debate emerged and forced departmental-level organisations to make their positions explicit.

The 2001 Chamber elections confirmed the CP's progress, gaining six points at the national level compared to the 1995 election and capturing 26% of the vote, with 30% of the vote or more in over 30 departments. However, while many department-level organisations saw their results climb, sometimes spectacularly, others stagnated or even regressed, especially in much of northwestern France. For example, in Finistère (a department of Brittany) the CP lost leadership of the Chamber of Agriculture, its score dropping nearly 20 points below the 1995 figure. The 2001 election thus put an end to the steady progression in heavily CP departments since the 1980s (leaving only the Chamber of the department of Loire-Atlantique in CP control), prompting activists to seek an explanation for the failure. From the departmental activists' perspective, members of the national board were partly responsible for the situation, and they expressed reservations over the capacity of 'the National [leadership]' to 'mobilise the peasants', because they were also concerned about how 'the peasant world' perceived media coverage based mainly on José Bové's actions. To their way of thinking, there was little doubt that farmers were more preoccupied with the 'mad cow' crisis than by what was happening at the WTO. Along the same lines, they also thought that some CP positions heavily covered by the media (such as the Israeli-Palestinian conflict) blurred the CP's image, and that a farm union should not allow itself to be associated with such political issues.

The media attention paid to the 1999 Millau action actually had the effect of promoting 'movementist' and 'internationalist' positions in the national-level organisation. It happens that four of the nine national board members in 2001 were *néo-ruraux* who had started farming in southern France after being involved in the extreme left in the 1970s. In addition to political convergences, a certain way of being an activist in the CP had settled into the national level of the organisation, particularly manifest in the imposition of a dominant line holding that 'only a significant agricultural policy change at the international level' would allow the union's objectives to be met (Confédération Paysanne, 2002). In parallel, these national-level representatives also worked to elevate the level of the union's engagement and join other forms of 'social movements', employees' unions, associations (such as ATTAC[3], which fights for the taxation of financial transactions), anarchists, and political parties of the extreme left.

Initially rather scattered and heterogeneous, criticism of the national board's actions would come to be expressed by and incarnated in a group of former national representatives. In 2002, at the General Assembly in Créteil (in greater Paris), these union 'figures' introduced a set of nine amendments to the activity report. The third amendment is particularly explicit:

> Agriculture is everyone's business, but the peasant remains the core actor ... But be careful, the peasant is at the heart of the system. He's the one who is subject to its constraints. He, then, is the main actor. He should remain the central concern of our union. The Confédération Paysanne's primary demand is that of its members – isn't that why it exists? ... Our

[3] Association for the taxation of financial transactions and citizen's action.

> peasant battles should keep their citizenship dimension. But it is not the non-peasants who can define, or even dictate, agricultural policy and what peasants ought to do ...

Members of the national board had a difficult time accepting this kind of challenge, some seeing it as a 'provocation', others 'a declaration of war' or 'an attempt at destabilising' the leadership team. Nevertheless, they agreed to open an internal debate to resolve the crisis, asking department-level organisations for their 'contributions'.

The positions of the conflicting parties were discussed and 'translated' in departmental groups. In January 2003, at a supplementary General Assembly, the delegates in attendance saw two conceptions of professional representation face off: one holding that the CP should mainly 'speak to' peasants and be led by the activists most able to convince them, the other preferring to join a movement for social transformation and honour the union's 'non-corporatist' tradition. And indeed, I observed the delegates of the two departmental organisations I studied, from the Orne and the Aveyron, react differently: the Aveyron CP backed the national board's actions, while activists from the Orne favoured 'recentring on the local' in both their prior text and internal debates. Although the Aveyron delegates' attitude may be explained by the presence of two of their own on the national board, since 2000 the Orne CP was also 'represented' in the national team by Jean-Pierre, who was also a fervent promoter of the anti-globalisation line. Consequently, other explanatory elements should also be taken into account, beginning with the 2001 Chamber election results. In the Aveyron, the CP received 44% of the vote, an advance of 15 points over its 1995 level, thus giving it a glimpse of the possibility of a future victory over the FDSEA. In contrast, in the Orne the CP garnered 32% of the vote compared to 33% in the previous election, and activists were trying to 'understand' the reasons behind this stagnation. This reading is not entirely satisfactory, however. For one thing, being based on the hypothesis that departmental delegates' votes at national meetings express the range of positions defended locally, it contributes to the maintenance of a reifying conception of activists groups. Secondly, by making the interpretation of electoral results (far from homogeneous in the two departments) autonomous, this analytical approach does not take account of the relationship between electoral interpretations and other mechanisms of differentiation.

The politicisation of social distinctions in the Aveyron

From the outset of my field research in the Aveyron (2002-2004), the group of about 50 participating activists was presented as being rather heterogeneous, split into two geographically based divisions opposing activists from the southern Aveyron to those from the centre, north and west. The formation and stabilisation of this 'trend' (to take up the term used by group members) was the result of several overlapping logics of differentiation that were maintained by the division of activist roles, the crossed perceptions of the political sensibilities of others, and the varying degrees of a sense of subjective belonging to the 'peasant world'.

Since the early 1980s, the Aveyron 'peasant left' has set itself apart by conducting spectacular actions and frequently calling on 'strategies of scandalisation' (Offerlé, 1998: 122). The targets were diverse – leaders of the majority union (FDSEA), directors of private businesses and cooperatives, departmental and national political representatives – and the palette of means was wide, including hunger strikes, 'locking in' people with political and economic responsibilities, and occupying official buildings. Unable to mobilise as many protesters as the FDSEA, their approach to opposition favoured forms of action whose effectiveness did not depend on the number of participants. These left-leaning peasants would always 'go into action' in front of (or even with) journalists from the regional media, in the light of day and with their faces exposed. The media presence contributed to the staging of the conflictual relations that the activists maintained with the representatives of power. It was primarily activists from the south of the department who practised and vaunted this confrontational unionism – the same ones who, with José Bové, ran the local *Travailleurs-Paysans* union (member of the CNSTP) between 1982 and 1987. In most cases they were *néo-ruraux* who had set themselves up on the Larzac plateau and in the Millau region in the late 1970s and early 1980s. Several of them participated in the Larzac struggle (1971-1981) by lending their support to peasants on the plateau who opposed the expansion of a military camp in the area (Alland, 1994). Until the mid-1990s, the southern Aveyron group easily formed a relative consensus on the virtues of this kind of collective practice.

The other 'trend' only began to emerge in the late 1990s, following the union's establishment in the northern part of the department, especially in the areas of the Aubrac and the Lot Valley. The CP recorded its first individual memberships there in the early 1990s before experiencing strong growth in the zone starting with the January 1995 Chamber of Agriculture elections. This electoral success would legitimise the practices of the northern activists, who gradually specialised in 'fieldwork', as they called it – that is, participating in official commissions such as those dealing with real estate transfer or production rights attribution.

Thus, over the years a theory of a division of skills and knowledge took root, holding that some activists – those from the south – were 'good' at carrying out actions, and others – to the north and west – had perfect mastery of technical subjects. However, these specialisations could also be related to opposing points of view. Northern and western activists reproached the southern 'militants' for not preparing their actions well enough, putting actions ahead of achievements, or even carrying out actions that might undermine their representational work in commissions. For example, Georges, a farmer near Rodez (in the centre-north-west zone), had often disagreed with 'the methods' of his comrades from the southern Aveyron:

> *The split between northern and southern activists, what do you think it's due to?*
> It's not a difference in objectives, it's a difference of method. That is, some think that as soon as a problem comes up, some people will say 'Let's do an action!' and then I don't agree with you, I lay into you… well I'm exaggerating a little bit but that's it. And others who say 'Before you lay into each other, let's try to see if we can straighten out the problem another way…' (…).
>
> *And, you've disagreed with a certain number of actions over the years?*
> Yes, always. Well, 'always'… Personally I disagree with the method, partly because the actions are often poorly planned, we don't come into agreement on how we're going to go about it, and we don't agree on the final objectives. I mean, today, all of a sudden, bang! something comes up and, well, let's go! We get there, and there, 'do we remove the desks, or not? C'mon!' Yes, we remove the desks, we put them through the windows – but it was never decided (…). Then we go back home, and after, we've got cops that come to investigate and then we see each other in court, we follow the whole business more or less well, you've got to get lawyers, you've got to pay the lawyers.

On their side, activists from the south may stigmatise their comrades' jitteriness and inability to react when the situation calls for it. For Gérard, a *néo-rural* living on the Larzac plateau since the late 1970s, northern activists 'are scared':

> *In the fight against the WTO, the danger is making people think that everything happens, like, far away… right?*
> Of course. So the problem, when they say that to us, we say 'ok guys, in the northern Aveyron there's a huge problem, a big plan for an irrigation lake for six farmers that's going to cost 6 million francs' [one million euros], I think that on top of that they'll drown 19 hectares of good land, for the little pals of the director of RAGT [a major agricultural supply company]. And so for that whole thing, people said: 'We've got to do something'. People from the south, we said to people from the north, 'Well, organise, we'll come if there's a demonstration'. People from the north, who are rather close, um, well, to the PS: 'yeah, don't worry about it, we'll organise something'. It's been a year we're waiting for something to happen, the lake will get built, there was never an action planned, not even a press conference, nothing… [he raises his voice]. They moan, but they'll never dare to do anything, uh, like, physical, I mean, to be seen.

> *And in the north, they are less able to do it?*
> Well they aren't capable, no, they're afraid, but at the same time, that's not how they do things.

This interview excerpt is particularly revealing of the associations and logic that activists use to make sense of where 'trends' are located, the division of roles between sub-groups, and the political tendencies of those who incarnate them according to a more critical or more supportive attitude towards the PS, the main political party on the left.

A little later in the discussion, Gérard even uses another register of distinction when he declares that 'the *néo-ruraux* are less PS than that'. It is this mix of differences that underlie the positions held in internal debates over the union's anti-globalisation engagement. For activists from the south, who are generally close to the extreme left without being formally affiliated with a party, the CP's involvement in a broader movement opposing 'neo-liberalism' is an additional sign of its 'independence' from political power. At the same time, activists representing the other 'trend' regret that the union's name is regularly associated with political organisations. In early 2003, Gilbert and Brigitte, farmers in the north of the department, were worried that the CP would end up losing its 'identity' by joining this political dynamic:

> Gilbert: Well for me, my feeling about the national organisation is that there's a few people at the level of the national board who want to make the Conf' into a vast so-called social movement, where me, I say it's a delusion because there's no social movement, it's a movement of ideas, and a movement of interesting ideas, rich even, it's true, it's really interesting everything that gravitates around the Conf', but to go so far as to want to build a social movement on it, I'm not for it, y'know. I think it's good when everyone comes together in the street at the same time, with their identity. (...) Me, I find the ordinary peasant has trouble making it today, he thinks – well, what I hear is that the Conf' does way too much politics.
>
> *Too political in what way?*
> Gilbert: A lot of things come down to making interventions and then, like, playing with the media...
> Brigitte: When there are declarations of support, it's either affiliated with the PS or the PC or the Greens, people can't stand that.
> Gilbert: So, uh, everyone knows that in 1987-1990 the Conf' had a very hard political line eh, it wasn't gentle with politics, left or right, and today there's a tendency towards absorption, a little bit.

During the interview, Gilbert and Brigitte emphasised their closeness to 'the ordinary peasant': they have farmers at their home on a daily basis and are called on to help with various problems, filling out official forms, giving advice on litigation over land, or consulting on financial problems. Gilbert has been an elected representative in the Chamber of Agriculture since 1995, and he works hard to represent this 'ordinary farmer' in official settings. However, because this 'invisible' work escapes media attention, those who live outside the agricultural world have little opportunity to hear about it. Under such conditions, it can be very difficult to escape the influence and visibility of the *néo-ruraux* on the Larzac plateau.

Other logics of differentiation also feed into this split, this time in the ways in which activists position themselves in the agricultural social space at issue. In the following excerpt from an interview, Gilbert and Brigitte try to explain what distinguishes them from their 'colleagues' in the union, successively evoking vacations, time spent working, their relationship with animals, and 'the relationship with the land':

> Gilbert: That too, we have a big debate in the Conf', when you talk about the quality of life in the country, well, when you talk about it with people from the south, actually I don't agree with them at all, they say 'Gotta create GAECs [groupement agricole d'exploitation en commun; joint farm ownership status], that way we can replace each other', where they're precisely in the process of boxing things in, that's not it at all (...). There's also difficulty living in the country, a difficult life has other origins, it's that people stick themselves in systems where they are no longer masters of their own choices, so already starting with that, it's hard to live with. Um, there's no-one left in the country, so there's no social life, no more of this mixing between the old, the young, the older generations, the new, no more people you talk to on the doorstep every morning, eh, and more specifically, in certain areas there's bachelorhood ... and GAECs aren't going to fix these three problems, eh? The GAEC is something conventional, (...) Me, it's enough for me to not take a month of vacation, we don't take any at all, I like to take vacation every day when I want to [laughs], I like to feel good the way I am, y'know. [Brigitte interrupts]
> Brigitte: Yeah, and also the approach to work is different. It's true that we are still attached to our critters, over there [on the Larzac] they are now only supports, it's evolved, there's no longer this ...
> Gilbert: ... this relationship with the land, basically ...
> Brigitte: Yes, exactly, I was looking for it. So we don't have the same relationship to the land, so ...
> Gilbert: Because actually everyone needs to recharge himself with the land, whether it's a city person or not. Everyone identifies with the land today, but it's not so simple as that, eh. Between old peasant roots three generations old, to really identify yourself, and the others.

These demarcating principles also structure how Chamber of Agriculture election results are interpreted. In January 2001, Gilbert headed the CP slate, while José Bové, who had been elected to the national board a few months earlier, only held the bottom position. The FDSEA received the majority of the vote (56%), while the CP scored 44%, capturing four of the 21 Chamber seats and confirming its progress. The CP had its best results in the north and west of the department, where in certain cantons it even won the majority. In the south of the department, with the exception of cantons containing Larzac farms, canton-level results were often below the departmental average. On the other hand, the CP also registered its highest local progression in cantons where right-leaning parties and candidates were historically largely in the majority. This is especially apparent in the comparison of canton-level results for the 2001 Chamber elections with those of the first round of the 2002 national presidential election. This apparent contradiction did not escape CP activists, who were keenly aware that their electorate was composed of a heavy proportion of farmers grounded in the political right. Nor did they hide the fact that they had trouble understanding the exact reasons why these farmers supported their union. The fuzziness around the analysis of electoral choices allowed them to make their own interpretations of the vote anyway, by attributing intentions to voters that were compatible with the positions they defended in the departmental organisation. Put another way, they equated their activist capital (Matonti and Poupeau, 2004) – and their specific resources and skills along with it – with their representation of the interests and 'mentalities' of the 'peasant world'. Gilbert interpreted the CP's good results in the north of the Aveyron as being the result of the quality of the 'fieldwork' of 'local activists':

> Gilbert: The whole north of the department, there were really, really good results, directly connected to the issues, peasants' problems, uh, when you fix a peasant's problem, you know, there's ten who know about it indirectly, family dinners, word of mouth happens and then there you are, you know. ...
>
> *And so how do you explain that it works so well around here, in the Chamber elections, in your opinion?*
> Gilbert: It's the fieldwork that's now connected with the local activists.
>
> *But fieldwork, what is it, then, exactly?*
> Gilbert: Well, fieldwork, it's uh, it's people at your place every night [laughs], for really different kinds of problems.

In contrast, for activists in the south of the Aveyron, the union's progress was proof that farmers in the Aveyron approved of the 'dismantling' of the McDonalds, adhered to their anti-globalisation discourse and understood the importance of the fight against the WTO. In January 2001, Gérard learned the results of the election while leaving the World Social Forum in Porto Alegre, Brazil:

> Gérard: I tend to say that peasants aren't all that stupid in the Aveyron. In the midst of Porto Alegre, we heard the results at the Sao Paulo airport, José [Bové] called the national office to find out how many the Aveyron got at the Chamber of Agriculture, we seriously got 45%, and the previous time it was 31 (...). And the time before that it was 20, so we're really developing, although we aren't known for having done a lot of things, uh, well like on the level of the everyday management of peasants' problems, we're more known for McDonalds, Roquefort and everything, but all the same, by defending Roquefort against the WTO, we really scored some points, it went over well locally, since people voted for us.

In 2002-2003, while the conflict begun by the former national representatives was in full swing, the Aveyron activists most invested in the anti-globalisation movement used this electoral increase to justify their support of the national board. At the same time, both groups exploited this resource locally as well. As a case in point, Gilbert and the other activists doing 'invisible' work used it as an argument to counter their comrades in the south of the department. Consequently, while this situation reveals the great variability in the extent of individual investments, it only reinforces previously established activist identities and does not lead to significant disengagement (Fillieule, 2005), since each 'trend' can appropriate the symbolic rewards arising from this electoral success for itself. In this department, unlike what was observed in the Orne, the expression of disagreement, authorised and made mundane by the collective recognition of a division, even tends to favour the maintenance of engagement over this period.

5.4 Concluding remarks

As this chapter has shown, the history of the CP is not simply characterised by opposition to agricultural policy and the orientations of the majoritarian unions. This collective history is also one of diminishing expectations that political authorities might respond, which is closely tied to the particularities of representing agricultural interests in the French system. Being situated on the left of the political spectrum is thus accompanied by a structural defiance of the dominant positions in the political field that in turn plays a central role in making individual engagements converge. A localised analysis of activism in the Aveyron, combined with a consideration of the national organisational context, allows for a refined understanding of how CP activists have positioned themselves vis-à-vis the political field. In department-level organisations there are manifest differences in political sensibilities that are related to other distinctions, be they activist skills or registers for attributing worth to professional careers.

Consequently, although the question of political party relations may have an influence on the intensity of engagement, it is only one of many dimensions of the work of organisational homogenisation, the effectiveness of which depends on the group's ability to offer symbolic rewards in keeping with their members' social trajectories. The union brings together farmers who have a distant relationship with 'their' profession, inseparable from their trajectories prior to farm ownership, and a propensity to a lifestyle inspired by the educated middle classes (Bruneau, 2006). Compared with other farmers, then, these activists' social trajectories appear rather homogeneous, but small differences in disposition, far more than a simple opposition between 'heirs' and the *néo-ruraux,* are still likely to be reactivated, as became so starkly clear in the anti-globalisation period.

References

Alland, A. and Alland, S., 1994. Crisis and commitment: the life history of a French social movement. Gordon and Breach Science Publishers, Yverdon, Switzerland, 190 pp.

Bernard de Raymond, A., 2010. Les mobilisations autour des OGM en France, une histoire politique (1987-2008). In: Hervieu, B., Mayer, N., Muller, P., Purseigle, F. and Rémy, J. (eds.) Les mondes agricoles en politique. Presses de Sciences-Po, Paris, France, pp. 293-335.

Bourdieu, P., 2000. Propos sur le champ politique. Presses Universitaires de Lyon, Lyon, France, 110 pp.

Bruneau, I., 2004. La Confédération paysanne et le 'mouvement altermondialisation'. L'international comme enjeu syndical. Politix 68: 111-134.

Bruneau, I., 2006. La Confédération Paysanne: s'engager à 'juste' distance. Doctoral dissertation in Political Science, Université Paris X – Nanterre, France, 627 pp.

Bruneau, I., 2008. Quand des paysans deviennent 'soixante-huitards'. In: Damamme, D., Gobille, B., Matonti, F. and Pudal, B. (eds) Mai–Juin 68. Les Éditions de l'Atelier, Paris, France, pp. 344-356.

Bruneau, I., 2010. Recomposition syndicale et constructions des collectifs militants. A partir d'une enquête sur la Confédération Paysanne. In: Hervieu, B., Mayer, N., Muller, P., Purseigle, F. and Rémy, J. (eds) Les mondes agricoles en politique. Presses de Sciences-Po, Paris, France, pp. 217-240.

Champagne, P., 1984. La manifestation. La production de l'événement politique. Actes de la Recherche en Sciences Sociales 52-53: 19-41.

Confédération Nationale des Syndicats de Travailleurs-Paysans (CNSTP), 1985. Combien de paysans avec quel revenu. Un choix qui interpelle toutes les forces sociales. Rapport d'orientation du Congrès de Villeneuve-sur-Lot, CNSTP, Villeneuve-sur-Lot, France.

Confédération Paysanne, 2002. Changeons de politique agricole. Éditions Mille et une nuits, Paris, France, 142 pp.

Cornu, P. and Mayaud, J.L. (eds), 2007. Au nom de la terre. Agrarisme et agrariens en France et en Europe du XIXe siècle à nos jours. La Boutique de l'Histoire, Paris, France, 462 pp.

Coulomb, P., 1990. Les conférences annuelles entre corporatisme et populisme. In: Coulomb, P., Delorme, H., Hervieu, B., Jollivet, M. and Lacombe, P. (eds) Les agriculteurs et la politique. Presses de la Fondation Nationale des Sciences Politiques, Paris, France, pp. 159-179.

Curade, M., 2011. Du syndicalisme à la politique. Ils sont passés de l'autre côté. Campagnes solidaires 260: I.
Dewalle, G., 1995. Il faut plafonner les aides. Libération January 12, 1995.
Duntze, N., 2011. Lucioles. Campagnes Solidaires 260: VIII.
Fédération Nationale des Syndicats d'Exploitants Agricoles, 1970. Rapport moral. Cited by Coulomb, P. and Nallet, H., 1972. Les organisations syndicales agricoles à l'épreuve de l'unité. In: Tavernier, Y., Gervais, M. and Servolin, C. (eds) L'univers politique des paysans dans la France contemporaine. Armand Colin, Paris, France, pp. 379-413.
Fillieule, O. (ed.), 2005. Le désengagement militant. Belin, Paris, France, 319 pp.
Gaïti, B., 1998. De Gaulle prophète de la Ve République (1946-1962). Presses de Sciences-Po, Paris, France, 378 pp.
Gaxie, D., 1977. Economie des partis et rétributions du militantisme. Revue Française de Science Politique 27(1): 123-154.
Gombin, J. and Mayance, P., 2010. Tous conservateurs? Analyse écologique du vote de la population agricole lors de l'élection présidentielle de 2007. In: Hervieu, B., Mayer, N., Muller, P., Purseigle, F. and Rémy, J. (eds) Les mondes agricoles en politique. Presses de Sciences-Po, Paris, France, pp. 193-216.
Keeler, J.T.S., 1987. The politics of neocorporatism in France: farmers, the state and agricultural policy-making in the fifth republic. Oxford University Press, Oxford, UK, 369 pp.
Lagrave, R.M., 1990. Les gauches syndicales (MODEF, CNSTP, FNSP). In: Coulomb, P., Delorme, H., Hervieu, B., Jollivet, M. and Lacombe, P. (eds) Les agriculteurs et la politique. Presses de la Fondation Nationale des Sciences Politiques, Paris, France, pp. 355-369.
Lambert, B., 1970. Les paysans dans la lutte des classes. Le Seuil, Paris, France, 191 pp.
Martin, J.P., 2005. Histoire de la nouvelle gauche paysanne. Des contestations des années 1960 à la Confédération paysanne. La Découverte, Paris, France, 312 pp.
Matonti, F. and Poupeau, F., 2004. Le capital militant. Essai de définition. Actes de la Recherche en Sciences Sociales 155: 5-11.
Mischi, J., 2013a, Contested rural activities: class, politics, and shooting in the French countryside. Ethnography 14(1): 64-84.
Mischi, J., 2013b. Les territoires ruraux, des espaces ouvriers en mutation. Métropolitiques, Paris, France. Available at: http://tinyurl.com/qcp7fxy.
Muller, P., 1984. Le technocrate et le paysan. Éditions Ouvrières, Paris, France, 173 pp.
Offerlé, M., 1998. Sociologie des groupes d'intérêt. Montchrestien, Paris, France, 158 pp.
Purseigle, F., 2010. La coordination rurale, un nouvel acteur sur l'échiquier syndical. In: Hervieu, B., Mayer, N., Muller, P., Purseigle, F. and Rémy, J. (eds) Les mondes agricoles en politique. Presses de Sciences-Po, Paris, France, pp. 241-272.
Reed, M., 2008. The rural arena: the diversity of protest in rural England. Journal of Rural Studies 24: 209-218.
Renahy, N., 2005. Les gars du coin. Enquête sur une jeunesse rurale. La Découverte, Paris, France, 285 pp.
Strijker, D. and Terluin, I.J., 2015. Rural protest groups in the Netherlands. Chapter 3. In: Strijker, D., Voerman, G. and Terluin, I.J. (eds.) Rural protest groups and populist political parties. Wageningen Academic Publishers, Wageningen, the Netherlands, pp 63-77.

Annex

Annex 1. Key events in the development of rural protest groups in France, 1945-2014

Year	Event
1946	Founding of the Fédération nationale des syndicats d'exploitants agricoles (National Federation of Farmers' Unions, FNSEA).
1947	Founding of the Cercle national des jeunes agriculteurs (National Young Farmers' Circle), which would formally become a union in 1956. It then changed its name into Centre national des jeunes agriculteurs (National Young Farmers' Centre, CNJA) in 1961.
1959	Founding of Mouvement de coordination et de défense des exploitations agricoles familiales (Movement for the Coordination and Defence of Family Farms, MODEF). This union, whose leaders are often members of or sympathisers with the Communist Party, denounce 'capitalist' policy that plans for the disappearance of 'family farms'.
1958-1962	Beginning of the Fifth Republic; new laws in 1960 and 1962, under the CNJA's influence, to facilitate farm expansion and 'modernisation'.
1968-1972	Emergence and coordination of internal conflict in the CNJA, leading to the formation of the Paysans-Travailleurs (Peasant-Workers).
1969-1981	The 'Golden Age' of agricultural policy co-management (the institutionalisation of consultation between the Gaullist executive branch and the FNSEA/CNJA pair): monthly meetings at the Ministry of Agriculture, annual conference established in 1971.
1981-1982	Founding of the Confédération nationale des syndicats de travailleurs paysans (National Confederation of Workers-Peasants Unions, CNSTP, 1981), and the Fédération nationale des syndicats paysans (National Federation of Peasant Unions, FNSP, 1982), two unions emerging from the 'peasant left' formed in the 1970s.
1981	Election of François Mitterrand as President of the French Republic. The Minister of Agriculture, Edith Cresson, acknowledges the existence of minority unions.
1986	Victory of the political right in legislative elections. Jacques Chirac, new Prime Minister, hands the Ministry of Agriculture to François Guillaume, president of the FNSEA since 1979.
1987	Fusion of the CNSTP and the FNSP, resulting in the creation of the Confédération Paysanne (Peasant Confederation, CP).
1989	By winning 18% of the vote in elections for the Chamber of Agriculture, the CP establishes itself as the second-ranking union organisation, but far behind the FNSEA-CNJA pair (61%).
1992-1994	Reform of the Common Agricultural Policy (CAP). An internal protest emerges in the FNSEA, especially in SW France, which in 1994 takes form with the founding of a new union, Coordination Rurale (Rural Coordination, CR).

Annex 1. Continued.

Year	Event
1999-2005	Following the 1992 reform of the CAP, the CP gradually joins a nebulous protest movement alongside non-agricultural organisations and radicalises its positions on international agricultural policies. The 1999 'dismantling' of the Millau McDonalds was part of this process, and its intense media coverage gave the CP a major role in founding an anti-globalisation movement in France.
2009-2010	Four former national-level leaders of the CP are elected European deputies or regional councillors from lists presented by the Europe Ecologie/Les Verts group.
2013	The FNSEA-CNJA pair remains largely in the majority following the Chamber of Agriculture elections (53% of the vote). However, the CP is no longer in second place, having been slightly overtaken by the CR (20% compared to 19%).

6 Austrian pastoral: rural protest at the margins of a hegemonic policy landscape

F. Seifert

Independent Researcher, Austria; fseifert@gmx.at

Abstract

Austria does not have an abundance of rural protest movements. Rural, understood as farmer, protests, are uncommon, as farmers are strongly integrated into – and thus domesticated by – a system of agricultural chambers, financial institutions and political representation through the conservative Catholic People's Party. The system is hegemonic and persistently involves internal conflict in representing the agro-sector's seemingly common interests. In post-war Austria we can identify only a few, small-scale dissident farmer movements, none of which ever managed to jeopardise the hegemonic system. In this contribution we take a closer look at two recent movements: the movement against agro-food biotechnology (1996-2005) and the movement against 'dumping prices' in the milk sector (2004-2012). The first movement is only a rural movement to some extent, and its rural agenda was quickly embraced by the hegemonic system. In contrast, the milk protest movement sought to established an economically viable alternative to the hegemonic system which, in turn, sought to marginalise it. Ultimately, neither of the movements managed to seriously challenge the hegemonic system.

6.1 Introduction

Rural protest movements are not abundant in Austria. Protest among rural communities and farmers is uncommon as farmers are strongly integrated into, and thus domesticated by, a system of agricultural chambers, financial institutions and political representation through the conservative Catholic People's Party. The system is hegemonic, persistently restraining internal conflicts while representing the agro-sector's apparent common interests. In post-war Austria, we can identify a few small dissident farmer movements; however, none ever managed to jeopardise the hegemonic system. In this contribution we take a closer look at two recent movements: the movement against agro-food

D. Strijker, G. Voerman and I.J. Terluin (eds.) **Rural protest groups and populist political parties**
DOI 10.3920/978-90-8686-807-0_6, © Wageningen Academic Publishers 2015

biotechnology (1996-2005) and the movement against low prices in the milk sector (2004-2012). The first movement was only a rural movement to a limited extent, and its rural agenda was quickly embraced by the hegemonic system. In contrast, the milk-protest movement sought to establish an economically viable alternative to the hegemonic system which, in turn, aimed to marginalise it.

6.2 The hegemony of the People's Party in agricultural policy

In the second half of the twentieth century, Austrian agriculture underwent the same landslide changes that transformed the rural world elsewhere in Western Europe: huge gains in productivity due to agrochemicals, mechanisation and industrialisation, a drastic loss of agriculture's relative weight in the economy, and a drastic shrinkage in agricultural employment, the numbers of farms and the rural population (Groier, 2006; Hanisch, 2002). In spite of the tensions entailed by this painful transformation, Austria has never developed a great tradition of rural protest. On the one hand, this could be due to historical reasons, such as a belated onset of modernity due to centuries of counter-reformation, conservative Catholicism, authoritarian Habsburg rule and, ultimately, integration into Hitler's Germany. On the other hand, the relative scarcity of open rural protest in post-war Austria also originates from its rather structural-institutional nature, which draws on a hegemonic system of political interests centred around the Austrian People's Party (ÖVP) (Hofreither, 1997). Austria is a country in which the organisational density of the two major political parties – the Social Democratic Party (SPÖ) and the People's Party – is traditionally greater, and their influence reaches deeper into social and economic life than in other comparable countries. This influence applies particularly to the agricultural sector. The major players in this sector are found in three areas, each of which is within the sphere of influence of the People's Party.

First, the Federal Ministry of Agriculture and Forestry, which is responsible for both the national orientation and the international coordination of agricultural policies, is typically dominated by the People's Party, with the only exceptions being the years of Socialist rule (1970-1986), when the People's Party did not participate in government. The archetypical form of government in post-war Austria has consisted of a grand coalition between the Social Democrats and the People's Party, combined with a political culture of consociationalism. This favours a politics of mutual recognition of vested interests, with the People's Party routinely claiming control of the agricultural agenda.

Second, in the Chambers of Agriculture – the farmers' representative body at the national level – the Austrian Farmers' Federation (Bauernbund), consisting of nine provincial organisations, always holds a clear majority (at least about 80% of the seats). As membership of this federation necessarily implies political membership of the People's Party, the party's reign over the agricultural and rural sphere is thus assured.

The Bauernbund has existed in its present form since the founding of the People's Party in 1945. Currently, it has about 300,000 members. Due to its strength in the chambers, the Bauernbund is also one of the People's Party's major sub-organisations, exerting a disproportionate influence on its policies. Political elites in the People's Party, for example, are often recruited from the Bauernbund; one of whom was Leopold Figl, the first Federal Chancellor in the crisis-ridden years after the Second World War (1945-1953). He had been chairman of the Lower Austrian Bauernbund in the early 1930s, and re-established and reintegrated the Bauernbund into the People's Party in 1945. Another prominent politician who held office in the Bauernbund at an earlier stage of his career was Franz Fischler, who was later appointed Federal Minister of Agriculture (1989-1994) and soon after, EU Agriculture Commissioner (1995-2004). After Austria's EU accession in 1995, the Bauernbund further strengthened its powerful position, as the chambers became the primary advisory institution guiding the way through the bureaucratic jungle surrounding EU subsidies and quality-control criteria.

Third, the Austrian Raiffeisen organisation is an unofficial, albeit significant, factor in the network of agricultural decision-making structures (Weiss, 2010). This former farmers' cooperative has been transformed into a financial giant, involved in construction, real estate, insurance, travel and media business. As a consequence, it has by far surpassed its initial agricultural agenda and its control by farmers. Due to its dominant position in many key agricultural markets and its strategic seats in major political organisations such as the Chambers of Agriculture, Chambers of Commerce and the Industrial Association, the Raiffeisen Group is instrumental in securing policy decisions.

6.3 Hegemony through adaptation

This hegemonic system of agricultural policy formation does not leave much space for farmer protests. Memorable protest events, such as demonstrations where hundreds of tractors besieged the Parliament in Vienna, only occurred in the years of Socialist rule in the 1970s. During this time, the People's Party found itself deprived of its traditional office, and these protests, which were organised by the Bauernbund, can be seen as reactions to what was perceived as the first serious threat to the People's Party's hegemony (Hovorka and Hoppichler, 2006). Yet, even in these years, the Social Democrats' attempts to exert an influence on agriculture, for example by introducing measures favouring the disadvantaged mountain farmers over the more productive farmers of the plains, failed in the long term. Ultimately, after the end of Socialist rule, the People's Party's reign over the agricultural sphere was seamlessly re-established.

A major discursive strategy employed by the agricultural elites when facing criticism is to deny inequality among Austrian farmers and the tensions arising from it, and denounce critics as enemies striving to undermine farmers' unity (Rohrmoser, 2001).

In an agricultural system where subsidies make up about half of farmers' income, the distribution of these subsidies is a natural matter of dispute. Until recently, the chambers kept quiet about this issue, barely hiding the fact that subsidies significantly favoured the larger producers and agro-industries (Weiss, 2010). Several groups outside the People's Party-Bauernbund have highlighted this non-transparent practice, such as the relatively small Workers and Farmers' Federation (Arbeitsbauernbund) (referred to as SPÖ Farmers below), other farmers close to the right-wing populist Freedom Party, and the progressive, ecologically minded farmers. These groups are usually discredited by the Bauernbund as 'peasant splitters' (Hovorka and Hoppichler, 2006; Krammer and Rohrmoser, 2012).

Yet another reason for the hegemonic agro-system's enduring success is its capacity to learn and adapt to crisis and design new policy solutions. These solutions are often envisioned by external critics from the ranks of a small agricultural opposition, lateral thinkers, teachers, researchers and contrarians in the chambers. Although the senior levels of the Bauernbund management remain reticent to consider alternative approaches, usually a pattern involving initial ridicule, rejection and marginalisation of dissenters is followed by an unaccredited emulation of their concepts and practices. In this way, critical impulses can be embraced and ultimately integrated into workable policies, with many agricultural innovations thus originating from external sources. Institutionalised farmers' representatives thus copy the innovations of developers who they earlier mocked and marginalised. This is illustrated by the fact that many innovative elements in Austria's agricultural policy that were first introduced under Socialist rule in the late 1970s, later became flagship programmes for Austria's agricultural policy, such as special subsidies for mountain farmers and programmes for regional development.

Adaptation of ecologically and socially sound market policy

Another example of the adaptation process refers to the way the ecological discourse was adopted by Austria's agricultural policy in the early 1990s (Hanisch, 2002). In the wake of key controversies over the environment in the 1970s and 1980s, the ecological discourse gained strength and the Green Party emerged as challenger to the traditional dual party system. The Green-alternative movement also brought about a new critical perspective on agriculture, and identified over-production and the destruction of nature as basic agricultural problems. While the goals of agricultural policy in the 1970s – growth, enlargement, intensification and specialisation – were still in the minds of most farmers, they increasingly suffered from their negative image as destroyers of nature. Most farmers resented the emerging Green Party for its 'unrealistic', 'left-wing' criticism. Only an often ridiculed minority of Bauernbund dropouts embraced ecological alternatives in agriculture such as organic farming.

However, with the imminent accession to the EU agreed upon by the political elites in the late 1980, great fears arose in agricultural circles. These fears concerned competitive pressures, the influx of foreign agro-products into the domestic market, agricultural price decreases and a drop of farming income. Hence new concepts for agriculture were sought, which resulted in the idea of the creation of a distinctive Austrian product identity. After the People's Party had regained control of the agricultural portfolio in the late 1980s, it put forward the outlines of an 'ecologically and socially sound market policy' (*Ökosoziale Marktwirtschaft*). This concept advocated economic performance combined with environmental conservation and social equity through direct payments and support for disadvantaged farmers and regions. In preparation for integration into the European single market, the Austrian agricultural budget was almost doubled. A package of measures was adopted which mainly consisted of a comprehensive environmental programme, including the strong promotion of organic farming. Public relations also played a role in embracing this new concept, as both the quality of the food and the fate of Austrian farmers were prominent issues in the debates prior to the country's accession to the EU. The promise to subsidise organic farming helped gain public acceptance. With slogans such as 'eco-land' or 'delicatessen shop Austria' (*Feinkostladen Österreich*) the government effectively communicated its new strategy to the public.

In 1995, the Austrian Agri-Environmental Programme ÖPUL (Österreichisches Programm für umweltgerechte, extensiven und den natürlichen Lebensraum schützenden Landwirtschaft) started. Today, ÖPUL accounts for around 50% of the annual payments under the second pillar of the Common Agricultural Policy (CAP) (Groier, 2011). About 17% of ÖPUL subsidies are reserved for organic farming. Due to this policy, the proportion of organic farmers increased dramatically after Austria's accession to the EU in 1995, today stabilising at 16.5%. In the European context, Austria thus became one of the European forerunners in both environmental policymaking and organic farming.

The eco-social approach illustrates the innovative and adaptive capacity of Austria's hegemonic agricultural system. By reconciling previously rejected elements with established policy goals, the concept tackled several major challenges: it adapted Austrian agriculture to the CAP regime and the EU's liberalised agricultural market; it combined the ecological with the social agenda, as small mountain farms and extensive grassland and dairy farms converted to organic farming; and it ensured general compliance.

6.4 Dissenting voices

In spite of the system's flexible stability, there are dissenting voices, both in the chambers and, probably more importantly, in the form of independent organisations outside the chambers. In relation to the opposition in the chambers, shortly after the Second World War, communist farmer associations attempted to challenge the People's Party's predominance in the chambers, but soon disappeared. Until the mid-1980s, SPÖ Farmers were the only rivals in the chambers worth mentioning. However, they were not able to mount a serious challenge to the Bauernbund's supremacy. In the 1980s, when discontent was on the rise in relation to Austria's accession to the EU, the Bauernbund gradually started to lose seats to new contenders (Table 6.1), who we briefly introduce below.

Freedom Party Farmers

The Freedom Party Farmers (Freiheitliche Bauernschaft) were the first to make inroads into institutional power, particularly in the province of Carinthia. As the name suggests, this farmers' organisation is related to the Freedom Party, and their rise was mainly a reflection of the rise of the charismatic right-wing populist Jörg Haider beginning in the mid-1980s. The flamboyant Haider was a many-sided character, with his political power base in Carinthia, where he held the office of provincial governor. As owner of a large Carinthian estate known as Bärental (Bear Valley), he also had roots in the rural world, which he emphasised when addressing a rural audience (Kalina, 2008). Haider's major political momentum came from his fierce attack on the established dual party system and its entrenched power, which prevailed in the agricultural sphere as in all sectors of public life. The Freedom Party Farmers' position reflected Haider's

Table 6.1. Overview of farmers' groups and their relationship with political parties in Austria.

Farmer's group	Founded in	Related to political party
Austrian Farmers' Federation (Bauernbund)	1945	Conservative, Catholic Austrian People's Party (ÖVP)
Workers and Farmers' Federation (Arbeitsbauernbund)	1945	Social Democratic Party (SPÖ)
Independent Farmers' Association (Unabhängiger Österreichischer Bauernverband UBV)	1952	-
Freedom Party Farmers (Freiheitliche Bauernschaft)	1956	Freedom Party
Austrian Mountain Farmers' Association (Österreichische Bergbauernvereinigung ÖBV).	1975	-
IG Milch	2004	-

right-wing populism: anti-EU resentment, denouncing vested interests, as well as the People's Party dominance in agricultural policy. By contrast, immigration – a core focus of Haider's populist mobilisation strategy – was not a major issue in the countryside or for the Freedom Party Farmers.

Independent Farmers' Association

The only dissenting group that has remained without direct party affiliation while running for office in chamber elections is the General Austrian Farmers' Association (Allgemeiner Österreichischer Bauernverband), known after 1999 as the Independent Farmers' Association (Unabhängiger Österreichischer Bauernverband, UBV). The association was founded in 1952, picking up on the traditions of the pre-war Rural Federation (Landbund), a farmers' party with a pan-Germanic, anti-clerical and anti-Socialist outlook from the period of the First Republic. Considering their ideological affinity, it is no surprise that the UBV and the Freedom Party (founded in 1956 as the successor to the Federation of Independents, VdU) appealed to an overlapping membership, particularly in the formative years after the war. In fact, founding members of both organisations had been previously engaged in the Nazi Party (Nationalsozialistische Deutsche Arbeiterpartei, NSDAP) (Kalina, 2008). However, as the UBV chose to retain its role as a non-partisan, purely professional farmers' representative body operating according to a trade union model, it became distanced from the partisan farmers' organisations, which also included the Freedom Party Farmers (although a certain affinity remained).

The organisation's main thrust was directed against the ubiquitous public influence of the dominant political parties, in particular the People's Party's domination of agricultural policy. The association denounced the acquiescence of the Bauernbund in the face of both the severe income disparities between farmers and the existential threat which structural change and agro-business posed to family farms. In the 1960s, the UBV staged a number of spectacular protest events which the Bauernbund conspicuously did not support (Kalina, 2008). Although less nationalistic than the Freedom Party Farmers, the UBV nevertheless advocates the roll back of the EU CAP and a liberalised world-market regime.

In recent years, the Green Party has also begun to run in chamber elections with little success, as the agendas of both groups widely overlap. For the small UBV, represented only in the chambers of the provinces of Salzburg, Upper Austria and Carinthia, this nevertheless means some competition.

Austrian Mountain Farmers' Association

As for the extra-institutional opposition, the longest standing player is the Austrian Mountain Farmers' Association (Österreichische Bergbauernvereinigung, ÖBV). Founded in 1975 by a group of young farmers, the ÖBV continues to challenge dominant agricultural policy, which, according to the ÖBV, preserves the vested rights of big farmers, bars egalitarian or redistributive policies, and has never renounced its productivist bias in spite of ecological reforms (Kalina, 2008; Rohrmoser, 2001). Its criticism of the eco-social reforms illustrates its critical edge: according to the ÖBV, the simultaneous promotion of ecological sustainability and peasant agriculture and competitiveness in liberal EU markets is a contradiction in terms. The ÖBV considers that eco-social agricultural policy is merely a very expensive measure tailored to cover up the internal contradictions of EU agricultural policy. Since additional funding still linearly correlates with holding size, the policy only benefits agricultural industry and large farms. Thus, according to the ÖBV, the national agricultural system with all its contradictions and imbalances has merely been transferred to the CAP system without major changes.

Unique in Austria and uncommon for any agricultural association, the ÖBV has strong international networks. It is a founding member of the European Farmers' Coordination (Coordination Paysanne Européenne, CPE) and the Vía Campesina, both transnational peasant and farmers' organisations who defend the right to food sovereignty and sustainable family farming, reject the privatisation of seeds and the concentration and industrialisation of agricultural production, and advocate sustainable agriculture based on solidarity rather than profit motives (Seifert, 2009).

The ÖBV works at the fringes of agricultural politics, mainly by envisioning and promoting alternative concepts for the rural world. In fact, a great number of innovations in agricultural policy were initially formulated by the ÖBV and only later taken up by either the SPÖ or ÖVP governments. Endowed with very scarce resources, the organisation's major assets are the commitment and visionary power of its few steady members and its readiness to create network links with other organisations. In the chambers, the UBV adopts the role of the ÖBV's institutional arm.

An example of ÖBV's ability to create social networks is the Alliance for an Environmentally and Socially Sustainable Peasant Agriculture (Agrarbündnis für eine bäuerliche, umwelt- und sozialverträgliche Landwirtschaft), which the ÖBV has coordinated together with the UBV since April 2001. Against the background of BSE and other food crises, the Alliance mobilised a variety of partners, such as environmental, Third World, consumer, animal welfare and church organisations, as well as the largest organic producers' association to push for a further strengthening of ecological and social reform of agricultural policy, including fair prices and an equitable distribution of public funds. The Alliance is also involved in a partnership between farmers and consumers, and in integrated rural development programmes on the greening of agriculture, aiming to avoid surpluses and protect the environment, as epitomised by the organic farming model.

6.5 Two protest movements in recent years

The fact that critical voices exist, albeit external to agricultural politics, suggests a potential for the organisation of discontent and perhaps even wider protest. In contrast to the way the hegemonic agricultural system represents itself, constant pressures to modernise and liberalise the rural sphere do not go without considerable internal tensions. How do protest movements play out in the constellation described? In the following sections, we review two major protest movements that have occurred in recent years: the movement against agricultural biotechnology and the protests against low milk prices. The two movements illustrate two diametrically opposed ways in which the hegemonic system copes with protest: inclusion and exclusion.

The protests against agricultural biotechnology

Today, Austria is one of the staunchest defenders of a GMO-free status in the EU. This policy has been portrayed as a consensual and state-supported 'NIMBY' (not-in-my-backyard) policy (Seifert, 2009). All sorts of pains are taken to prevent the cultivation of GMOs on Austrian ground in order to 'protect' Austria from the alleged dangers of GMOs, although this does not accord with EU legislation and therefore requires complex political, legal and scientific strategies. Public and political support for this stance is ubiquitous and virtually unanimous. GM resentment is backed by the tabloid *Kronen Zeitung*. Due to its high reach and incontestable reputation as a 'voice of the people', this tabloid has a crucial influence on Austrian politics. Furthermore, all major social groupings in parliamentary opposition concur with, or at least acquiesce to, the denouncing of GMOs. There are, at best, minor divisions between political parties on the issue. In other words, any government responsible for a GMO release 'in-our-backyard' would appear to act against the public will (Seifert, 2009).

This state of affairs has evolved over time, and it is informative to look into this process as it provides insights into the workings of the hegemonic system when confronted with pressures for change. Ultimately, the Austrian GMO-free policy was the result of a massive social debate, which occurred in the second half of the 1990s. Triggered by coincidental events, broad anti-GMO mobilisation resulted in a highly successful popular initiative in 1997, demanding the ban of GMOs and GM food from Austria. The major actors who mobilised the public were international environmental NGOs – Greenpeace and Global 2000 (the Austrian chapter of Friends of the Earth).

Farmers, by contrast, kept a low profile and rarely staged public protests. As a result, the controversy over agro-food biotechnology did not appear as a rural protest movement. Nevertheless, agriculture was a major issue in the critical debate, as biotechnology was seen as incompatible with organic farming, and one political organisation dealing with the agricultural agenda, the ÖBV, was crucially involved in the campaign. The ÖBV was a member of the organising board of the public initiative; in fact, it was the first actor to put forward the idea of a plebiscite, at a time when the suggestion was of no interest to other critical groups. On its own, the ÖBV lacked the public standing required to run such a campaign effectively. Only later, when the yellow press became involved and created the opportunity, did the other organisations step in to mobilise the public.

Inclusion by the hegemonic system

The masters of the hegemonic system – the People's Party, the Bauernbund and the Chambers of Agriculture – by contrast, needed several years to arrive at a workable position regarding GMOs, as they kept struggling to balance the interests of organic and conventional farmers with the interests of preserving the option to resort to GMOs. It took until 2000 and the appointment of Josef Pröll (ÖVP) as Minister of Agriculture to formulate a common GMO-free position. Since then, the Austrian Ministry of Agriculture has renounced agro-biotechnology in general and, thus, has become the domain where the country's GMO-free policy is played out. The role of organic farming became key in this policy. As organic farmers are obliged by their statutes not to resort to agro-biotechnology, they are the main beneficiaries of a national GMO-free policy. The Austrian co-existence policy most clearly demonstrates the prominence of organic farming. In 2005, the EU Commission began to develop a policy to ensure the workable co-existence of both GMO and non-GMO production in Europe, leaving it up to member states to elaborate detailed regulations. Austria used this initiative to pre-empt GMO cultivation: exceedingly high separation distances were set to prevent the intermingling of crops, and state sponsored 'GMO-free' campaigns were run to create a prohibitive environment for agro-biotechnology (Seifert, 2006).

The reason why this policy ultimately could be integrated into the hegemonic system is that it could fit into the environmental policy trajectory that took off in the late 1980s. After Austria's accession to the EU, organic farming became the 'sacred cow' of Austrian agricultural policy, and therefore anything which might corrupt the sector tended to provoke aversive reactions from policymakers. When, in the wake of the European anti-GMO movement in the late 1990s, 'GMO contaminated' food became the bogey for European food markets, organic products guaranteed not to contain GMOs came to enjoy a competitive advantage over conventional products, which elevated the status of organic farming even more. From this it was a small step for Austrian agricultural policymakers to sponsor a 'GMO-free' policy for the entire farming sector, arguing that all Austrian producers might benefit from its rigorous policy, which gives Austrian products a competitive edge irrespective of whether they are organic or conventional. Thus, while in the first instance the traditional leading groups in the agricultural domain had considerable difficulties with the anti-GMO movement, they proved themselves able to learn and adapt.

The milk price protests by IG Milch

In contrast to the movement against agro-biotechnology, the Austrian milk protests were a genuine farmers' protest movement. Such farmers' protests are extremely rare in the shepherd idyll of Austrian agricultural politics, as they constitute a revolt against this very idyll. Significantly, the movement was not only directed against industry, government and the EU's supranational regime, but also against the hegemonic system controlling domestic agricultural policy.

The main proponent of the milk protests was IG Milch (http://www.ig-milch.at/). Its central concern was to achieve a fair price (of 40 euro cents per kg) for the structurally disadvantaged milk producers. The acronym IG has a double meaning: it can be read as 'Interessengemeinschaft' in the general sense of 'shared interest' or 'community of interest'. However, the name is also a play on established trade union names, such as IG Metall, thus emphasising the contrast to the prevailing form of organisation in the agricultural domain, the federation (Bund). The IG was founded in 2004, in the wake of spontaneous protests in the province of Upper Austria against the ever-decreasing milk price (Kalina, 2008). Upper Austria was not a random starting point: there is a very high proportion of grassland and dairy farmers in Upper Austria and neighbouring Salzburg. There was also a clear affinity and overlap between the IG and UBV, who had also made the quest for a fair milk price one of its major campaign issues. In 2011, the IG had about 4,000 members, with 43,000 Austrian dairy farmers in total (Gmeiner, 2011).

IG Milch associated with European Milk Board

The syndicate-like association of IG continues to fight for a better milk price: a price that covers the cost of average milk production. IG has protested in multiple ways: organising and engaging in media-effective protest action; using direct political lobbying at the domestic as well as EU levels; creating alliances with supportive political and industrial actors; engaging in European networking and coordination through the European Milk Board (EMB); and by creating independent product labels and marketing concepts (Halbmayr *et al.*, 2006).

A review of the IG's eventful history illustrates the various actions undertaken as well as the various contextual constraints and opportunities, and its achievements and drawbacks. The IG grew very rapidly after being founded because of the discontent of dairy and grassland farmers. Milk prices had steadily decreased for the past ten years and so had the number of dairy farms. Since Austria's accession to the EU, for example, 3,000 dairy farms have gone out of business each year (Halbmayr *et al.*, 2006). A great number of farmers felt exposed to an existential threat, as revenues dropped beneath a sustainable threshold. The IG's claim to lift the base price to a level covering production costs quickly found followers. Creameries engaging in dumping price competition were identified as primarily responsible for the bad situation. Dairies and retailers offering milk products at dumping prices became targets of various spectacular protest actions with the result that overly cheap prices were soon corrected. Representatives of the agricultural establishment and local People's Party politicians, however, immediately denounced the 'rebellious' protests, which in a way also questioned their authority (Kalina, 2008). From the IG's point of view, political decisions were desperately needed. Milk prices needed to be regulated and overproduction by some highly efficient producers needed to be banned. For representatives of the hegemonic system, the issue was not to be mingled with politics, as milk prices were considered to be determined by the market. In any case, the IG's call for 40 euro cents was ridiculed as being utterly unrealistic.

As the milk crisis was felt across Western Europe, and its causes were considered by IG to be rooted in EU dairy policy (currently, milk supply is steadily increasing as milk quotas will be abolished in 2015), the IG aligned with its counterparts in eight EU member states (Italy, Belgium, Denmark, France, the Netherlands, Sweden, Spain and Germany). Together they established the EMB in 2005. They agreed that the EMB should primarily work towards lifting the milk price to 40 euro cents per litre (a goal still held today). To apply pressure on politicians and industry, preparations began for a European milk strike.

The Austrian IG thus became part of a trans-European network and transnational movement. In fact, the IG arguably emerged as part of a transnational dynamic, as the organisation's creation in 2004 had drawn on the model of the Federal Union of German Dairy Farmers (Bundesverband Deutscher Milchviehhalter, BDM). This organisation, founded in 1998, unites about one third of all German dairy farmers and is considerably larger than the Austrian IG. Particularly, the milk strike actions in 2008 and 2009 can be seen as part of a general European campaign. The same can be said of those concepts advocated by IG that pertain to the CAP as an alternative to the fading out of the milk quota system. The EMB, for example, advocates the introduction of a flexible market control by means of an agency that monitors production costs and prices.

IG Milch's successful marketing activities

In 2006, another important achievement of the IG followed at the domestic level as it succeeded in establishing its own brand of milk, 'A faire Milch' (Fair Milk), sold by some large retail chains. The extra cost of 10 euro cents per litre of milk would be transferred to a trust which would proportionately redistribute the revenues among suppliers. The membership fee for the trust was set at 50 euros. 'Faironika', a plastic dairy cow painted in the likeness of the Austrian flag, was created in order to serve as the promotional symbol for Fair Milk. In the following year, the Fair Milk campaign was honoured with a national prize for best marketing. Later, the EMB adopted Faironika as a marketing and political tool, in which member state organisations used national names for Faironika (e.g. 'Justine' in France) and painted the cow in national colours.

IG established the direct marketing system 'Free Milk Austria' (Freie Milch Austria) in 2009. The setting up of a system for collecting and marketing milk independently of dairies (accused of reckless price competition) is thought to generate higher income for dairy farmers and shatter the monopoly of dairy cooperatives. In the past three years, the company has grown rapidly, producing 740,000 tons of milk annually that is collected from 580 farmers (mostly IG members) and marketed mainly to Germany and Italy. This amount is comparable to a small local dairy in Austria. Nevertheless, competition with dairies is tight and prices currently paid to farmers are not above, but are in fact below, the average.

In 2008 and 2009, IG was involved in major protests. The strike in May and June 2008 took place in the context of the European milk strike organised and coordinated by the EMB. The boycott in September 2009 was triggered by the earlier collapse of the European milk market. In both cases, IG called for a nation-wide strike and protest action in response to both European price decline and European mobilisation and coordination, albeit less successfully than its counterpart movements in Germany or France.

Exclusion of IG Milch

The IG is in a constant struggle with the hegemonic system governed by the People's Party. Representatives of the latter seek to impede the advance of the milk syndicate in various ways. Provincial People's Party politicians, for example, have prohibited IG billboards from being attached to the walls of public buildings. Local governments have banned Faironikas from roadsides for insubstantial reasons. Office holders in the Bauernbund or provincial governments have publicly denounced the syndicate's actions. Mutual recriminations have ended up in lawsuits. Officials and functionaries typically ignore the IG's invitations to hold talks. At the 'milk summit' talks in Vienna during the milk strike in the summer of 2008, which was held in response to the strike organised by IG, IG itself was not allowed to participate, while retailers, dairies and the Chamber of Agriculture, all of whom had censured the strike (and arguably had their part in creating the situation leading to it), sat around the table. At the peak of the milk crisis in 2009, IG Milch managed to have talks with Federal Chancellor Werner

Faymann (SPÖ). However, the Minister of Agriculture, Nikolaus Berlankovic (ÖVP) called off a similar meeting because of prior commitments. To date, direct talks have been steadfastly avoided.

The People's Party-governed agricultural establishment is a major target of criticism by the 'milk rebels'. Milk protests involve protests against the chambers, the Bauernbund and national agricultural policy, all accused of complicity with agro-industry, processors, retailers and EU market ideology. The criticism clearly strikes a cord in traditional Austrian farmers, many of whom feel that their vital interests are threatened. They did not appreciate the stance taken by the chambers and the Bauernbund at the height of the milk crisis in 2009. Not only did the latter oppose strike measures and censor confrontational actions, particularly when directed against policymakers, but they also sought to appease their constituency, arguing that the Austrian price level was still well above the German average and, in the usual way, chose to make a gentlemen's agreement with industry and retailers. Farmers' discontent with their long-established representatives have made some minor inroads during chamber elections, to the detriment of the Bauernbund, but are yet to mount a serious challenge. Thus, at the recent chamber elections in 2009, the Bauernbund still won 79% of the votes and 206 of the seats, while the UBV (only in Upper Austria and Salzburg) obtained a little over 3% of the votes and 9 seats. The respective numbers for the SPÖ Farmers were 30 seats (9.4%); the Freedom Party Farmers, 11 seats (6.4%); and other groupings, 4 seats (1.8%) (Landwirtschaftskammer Österreich, 2010).

Additional cases illustrate the competitive relationship between the 'milk rebels' and the established system. The Bauernbund, for example, had plans to introduce an 'ethical label' for milk products, as the Austrian market had signalled some demand for this, but the plans were dropped when 'A faire Milch' successfully occupied the 'ethical' niche market. Another illustration is the harsh reaction of the Bauernbund to the initiative of one of its office holders to found an 'IG Fleisch' (Interest Group Meat) inspired by IG Milch: the office holder was immediately fired.

Meanwhile, IG is not completely alone in its struggle against the People's Party hegemony. At least morally it can rely on the sympathy of all rival parties (including the Freedom Party Farmers), while in chamber elections the non-partisan UBV constitutes its institutional arm. Another ally is the ÖBV and its Alliance for an Environmentally and Socially Sustainable Peasant Agriculture, which, together with IG, has staged protests against current agricultural policy. None of these allies, however, has the clout to seriously threaten the current system.

Downturn of IG Milch

Ultimately, the instability of the membership base of any 'rebellious' organisation and the realities of the agricultural market might be more of an obstacle than the strong political and institutional headwind constantly encountered by the IG. Recent years, for example, have proved particularly difficult for IG because of internal rivalries and legal disputes over budgetary irregularities. Coupled with a lack of economic success, these mishaps have brought about a dramatic decline in membership, from 4,000 to 2,000.

6.6 Conclusion: the pastoral conserved

The milk protests constitute a rare example of a genuine farmers' protest movement that stands directly opposed to the hegemonic system of Austrian agricultural policy. For other examples we must go back to the UBV protests in the 1960s and the time of the one-party government of the People's Party. However, the milk protests caused little more than a tiny scratch on the political surface, even though they made a lot of noise. In contrast, the case of the anti-GMO movement, agricultural elites soon dropped their initial reservations against the quest for a GMO-free Austria. In the case of the milk protests, the hegemonic system did not address the problem by learning and incorporating an external policy concept into already existing adaptive strategies but by excluding the contender.

Several reasons account for these different responses. The Austrian anti-GMO movement was a movement among the people, not just farmers, although a critical farmers' organisation was centrally involved and the protection of organic farmers' interests was key to legitimising a nation-wide GMO-free policy. The overwhelming popular support for this policy, the relatively low cost to small-scale Austrian agriculture of renouncing agro-biotechnology, and its logical integration into an ecologically sound policy narrative adopted in the course of the country's integration into the EU, lent itself to the adoption of such a policy in the long term. The milk protests, by contrast, grew out of the disgruntled peasantry whose demand for cost-effective milk prices went along with a bitter reproof of the monolithic Austrian agricultural establishment, involved in balancing the interests of its various clientele. Thus, the milk protests were an attack on the People's Party's hegemony over agriculture. Fighting back by excluding the 'peasant splitter' was therefore the logical choice. Because self-organised producers are considered to be rivals to the hegemonic system, the latter cannot emulate their organisational form, at least for the time being. The only option is to outcompete their rivals in the economic field by making use of enormous advantages in terms of organisation, money and political power.

At the height of the 2009 crisis, this primarily meant playing the European card. The Minister of Agriculture, Berlakovic (ÖVP), was active in rallying his European counterparts to put pressure on the European Commission. This was instrumental in moving forward the release of direct payments, which served as one immediate rescue measure among others in the face of tumbling milk prices. When the milk price rose again in late 2009, it was presented as the result of these efforts. The pressure on agricultural policy diminished due to higher market prices in 2010 and 2011. Conversely, the marketing system built up by the 'rebels' could not realise the higher farmer incomes that had been hoped for, and IG membership tailed off. Today, the IG has around 2,000 members, less than 5% of all Austrian dairy farmers. The larger part of Austrian farmers prefer to stick to their regular processors and distributors.

People's Party hegemony likely to continue

What are the hegemonic system's chances of remaining intact in the future? Based on the discussion in this chapter, we judge it to have the brightest prospects of survival. The dominance of the People's Party is still overwhelming, even after the recent unrest. Its tangled power base remains unbroken. The Bauernbund remains dominant in the chambers, thus ensuring access to and coordination of subsidies. The Federal and provincial administrations continue to be a strong base, together with the Ministry of Agriculture and the Raiffeisen bank system. For the time being, no serious contender is in sight: no other party organisation in particular has shown any potential to challenge the hegemony. Nevertheless, there is potential for protest, even for protest votes in the chambers. The UBV has particularly benefited from the milk crises and won a couple of seats in recent chamber elections while pragmatically coalescing with local Green or Freedom Party candidates. Green Party groups struggle to gain a foothold in the countryside, while SPÖ Farmers and Freedom Party Farmers stagnate. Protest can either be addressed by incorporating criticism into 'ecologically and socially sound' and high-end niche-market policies, or rivals can be kept at bay simply by disregarding them. Nevertheless, precisely because of its dominance, it is remarkable how sensitive the system is to rival currents. This testifies to its readiness to learn and adapt to challenges emerging from within.

These challenges are likely to recur in the future. The conflicts dealt with here are symptomatic of the enormous structural pressures (in our examples, technological productivity enhancement and market liberalisation) on Austria's still largely family-farm-based agriculture, which owes its survival to a costly system of subsidies. Today, market liberalisation and structural change appear as agriculture's inescapable fate in the long term. The double-layered protective shield consisting of the CAP and Austria's domestic policy is arguably based on this premise and appears designed to slow down and moderate this painful transformation rather than to reverse it. As the forces unleashed by world-market integration increase, it is likely that new protest movements that challenge the dominant system will emerge. This adds to the observation that both protest movements studied here appeared within wider transnational movements and networks. Given the transnational nature of emerging problems and grievances in the agriculture sector, it is plausible to predict the rise of further transnational protest movements in the rural sphere that are also likely to impact on Austria.

The ability of the hegemonic system to assert itself against new contenders, either by incorporating them or fighting them off, ultimately depends on how it performs its role as the Austrian farmers' 'good shepherd'. We might suppose that individual farmers do not base their choices on the restricted rationality of the client-patron relationships prevalent in this system alone, but above all on economic parameters. In other words, does the reigning system provide enough planning reliability and income for them? As long as the great majority of farmers answer this question in the affirmative the Austrian pastoral sector will be preserved.

References

Gmeiner, H., 2011. Mühsamer Kampf um faire Milchpreise. Salzburger Nachrichten November 28, 2011.

Groier, M., 2006. Tabuthema Betriebsaufgabe. Eine Darstellung aus Sicht der Betroffenen. In: AgrarBündnis e.V. (ed.) Der kritische Agrarbericht 2006. AbL Bauernblatt Verlags-GmbH, Hamm, Germany, pp. 63-68.

Groier, M., 2011. Wohin fließen die ÖPUL-Gelder? Ländlicher Raum Online-Fachzeitschrift des Bundesministeriums für Land- und Forstwirtschaft, Umwelt und Wasserwirtschaft 11-1. Available at: http://tinyurl.com/pwod73e.

Halbmayr, E., Grünzweil, E. and Bauer, J., 2006. 40 Cent pro Liter Milch – mindestens! Zur Situation der Österreichischen Milchwirtschaft und den Forderungen des European Milk Board. In: AgrarBündnis e.V. (ed.) Der kritische Agrarbericht 2006. AbL Bauernblatt Verlags-GmbH, Hamm, Germany, pp. 134-138.

Hanisch, E., 2002. Die Politik der Landwirtschaft. In: Bruckmüller, E., Hanisch, E., Sandgruber, R. and Weigl, N. (eds.) Geschichte der Österreichischen Land- und Forstwirtschaft im 20. Jahrhundert. Politik, Gesellschaft und Wirtschaft. Verlag Ueberreuter, Vienna, Austria, pp. 15-187.

Hofreither, M.F., 1997. Agrarpolitik. In: Dachs, H., Gerlich, P., Gottweis, H., Horner, F., Lauber, V. and Müller, W.C. (eds.) Handbuch des politischen Systems Österreichs. Die Zweite Republik. Manz, Vienna, Austria, pp. 641-651.

Hovorka, G. and Hoppichler, J., 2006. Agrarpolitik. In: Dachs, H., Gerlich, P., Gottweis, H., Horner, F., Kramer, P., Lauber, V., Müller, W.C. and Talós, E. (eds.) Politik in Österreich. Das Handbuch. Manz, Vienna, Austria, pp. 701-710.

Kalina, W.F., 2008. Auf Bauer – wehr dich! Der Allgemeine Österreichische Bauernverband – die Agrarrebellen der Zweiten Republik. Leopold Stocker Verlag, Graz, Austria, 522 pp.

Krammer, J. and Rohrmoser, F., 2012. Im Kampf um ihre Rechte: Geschichte der Bauern und Bäuerinnen in Österreich. Promedia, Vienna, Austria.

Landwirtschaftskammer Österreich, 2010. Agrarpolitik 2009/10. Leistungen und Forderungen der Bauernvertretung. Available at: http://tinyurl.com/laaawob.

Rohrmoser, F., 2001. Erschließung neuer Formen der Konfliktbearbeitung in bäuerlichen Strukturfragen. SWS Rundschau 41: 497-515.

Seifert, F., 2006. Regional GM opposition as multilevel challenge? The case of Upper Austria. Tailoring Biotechnologies 2: 11-36.

Seifert, F., 2009. Consensual NIMBYs, contentious NIABYs: Explaining contrasting forms of farmers' GMO opposition in Austria and France. Sociologia Ruralis 49: 20-40.

Weiss, H., 2010. Schwarzbuch Landwirtschaft. Die Machenschaften der Agrarpolitik. Deuticke-Verlag, Wien, Austria, 176 pp.

7 Farm protest and militancy in Australia: supporting or under-mining interest-group politics?

D.R. Halpin

Research School of Social Sciences, Australian National University, Canberra, Australia; darren.halpin@anu.edu.au

Abstract

This chapter approaches the question of rural protest in Australia. It does so by charting the professionalisation of farmer interest groups and the implications this had on political participation by farmers. The value of such a perspective is that it makes explicit the interconnection between 'outside' strategies of organised political advocacy and more acceptable and institutionalised forms. The chapter offers a historical account of the ways in which protest – and other forms of collective political response – have been used by Australian farmers.

7.1 Introduction

The political representation of primary producers in Australia is currently achieved through a series of State Farmer Organisations (SFOs) and Commodity Councils (CCs) under the umbrella of the National Farmers' Federation (NFF). This 'family' of interest groups is the recognised voice of Australian primary producers, although membership for farmers is not compulsory. While a relatively recent addition to the political landscape, the NFF family structure has its roots in more than a century of political organisation. In the quest for political representation, primary producers have moved from local voluntary associations, parliamentary parties (specifically, the National Party), state and commodity-based sectoral interest groups to, finally, a federated national interest group. This change seems to have come about because of changes in the socioeconomic conditions of primary producers and the political conditions enforced by the prevailing political-administrative system in bringing about transitions in representative political formations. However, change has not been seamless. This chapter focuses principally on the period where professional groups sought to establish themselves and thence to assert dominance over farmers' political representation.

DOI 10.3920/978-90-8686-807-0_7, © Wageningen Academic Publishers 2015

Between the 1960s and mid 1980s, militancy and farm protest were common, serving both to threaten and strengthen the emerging NFF structures. This chapter addresses the place of protest in the context of the organisation and representation of farming interests through interest groups. While there is a tendency to see protest as an anathema to institutionalised politics, research has established that for the most part these outsider or indirect strategies often form a part of the repertoire of most interest groups or social movement organisations. For the representation of farmers' interests in Australia, militant protest was an important contributor to the emergence of a powerful and professionalised interest group, but at the same time such forms of protest eventually became a threat to the pursuit of a highly professionalised form of institutionalised representation with a (neoliberal) reforming government.

Complex and detailed histories have been written on the development of primary-producer political formations (Connors, 1996; Graham, 1966; Harman, 1968). The task of documenting these histories is the sole focus of many books and doctoral theses. To attempt even a credible summary of these works would require a book dedicated solely to this task (see, e.g. Connors, 1996). Therefore, this section will provide a 'potted' history of the political activity of primary producers in the lead up to the formation of the longstanding peak structure, the NFF, with particular emphasis on New South Wales (NSW). While the pattern of development of primary-producer political strategies – from local group to pressure groups, to parliamentary party – were consistent throughout Australia, there were qualitative differences between states (Graham, 1966). Here, we will plot the changes or transitions in formations used by primary producers to represent their interests.

7.2 The amalgamation of farmer organisations in New South Wales

The economic hardship experienced by Australian agricultural producers in the early 1960s was reflected in the financial position of primary-producer political organisations. Given that organisations would often duplicate work, despite many of them maintaining offices in close geographical proximity to one another, amalgamation was advocated as one way to cut costs and ensure organisational effectiveness. Financial pressure was reinforced by political pressure. The Federal Minister for Agriculture, McEwen, had insisted that farmers speak with one voice in policy representations (Campbell, 1980). A lack of coherence and inconsistency among farm groups, it was argued, would leave government with no other choice but to act on their departmental advice or the advice of other more professional and cohesive sectors.

The word most often used to characterise the joining together of various farm organisations is 'unity'. However, 'amalgamation' is used here for the reason that it was purely an organisational measure motivated by financial considerations, rather than a measure aimed at communicatively deriving a more homogenous set of interests from which to determine the proper direction for agricultural policy.

With the financial and political cost of parallel organisations increasingly apparent, primary-producer organisations commenced the task of amalgamation in the late 1960s, a project largely completed by the end of the subsequent decade. Given that federal organisations were beholden to their state organisations, it is with state-based organisations that analyses of amalgamation should start. It is relatively clear that the precedent for organisational amalgamation came from New South Wales (the most populous state of Australia), so it is on developments in this state that most emphasis is placed.

Amalgamation in New South Wales

In 1962, the Wheat and Woolgrowers' Association[1] and the Farmers and Settlers' Association (FSA) merged to form the United Farmers and Woolgrowers' Association (UFWA) (Harman, 1968). This merger left the UFWA and the Graziers' Association as the two major primary-producer organisations in NSW. Despite considerable external political pressure to forge a united position, fundamental structural changes in the agricultural sector made unity increasingly harder to achieve between farmers and graziers.

By the 1960s many of the Graziers' Association members were producing wheat and wool: over 50% of properties with sheep were producing wheat, while nearly all wheat farmers had some sheep or cattle (Harman, 1968). Harman claims that the emergence, and subsequent dominance, of mixed farming resulted in the two bodies having their own policies on the same industries, bringing the Graziers' Association and the UFWA into conflict. Wool marketing was a particularly contentious issue. The Graziers' Association resisted the implementation of a wool-stabilisation scheme, similar to that operating in wheat, while the UFWA was generally supportive. While able to thwart a 'yes' vote in two referendums on wool assistance, the Graziers' Association ultimately acceded and the Reserve Price Scheme was instituted.

It is worth noting that the Graziers' Association's determination to disprove their image among the public and politicians as representing the rich rural elite may have eroded its capacity to resist government assistance for woolgrowers. The Graziers' Association actively promoted the cause of smaller woolgrowers in its literature and recruited them in such numbers that they constituted the bulk of its membership (Harman, 1968). It seems ironic that the tactics employed to recruit members and enhance its political

[1] This was a breakaway group of NSW wheat farmers who split from the FSA in 1934.

palatability may ultimately have reduced its ability to influence policy in the direction its core membership originally intended.

Concern was often expressed that without amalgamation politicians would play off one section of an industry against the other. However, some questioned the conventional wisdom regarding amalgamation, prompting one commentator to ask (Anderson, 1972: p. 122): 'Do farmers, when united, really 'speak with one voice' or merely 'listen with one ear'?' In his presidential address to the 1966 NSW Graziers' Association Annual Conference, Bruce Wright warned that unity would benefit politicians more than primary producers. He said (*op. cit.* Connors, 1996: p. 121):

> 'In my view we should not be concerned about the convenience of government or political parties ... to force into a common organisation interests which are not ideologically homogenous must – and does – lead to the suppression of ideas, to intolerable delays in evolving policy and the emergence of watered down compromise lacking in the vigour of conviction. I fear the regimented suppression of opinion far more than I do the consequences of government being confronted with conflicting policies.'

These sentiments were more than politically convenient rhetoric. Differences between farmers and graziers in terms of production systems also translated into significant differences between the internal operation of both the Graziers' Association and the UFWA and their conceptions of democracy (Connors, 1996). The Graziers' Association leaders were selected by a small group of senior members, while the UFWA left the Annual Conference delegates to elect office bearers. The UFWA had a reputation for being antagonistic towards bureaucrats and academics, while the Graziers' Association's members were largely well educated and employed university-trained staff to act on their behalf. Connors (1996) notes that the Graziers' Association felt comfortable with its staff publishing reports and press releases under their name, while the UFWA released information drafted by staff from the President. In terms of influencing government, the Graziers' Association preferred reasoned argument and written submission, while the UFWA preferred rallies and marches. In 1967, the Graziers' Association set up a Unity Study Committee to canvass the idea of uniting with the UFWA. This committee found sufficient differences to recommend against unification.

While structural, philosophical, political and social differences between farmers and graziers were not resolved, the effects of drought, low commodity prices and rising farm costs were felt by all primary producers. This adversity fostered a climate conducive to amalgamation. The task of returning prosperity to agricultural Australia became a rallying call for the movement towards amalgamation. A further rallying point was the live sheep export issue (Connors, 1996). In response to the union movement's refusal to load sheep at ports unless a proportion of them were slaughtered by union labour in Australia, primary producers moved sheep to a rural port and loaded ships themselves. Given that the last time primary producers opposed unions was when the pastoralists formed the Pastoralists' Federal Council in 1890 to oppose the Amalgamated Shearers Union, the symbolism in terms of amalgamation was immense.

The adverse economic conditions also created a climate of militancy, and initially producers acted independently of their organisations. In NSW, the Rural Action Movement (RAM), which demanded higher prices and marketing reforms, was established. Despite the UFWA's early opposition to the RAM, only two years later it was party to a joint petition to NSW Parliament for the setting up of a sheep meats marketing board (Connors, 1996). While the RAM reportedly faded from view,[2] it was indicative of the turbulent times. According to Campbell (1971), three different types of organisations emerged in opposition to established primary-producer organisations. Groups proposed the formation of cooperatives to reduce the cost of agricultural inputs, other organisations had the aim of lobbying for a particular marketing reform, while others were opposed to the established positions of farm organisations. All these groups placed established farm organisations under increased pressure to act.

Rural downturn in the 1970s placed significant financial pressure on the organisations themselves, prompting the Graziers' Association and the UFWA to make joint submissions and representations to government on several occasions (Richmond, 1980). The widespread economic hardship amongst primary producers, coupled with the lack of funds for both representative organisations, was enough to catalyse a serious move towards amalgamation. Of course, the removal of key policy disagreements also assisted. Connors (1996) suggests that the resolution of the wool marketing issue in the early 1970s paved the way for amalgamation. The establishment of a reserve price scheme ended years of division between the pro-protectionist small woolgrowers represented by the UFWA, and the free-trading graziers represented by the Graziers' Association.

In September 1975, an amalgamation committee was formed between the Riverina Graziers' Association (a regional branch of the NSW Graziers' Association), the NSW Graziers' Association and the UFWA, to consider plans for a merger (Graham, 1978). The report released by the committee listed the positives as reduced costs, ability to resolve conflict within the organisation and the reduction of farmer and grazier divisions. The negatives were that the reduction in competition between two bodies could leave some important views off the agenda and, hence, facilitate the emergence of splinter groups. The tone of the report gave a clear indication that the advantages outweighed the disadvantages. The plans were put to a referendum of members of the organisations in July 1977, which returned an overwhelming 'yes' vote for amalgamation.

[2] The RAM seems to have re-appeared in the 1990s (Meeting notice in *The Land*, 1997) in NSW along with other militant groups. However, these all receded into the background after initial signs of popularity.

According to Trebeck (1990), the response to the referendum was 93% in favour (for the referendum to be valid, over 50% of the financial membership had to vote). In 1978, the Livestock and Grain Producers' Association (LGPA) was established. The title of the new organisation illustrates the level of compromise necessary to bring together the two organisations; neither farmer nor grazier mentioned.

The mood at the time of amalgamation can be captured by a panel discussion regarding the future of farm organisations and militancy printed in *The Land* newspaper during 1977. One said that the bickering between farm organisations 'probably played right into the politicians hands. I don't think this should happen. They should go with one voice to debate these issues' (Hardy, 1977: p. 14). A fellow-panellist said, regarding professionalism (Hardy, 1977: p. 14): 'The leaders we have at present are more or less on an honorary basis, aren't they. This is the point we've got to look at really. We've got to have a man who's getting a pretty good salary, probably in the $40,000-$50,000 bracket. You know one who is going to be there all the time'. The appointment of professional salaried staff was perceived by some as entailing a loss of control by hands-on farmers of their organisations (Hood, 1977: p. 14): 'I think you still have got to have a fellow that's got to go back to the farm and get his hands dirty and sow his crop'.

It is clear that amalgamation had not, in itself, resolved diverging interests, ideologies or identities. Rather, financial necessity from an organisational perspective and political pressure from an external perspective had set aside rather than resolved the heterogeneous interests of the farming constituency.

7.3 Amalgamation of farmer organisations in the rest of Australia

The pattern of amalgamation was replicated in most states, with the exception of Western Australia and Queensland, where regional differences in agricultural production systems made it more difficult (Trebeck, 1990). At federal level the amalgamation process proceeded parallel to that which occurred in NSW. A working group was established in 1977 between the Australian Woolgrowers' and Graziers' Council (AWGC), the Australian Wool and Meat Producers' Federation (AWMPF) and the Australian National Cattleman's Council (ANCC). Following preliminary reports in 1978, the group was widened to include the Australian Farmers' Federation (AFF),[3] the Australian Wheatgrowers' Federation (AWF) and the Cattleman's Union of Australia (CUA).

[3] In 1969 the Australian Primary Producers Union (APPU) amalgamated with the National Farmers' Union (NFU) to form the Australian Farmers' Federation (AFF) (Campbell, 1980; Connors, 1996).

The formation of the NFF, a federation of commodity councils and state organisations, was originally set down for 1 January 1979. However, its commencement was delayed by infighting regarding the level of voting entitlements on commodity councils. State organisations such as the NSW Livestock and Grain Producers' Association (NSWLGPA) threatened to withdraw financial support for the new organisation if it was not sorted out promptly (The Land, 1979). This delayed commencement until 20 July 1979.

The NFF structure reflects the pattern of agricultural policy development in Australia. For the first half of the century, the states held most power in relation to agriculture. State primary-producer organisations, often representing single commodities, emerged as subscription-based organisations to influence state governments. However, post-Second World War, the Commonwealth played an increasingly important role in controlling prices, regulating the supply of commodities and funding research and development in order to increase post-war production and, hence, increase export earnings. State organisations, still commodity-based, then formed commodity federations as conduits for channelling their demands regarding commodity marketing to the Commonwealth Government. While state organisations had amalgamated into general farm organisations, federal bodies remained commodity-based. Federal primary-producer organisations, apart from the APPU, were always federations of existing state and commodity groups rather than a discrete subscription-based organisation. The NFF was bound to a structure that reflected this history.

The formation of the NFF family in 1979, as the peak 'encompassing group' for farmers[4] in Australia, is the exception rather than the rule for 'employer associations' in this country. That the NFF family defies broader organisational trends in business representation suggests that farmers are not a typical case.

7.4 Cementing a professional interest-group system: containing militants and embracing protest

By the end of the 1970s there was an emerging acceptance by farmers of the need to adopt more conciliatory tones with governments of both persuasions rather than merely aligning themselves with the Country/National Party. A letter to *The Land*, entitled, 'Forget the Parties', expresses the sentiment of this period in rural politics (Vallance, 1977: p. 6):

[4] From this point on, the term 'farmers' will be used in preference to 'primary producers'. The establishment of the NFF illustrates that the farmer versus grazier distinction has less contemporary significance.

> It is patently clear that the only course open to electors in rural Australia is to forget the parties, as all they have achieved in the past two decades is the further destruction of rural Australia.

Farm organisations, such as the NSWLGPA, put this resolution into effect by banning donations to any political party, ending the practice of distributing National Party electoral material through its branches, and passing motions declaring the organisation non-political. In the context of farmers' disillusionment with parliamentary means of influencing policy, it was significant that the NFF had its initial conflict with a coalition government, led by Fraser, and of further significance that its most heated exchanges were with the National Party's leader Doug Anthony.

The farm militancy that preceded the amalgamation continued under the new NFF family. While the amalgamation resolved many financial issues from an organisational perspective, it had not addressed the frustration expressed by many farmers who were unable to find space for the expression of their points of view. Larger farmers were frustrated by their inability to change traditional farm policy. For example, in 1985 a group of grain producers in western NSW formed Graingrowers of Australia Ltd to oppose the policies of the NSWLGPA. They were agitating for deregulation of the wheat industry. They originated in the town of North Star, in northwest NSW. They were a single issue organisation and did not seek to take over the existing farm organisation (Campbell, 1985; Graham, 1985). Many small farmers, disillusioned with the direction of the NFF family – specifically its support of tax measures which favoured the use of farms as tax shelters (Gerritsen, 1987b) – formed their own organisations, both state and federal. These include the Canowindra Reform Group and the Women's Rural Action Group. In combination with other groups, these two formed the Union of Australian Farmers, which directly challenged the NFF. Its agenda was largely to criticise the NFF for pursuing macro-economic reform, the results of which were taking too long to materialise. They pointed out that while the lobbying occurred, farmers were being forced off the land.

That groups were attacking the NFF family for being both a supporter of traditional price-support policies associated with the wishes of small farmers and a supporter of big farmers who were surviving the farm crisis, illustrates the extent of the equivocation over the direction of farm policy. For example, the Canowindra Reform Group asserted that the NSWLGPA supported larger farmers' interests, while the Graingrowers of Australia asserted the opposite. Most of these militant groups failed within two years, largely due to the lack of financial resources.

Recognising the heightened level of farmer militancy and the increasing disillusionment among farm organisation members, the NFF family responded. Rather than address the militancy internally, it focused farmers' energies on the policies of the newly elected

Hawke Labor Government. This culminated in the 1985 Canberra farm rally. The strong crowd of farmers marching on Federal Parliament, estimated at 50,000, established the credibility of the NFF family with its constituency. A 'fighting fund' was established on the day of the rally, which raised somewhere between Australian $10 million (Trebeck, 1990) and Australian $14 million (Connors, 1996). Whatever the exact figure, the farming community, supported by the broader business community, had established a significant fund from which its campaigns could be financed. The NFF's use of direct action as a political tactic had its origins in conflict with the union movement over the live sheep export issue, the Mudgenberi Abattoir dispute and the Wide Comb dispute.

The rally was also significant in that it further eroded the relationship between the NFF and the National Party. The leader of the federal National Party, Ian Sinclair, was denied permission to speak to the rally. This action effectively, as well as literally, removed the National Party from the role of agenda setting for rural Australia (Connors, 1996). In turn, the NFF established a reputation with the new Labor Government as being 'ready for business'. Credibility with the farming constituency flowed from the rally, with the NFF touted as a body willing and able to act to change what the Nationals could not. The NFF held the hopes of an angry and militant rural constituency.

Some analysts have suggested that the farm militancy of the mid-1980s was a tactic by the NFF to divert rank and file attention away from internal organisational tensions. They argued that they developed an anti-Labor ideology to achieve this diversion; however, it was so successful that it threatened the control of the NFF leadership (Gerritsen, 1987a). Yet, this view has to be balanced by recognising the various militant acts that were undertaken independently of the farm organisations. These actions may have drawn on the same ideology but were not connected and in fact opposed the NFF leadership.

The militant activity not only raised the NFF's stature with its members but also raised its profile with the state. Due to its capacity for militancy the NFF had become an organisation that government could not ignore. Its well-researched and reasoned economic arguments gave its policy position a certain level of respectability. Most significant of all was the NFF's willingness to exchange some agricultural assistance measures in return for reductions in other areas of the economy (Martin, 1989). This willingness became the basis for the NFF's participation in subsequent economic policy formulation.

The Hawke Government recognised the NFF's willingness to make concessions, and in its 1986 'Economic and Rural Policy' statement challenged the farm lobby to get on board and assist the process of reform (Commonwealth of Australia, 1986: p. 50):

> If the rural sector really wants its off-farm costs to be reduced, the farmers must be prepared to support the Government's efforts in this area. While the Government can provide leadership in the process, major changes have an essentially self-help character.

In its introduction (Commonwealth of Australia, 1986: p. 1) this document states:

> In its review of economic and rural policies, the Government has identified many areas where little progress can be achieved by government alone. We are looking for a full commitment from rural groups generally, and farm organisations in particular to join us in developing specific courses of action in these areas. We have suggested machinery to facilitate this process.

The formation of the Rural and Allied Industries Council (RAIC) was part of the new corporatist 'machinery' which the Federal Labor Government established in order that its reformist agenda be successfully implemented. The NFF and its affiliate organisations took up the corporate role, as articulated by Hawke. Gerritsen and Abbot (1988) partly attribute the change in tactics to Kerin's action prior to the 1987 election, when he bypassed the NFF and went straight to state farm organisations. According to Gerritsen and Abbott (1988: p. 24):

> This reinforced the image of a government interested in constructive dialogue with the farm sector, and reminded the NFF that its authority amongst farm groups was dependent upon its representative function with government. Consequently, after the election a more cooperative relationship with government was sought by the NFF.

The approach adopted by the Hawke Government with respect to the rural sector was part of a broader policy of bargained consensus (Keating and Dixon, 1989); that is, the education of narrow interest groups as to the broader agenda, such that they were willing to trade specific measures for the pursuit of policy change which would assist the whole group. They concluded (Keating and Dixon, 1989: p. xii):

> ... that the Government's ability to achieve significant changes of direction in a range of economic policies has been aided by the growing influence of more encompassing interest groups. This has not, however, brought with it the disappearance of narrower groups with vested interests in increasing their share of national income (or at least preserving their existing share).

7.5 At the crossroads: militancy versus corporatism

The agricultural policy process changed significantly under Hawke (Gerritsen, 1987a). The early militancy had prompted the Hawke Government to provide NFF leaders with an opportunity to make policy gains through de-politicised activity. However, the NFF family's sponsorship of the 1985 rally and subsequent direct actions left them in a contradictory position. The rally had at once delivered increased access by the NFF to government and left sections of their membership with the belief that political change was best achieved through grassroots activity. The NFF leadership's corporatist role now required them to convince members to cease militancy. Given that state farm organisations had direct members, they were left to resolve the contradiction.

The NSWLGPA captured the contradiction better than most farm organisations – it was simultaneously advocating both militant and de-politicised policy approaches. In the same year as the Canberra farm rallies, LGPA members dumped wheat on the steps of Parliament House and presented the Government with a claim for Australian $3.6 million in compensation on behalf of Australian farmers – the cost per year of its economic policies (Lawrence, 1987). In 1986, it called for its members to be prepared, at only 48 hours' notice, to form roadblocks around Canberra if their demands were not met by the Hawke Government (Cribb, 1986a). At its February 1986 meeting, the General Council of the LGPA canvassed the range of strategies in which members could participate and which would advance their cause. These included withholding rates, letter-writing, applying for various forms of government assistance en masse, refusing to pay government rates and blockading Canberra and major roads (NSWLGPA, 1986b).

Yet, amidst this militant rhetoric, the NSWLGPA proposed that an 'Agricultural Accord' be entered into between the agricultural sector and the Commonwealth Government. The NSWLGPA (1986a: p. 2) asserted: 'An Accord has the advantage that it provides both parties with a single forum in which to consider the totality of circumstances affecting the sector'. The NSWLGPA saw a need for the NFF to create a similar relationship with the Federal Government that it afforded the Australian Council of Trade Unions (ACTU). It argued (NSWLGPA, 1986a: p. 2): 'The precedents for an Agricultural Accord exist not only in the form of the Government-ACTU Accord but also in the presence of omnibus policy vehicles such as the US Farm Bill and the EEC Common Agricultural Policy'. The reference to these overseas vehicles of farm policy implied that the NSWLGPA was advocating a return to high levels of government intervention in agriculture.

The NFF was furious at the public release of the proposal and opposed its content on a number of grounds. Correspondence from NFF Secretariat staff indicate its opposition centred on the proposal's implicit acceptance of the ACTU-Government accord, an arrangement which the NFF had opposed in its Interest Rate Taskforce paper released on the same day (Campbell, 1986; Trebeck, 1986). Trebeck claimed that the approach advocated by the proposal locked the NFF into 'a series of inane consensus type policy statements', as well as arguing for ongoing government-consultative arrangements that would lead to the further expansion of government. Apart from the substantive disagreement on policy principle, the submission was criticised as economically inaccurate, hence undermining the credibility of the farm lobby (Campbell, 1986; Trebeck, 1986). David Trebeck (1986) – NFF Director – signed his letter off with 'Hawke just learnt a lesson about the risks of making policy on the run. LGPA is yet to learn it'.

While both the NFF and NSWLGPA were committed to some form of corporatist relationship, there was disagreement on how it should be secured. The NFF was convinced it could achieve its economic rationalist policy direction without committing farmers to concrete targets or formal agreements. This position prevailed and instead of pursuing further government assistance the NFF adopted an economic rationalist stance and pursued cost reduction measures, identifying (Lawrence, 1987: p. 201), 'individual inefficiency amongst farmers and wage inflexibility in the wider community as two of the main problems facing agriculture'. However, the pace of reform was often too much for some NFF member organisations, in particular those commodity groupings still receiving assistance (Connors, 1996).

7.6 Controlling the militants

Despite the leaders having returned to the bargaining table from the barricades, it took until 1987 for the NSWLGPA to finally wrest back control of the agenda from militants. Farm polls in the 'National Farmer', whilst hardly a scientific instrument, consistently illustrated a disillusionment with state farm organisations, support for the NFF and a growing enthusiasm for militancy (Cribb, 1984, 1985, 1986a). This placed increasing pressure on the NSWLGPA to reinvent itself and, in particular, to tie itself more closely to the popularity of the NFF.

In 1987, the NSWLGPA changed its name to the NSW Farmers' Association (The Land, 1990). The name change signalled a new orientation towards a more tightly run and professional interest group. The NSWFA was able to deny its competitors financial and political recognition by virtue of its own economic strength and political palatability. This tactic effectively robbed groups such as the RAM and the Canowindra Reform Group of popular support. Ultimately, 'rebel' leaders had no choice but to rejoin the NSWFA, with the necessary assurances that their grievances would be given more attention in future.

While the NSWFA backed away from militancy, it still encouraged member participation in political activity. Many of its major campaigns in the 1980s relied on member participation for their success. Campaigns against fringe benefits tax and pay increases for farm workers included threats that if policy was not changed members would refuse to pay both tax and wage rises. In essence, militancy was replaced with coordinated member involvement in implementing sanctions. This policy of member participation and involvement was not carried through into the 1990s.

Series of organised local or regional meetings replaced the large-scale rally that was the hallmark of the 1980s. The NSWFA response to the 1990 rural crisis comprised endorsing unspecified 'direct action' and then, as a NSWFA publicist described, using local meetings to enable 'members to let off steam' (Stapleton, 1990). Needless to say, the final 'direct action' did not eventuate. This tactic helped to maintain pressure on governments while placating member's calls for prompt action, maintaining their focus on non-political matters and keeping them from endangering the formal political 'bargaining' process. This approach supported the 'myth', promulgated by the NFF, that the solution to the crisis in the agricultural sector lay at the feet of the individual farmer and that they must increase productivity/efficiency or leave the industry (Lawrence, 1987).

7.7 Corporatism and the professionalisation of the National Farmers' Federation

Throughout the late 1980s the Hawke Government pursued the formation of 'Policy Councils' for each agricultural commodity. These were established to replace the old Statutory Marketing Authorities (SMA) that had become a repository for many farm organisation leaders during the McEwen era. A selection committee made up of farm representatives and other notables appointed members to the new councils. This was done to ensure those with the appropriate expertise obtained positions of influence. The minister referred day-to-day policymaking to these councils, leading to farm organisation criticism that he was abdicating his ministerial responsibility and deflecting their policy concerns (Gerritsen and Abbott, 1988).

Ultimately, the aggressive policy position of the NFF was accompanied by active participation on, and detailed representations to, various quasi-governmental committees and commissions, such as the Industries Assistance Commission (IAC), the Australian Agricultural Council (AAC) and the Economic Planning Advisory Council (EPAC). The commencement of Senate Standing Committees provided additional forums at which farmers needed to be represented (Campbell, 1985).

Further initiatives such as the Agrifood Council and the recent 'Supermarket to Asia' strategy[5] created formal links between business, farmers, bureaucrats and politicians.

While the NFF's early participation in corporatist forums was often marked by antagonism, on balance, the NFF and the Labor governments of both Hawke and Keating had a close relationship (Connors, 1996). However, the close relationship did not extend across all policy areas. While on some issues, such as trade, they were close allies, on others, such as labour-market reform, they were enemies (Gruen and Grattan, 1993). Consequently, the NFF had to adopt a strategy of lobbying on a policy-by-policy, issue-by-issue basis. This approach is reflected in the NFF's participation in temporary coalitions with other peak groups around specific policy issues. For example, the Australian Conservation Foundation (ACF) and the NFF put a joint proposal to the Federal Government that led to the development of the National Landcare Program. Again, in 1994, the NFF and the Australian Workers' Union (AWU) put together a three-year deal agreeing on the level of payment for shearers, cooks and shed hands, thus avoiding the need for costly recourse to the arbitration commission. This strategy also reflected a realisation by the NFF that the bush was a declining force in politics and needed allies to exact maximum influence (Connors, 1996).

The trend toward a corporatist relationship with government continued until the end of the Keating Labor Government. In his address to the 1995 Annual Conference of the NFF, the then Prime Minister congratulated the peak interest group for its approach to policy. He revealed that from a government point of view, there was 'not much more they could ask for' from an interest group. The NFF was praised for (Matthews and Warhurst, 1993: p. 94) 'being able to look beyond sectoral interest ... [and having a] ... comprehensive approach to policy'. The NFF had changed farmer producer groups from a 'rent-seeking distributional coalition' to an encompassing organisation.

Given the non party-political approach of farm organisations since the early 1980s, one would have expected a level of continuity between the rural policy process of Hawke/Keating and that of the Howard Coalition Government. As with the previous government, areas of conflict with the NFF were in those instances where it watered down commitments to a reduction in tariff protection to the Automotive and Textile, Clothing and Footwear sectors. This represented an unravelling of the NFF's bargaining power of withdrawing assistance in return for reform elsewhere in the economy and, as such, will always be a source of tension.

However, as Matthews and Warhurst (1993) maintain, for producer groups 'party politics does matter'. The expectations of positive and rapid change were far greater among the farming constituency when a National Party member rather than a Labor

[5] The latter initiative brings together agribusiness in the form of Woolworths, with government and primary producers to develop an integrated approach to creating sustainable export industries.

member is Minister for Primary Industry. Consequently, this raised the likelihood of tension. Further tension will likely arise if the Nationals, when in government, attempt to once again take up the agenda-setting role that the NFF had taken since the demise of the Fraser Government in 1983. The inability of the government to reach a consensus solution on the Wik peoples through deliberations with interest groups was in part due to the National Party's unwillingness to have the NFF seen as 'delivering' results.

7.8 Conclusion

The history of the political representation of farmers in Australia is a story of transition from one formation to another; from local grassroots groups, to local independent Members of Parliament, to parties and to interest groups. The movement from one form to another reflected the needs of primary producers to extend influence beyond their locality to the state and federal political scenes. Until the 1960s, primary producers exercised considerable influence through the parliamentary system, first through a dedicated party and then through associations. When primary producers could no longer maintain a dedicated political party, emphasis swung back to interest-group politics. This shift in turn catalysed a major set of reforms to the gaggle of primary-producer groups then involved in representational activities.

Within this broader transition, the refinement of the farmer interest-group structure involved a transformation from organisational fragmentation to amalgamation (Connors, 1996). While in most cases groups worked together (especially under the joint farmer and graziers' committee which oversaw the running of the Country Party), organisations resisted relinquishing their discrete identities and autonomy. Crucially, until the NFF's formation, producers organised around divisions that correlated with landownership, wealth and social status. The amalgamations of the late 1970s were, therefore, hard won.

An important observation from this historical analysis is that the organisational amalgamation necessary to form the NFF family was forged in order to meet the needs of the policy process. The political costs of organisational disunity and of poorly researched policy propositions were great enough for leaders to advocate amalgamation. Clearly, it was not a decision driven by the emerging homogeneity of farmers' interests. While the widespread economic downturn of the late 1970s provided a useful point of reference for all primary producers, the social, economic and political divergences remained.

The leadership's rush to produce the united voice that the political environment demanded, even though no such voice could be said to exist amongst the grassroots members of the NFF, caused problems. It is hard to argue against the idea that the social and economic conditions of many farmers did not foster a unity of interests in the period after its formation. The extent of organised protest and militancy, and the existence of 'splinter groups', is evidence of the fragility with which the new organisations exercised representative dominance during the mid-1980s. Ironically, however, this militancy itself was rather productive for the more professionalised organisations. The involvement of members in policy activities – even ones that contested the policy line of the NFF – allowed divergences to be in some way vented and discussed. However, professionalism, financial security and recognition by government became paramount, while direct participation by members, transparency and internal democracy slowly became secondary. Significantly, the NFF family was designed at a time when the farm sector was battening down the hatches for a Labor onslaught. It was designed to facilitate the autonomy of those elected to lead, under the guidance of the best experts money could buy. The development of a family of hierarchically organised professional interest groups, with the support of successive federal governments, has generated the necessary authority to secure control over militant elements and see to the dissolution of splinter groups.

The National Farmers' Federation family of organisations is now ensconced as the dominant avenue for the political representation of farmers' interests. According to Connors (1996: p. 61): 'Today, a farm organisation's reputation is judged more on the quality of its research and presentation than the noise it can make in the media or its influence with the National Party'. It appears that the NFF came to concur with these remarks and has conducted itself accordingly.

References

Anderson, R., 1972. Crisis on the land: weighing the farmers' Worth. Sun Books, Melbourne, Australia.

Campbell, K., 1971. Australian farm organisations: the unity issue. Politics 6: 148-160.

Campbell, K., 1980. Australian agriculture: reconciling change and tradition. Longman Cheshire, Melbourne, Australia, 274 pp.

Campbell, K., 1985. Changing institutions, processes and issues in the formation of Australian agricultural policy. Australian Journal of Agricultural Economics 29: 210-224.

Campbell, R., 1986. Letter from Robert Campbell (NFF Economist) to NFF President on January 28, 1986. NBAC/ANU NFF, N143/ 93 M1.

Commonwealth of Australia, 1986. Economic and rural policy: a government policy statement. AGPS, Canberra.

Connors, T., 1996. To Speak with One Voice. NFF, Canberra, Australia.

Cribb, J., 1984. Farm organisations: where they are letting you down. National Farmer March 20, 1984.

Cribb, J., 1985. Raking broadside for State farm leaders. National Farmer June 27, 1985.

Cribb, J., 1986a. Farm chiefs fail the performance test. National Farmer April 3, 1986.

Cribb, J., 1986b. Kerin outflanks farm militants. National Farmer March 6, 1986.

Gerritsen, R., 1987a. Making policy under 'uncertainty': the Labour Government's reaction to the 'rural crisis'. Discussion paper no. 3, graduate program in public policy. Australian National University, Canberra, Australia.
Gerritsen, R., 1987b. Why the 'uncertainty'? Labor's failure to manage the 'rural crisis'. Politics 22: 47-59.
Gerritsen, R. and Abbott, J., 1988. Shifting to certainty? Australian rural policy in 1987. Review of Marketing and Agricultural Economics 56: 9-26.
Graham, B.D., 1966. The Formation of the Australian country parties. Australian National University Press, Canberra, Australia.
Graham, V., 1978. LGPA pledges to fight for equity of farmers. The Land January 3, 1978.
Graham, V., 1985. Generals of wheat free trade assault. The Land February 3, 1985.
Gruen, F. and Grattan, M., 1993. Managing government, Labor's achievements and failures. Longman Cheshire, Melbourne, Australia.
Hardy, D., 1977. Farm leaders new militancy and its effect. The Land September 15, 1977.
Harman, G., 1968. Graziers in politics: the pressure group behaviour of the Graziers' Association of New South Wales. PhD thesis, Australian National University, Canberra, Australia.
Hood, W., 1977. Farm leaders new militancy and its effect. The Land September 15, 1977.
Keating, M. and Dixon, G., 1989. Making economic policy in Australia: 1983-1988. Longman Cheshire, Melbourne, Australia.
Lawrence, G., 1987. Capitalism and the countryside. Pluto Press, Sydney, Australia.
Martin, W.J., 1989. Lessons from Australian rural policy reform. Working paper no. 89/7, National Centre for Development Studies, ANU, Canberra, Australia.
Matthews, T. and Warhurst, J., 1993. Australia: interest groups in the shadow of strong political parties. In: Thomas, C. (ed.) First world interest groups: a comparative perspective. Greenwood Press, Westport, CT, USA, pp. 81-95.
New South Wales Livestock and Grain Producers' Association (NSWLGPA), 1986a. A Proposal for an Agricultural Accord, January 1986. Noel Butlin Archives, NBAC/ANU.NFF, N143/93 M1.
New South Wales Livestock and Grain Producers' Association (NSWLGPA), 1986b, Minutes of the General Council Meeting 4-6 February 1986. Noel Butlin Archives, NBAC/ANU NFF, N143/93 MB1.
Richmond, K., 1980. The major rural producer groups in New South Wales. In: Scott, P.R. (ed.) Interest groups and public policy. Macmillan, South Melbourne, Australia.
Stapleton, J., 1990. Farmers to block Hume highway. Sydney Morning Herald November 19, 1990.
Streeck, W., 1992. Social institutions and economic performance: studies of industrial relations in advanced industrial economies. Sage, London, UK.
The Land, 1979. LGPA threatens NFF pull-out. The Land, July 5, 1979.
The Land 1990. Single voice a long time coming. The Land, July 26, 1990 p.21.
Trebeck, D., 1986. Letter to Gary Bond (NSWLGPA) economist from David Trebeck (NFF). Noel Butlin Archives, NBAC/ANU NFF, N143/93 M1.
Trebeck, D., 1990. Farmer organisations. In Williams, D.B. (ed.) Agriculture in the Australian economy. Sydney University Press, Sydney, Australia.
Vallance, J., 1977. Forget the Parties. The Land, October 6, 1977.

8 Contemporary populism, the agrarian and the rural in Central Eastern and Western Europe

S.L. de Lange and M. Rooduijn**

Department of Political Science, University of Amsterdam, Amsterdam, the Netherlands; s.l.delange@uva.nl; m.rooduijn@uva.nl

Abstract

Populist parties have been successful in many Central Eastern and Western European countries. Although populist parties differ from each other in many respects, they also share a common message, namely that the 'bad' elite is no longer, or not acting in the interests of the 'good' ordinary people. In this chapter we examine the characteristics and subtypes of populism (e.g. agrarian populism, conservative or neo-liberal populism, national populism and social populism), providing a comprehensive overview of contemporary populist parties in Europe. We also study possible explanations for the electoral successes of these parties and explore the relationship between the agrarian, the rural and the populist. We conclude that although agrarian populism is no longer a widespread phenomenon in Central Eastern and Western Europe, most populist parties have a clearly defined stance on agrarian issues and attract considerable electoral support from the countryside.[1]

8.1 Introduction

Since the 1980s, Europe has witnessed an upsurge of populist movements, parties and politicians. In Italy, for example, the media-mogul Silvio Berlusconi and his party Forza Italia (FI) won the Italian parliamentary elections in 1994, 2001 and 2008, while in the Netherlands the eccentric Pim Fortuyn obtained 26 of the 150 Dutch parliamentary seats in 2002 with his Lijst Pim Fortuyn (LPF). Another example is France, where the nationalist Jean-Marie Le Pen and his Front National (FN) managed to reach the second round of the 2002 French presidential elections. Although these examples might suggest that most populist parties can be found on the right or far right of the political

[1] This chapter was finalized in 2013 and as a result the analyses, classification of parties, and literature review are based on the information available in that year.

DOI 10.3920/978-90-8686-807-0_8, © Wageningen Academic Publishers 2015

spectrum, populist movements, parties and politicians come in different forms and shapes. They include left-wing populist parties such as Die Linke in Germany, Sinn Féin (SF) in Ireland, Smer in Slovakia and the Socialistische Partij (SP) in the Netherlands, and right-wing populist parties such as the Dansk Folkeparti (DF) in Denmark, the Freiheitliche Partei Österreichs (FPÖ) in Austria, Fidesz – Magyar Polgári Szövetség (Fidesz) and Jobbik Magyarországért Mozgalom (Jobbik) in Hungary, and the Vlaams Belang (VB) in Belgium. They also include grassroots movements such as José Bové's movement against the Americanisation and globalisation of the French food industry (Chapter 5; Bruneau, 2015), or the Occupy Wall Street movement, which campaigned against financial institutions.

At first sight, these movements, parties and politicians might not seem to have any characteristics in common. Some have emerged in the consolidated democracies of Western Europe, while others have gained representation in the unconsolidated systems of Central and Eastern Europe. Some have an egalitarian ideology, rooted in socialism, while others have an anti-egalitarian ideology founded upon the principles of nationalism or neo-liberalism. Some are established elements of the party system and have been represented for decades (e.g. the FN and FPÖ), while others have been nothing more than a flash in the pan (e.g. the LPF). At the same time, however, the movements, parties and politicians also express an identical message: they accuse the elite of having lost touch with the customs, desires, interests and preferences of the 'ordinary' people. In the literature, they have therefore been described as populist.

This chapter serves two purposes. First, it examines the characteristics of populist parties in Europe. It thereby provides a comprehensive overview of left and right-wing populist parties that have emerged in Central Eastern and Western Europe since the early 1980s and focuses on possible explanations for their upsurge. Second, it explores the relationship between populism on the one hand, and the agrarian and the rural on the other. To this end, it addresses three related questions: (1) Does an agrarian populism exist, both in theory and in practice; (2) What role do agrarian and rural questions play in the ideologies and programmes of populist parties; and (3) Is there an agrarian or rural component in explanations for the success of populist parties and politicians?

8.2 Defining populism

Although many movements, parties and politicians have been labelled populist throughout history, an accepted and clear definition of populism has for many years been lacking. The term 'populism' has been applied to different phenomena on different continents and in different contexts (Canovan, 1981: p. 3). In the 1960s, Ionescu and Gellner (1969: p. 1) argued that 'There can, at present, be no doubt about the importance of populism. But no one is quite clear just what it is ... It bobs up everywhere, but in many and contradictory shapes. Does it have any underlying unity, or does one name cover a multitude of unconnected tendencies?' Also, more recently, scholars have concluded that the concept deals with 'an essential impalpability, an awkward conceptual slipperiness' (Taggart, 2000: p. 1).

Despite the conceptual confusion, three different approaches towards defining populism can be distinguished (e.g. Jagers, 2006; Pauwels, 2012). First, populism can be conceived of as a particular form of political organisation. In the literature on Latin America, for example, it is often emphasised that a populist movement is characterised by a strong leader who receives uninstitutionalised support from a heterogeneous group of people (Di Tella, 1997; Germani, 1978; Weyland, 2001). Similarly, in Western Europe, populism has been associated with loose movements led by a charismatic leader (Taggart, 1995, 2000). Second, populism can be conceptualised as a political style. Canovan (1999: p. 5) has argued that 'Populist appeals to the people are characteristically couched in a style that is 'democratic' in the sense of being aimed at ordinary people. Capitalising on popular distrust of politicians' evasiveness and bureaucratic jargon, they pride themselves on simplicity and directness' (see also Bos *et al.*, 2011; Jagers and Walgrave, 2007; Mazzoleni, 2003; Taguieff, 1995). Finally, populism can be defined as a particular 'thin-centred ideology' (e.g. Albertazzi and McDonnell, 2008; Canovan, 2002; Mudde, 2004, 2007) or 'discourse' (Hawkins, 2009, 2010; Laclau, 1977, 2005). Scholars employing this approach conceive of populism essentially as a set of ideas (Hawkins *et al.*, 2012). They equate populism with a Manichaean worldview, in which the people are described as Good and the conspiring elite as Evil.

In this chapter, populism is defined in accordance with this third approach – as a set of ideas.[2] More specifically, we employ Mudde's (2004: p. 543) definition of populism as 'an ideology that considers society to be ultimately separated into two homogeneous and antagonistic groups, 'the pure people' versus 'the corrupt elite', and which argues that politics should be an expression of the *volonté générale* (general will) of the people'. This definition of populism as a thin-centred ideology has been employed in various empirical studies of the concept (Pauwels, 2010; Rooduijn and Pauwels, 2011; Rooduijn *et al.*, 2014). It is important to emphasise that Mudde does not conceive of populism as being a 'full' ideology such as conservatism, liberalism or socialism. Following Freeden (1998) and Canovan (2002), he argues that populism is a 'thin-centred ideology': it does not offer an all-inclusive worldview, but only focuses on the relationship between the people and the elite. It can therefore be combined with other ideologies, both on the left and the right of the political spectrum (Mudde, 2004; Taggart, 2000).

[2] That populism is defined as a set of ideas means that populism is not understood as a feature of a specific political actor (such as a political party, a movement, or an individual politician) but as a characteristic of the message of such an actor. Because actors can endorse the populist set of ideas to a larger or smaller extent, populism then becomes a matter of degree. Actors who send out many populist messages are more populist than actors who send out only a few such messages.

According to Mudde's definition, populism essentially consists of two elements: people-centrism and anti-elitism. Populism is people-centrist because it emphasises the centrality and the sovereignty of the homogeneous people, and it claims that the general will of the people should be the point of departure for political decision-making (Ionescu and Gellner, 1969; Mény and Surel, 2002). The term 'the people' can have different meanings in different circumstances. It could refer, for instance, to the electorate, to farmers, to 'hardworking men' or to the nation (Canovan, 1981; Pasquino, 2008; Taggart, 2000). Often, however, it is rather unclear what populists are referring to when talking about 'the people'.

Populists are anti-elitists because they accuse the elites of standing in the way of the centrality of the people (Mény and Surel, 2002). They claim that the elites are arrogant, incompetent and selfish. Moreover, the elites have no idea of what ordinary people find important or what policy reforms they would like to see implemented (Barr, 2009; Canovan, 2002). According to Mudde (2004: p. 546), 'In an often implicitly Rousseauian fashion, populists argue that political parties corrupt the link between leaders and supporters, create artificial divisions within the homogeneous people, and put their own interests above those of the people'. Anti-elitism often concerns a political elite, but it could just as well be about a cultural, economic or judicial elite (Jagers and Walgrave, 2007).

Some scholars have claimed that populism opposes not only the elite but also 'dangerous others' – groups who are not considered part of 'the people', such as immigrants or people of another race (Albertazzi and McDonnell, 2008; Panizza, 2005; Taguieff, 1995). In this chapter, we follow Mudde (2004) and others (Canovan, 1981; Taggart, 2000) who have claimed that although this exclusionist feature might be a defining characteristic of national or right-wing populism, it is not an intrinsic property of populism as such. In fact, one of the essential properties of populism is that it can be combined with various ideologies. Including the exclusionist feature in our definition would introduce a bias towards right-wing populism.

8.3 Types of populism

Given that populism is a thin-centred ideology, it manifests itself in different shapes and forms. It can be exclusive or inclusive, left-wing or right-wing, movement-based or party-based, organised bottom-up or top-down. Consequently, many scholars have sought to differentiate between types of populism (e.g. Canovan, 1981; Mudde, 2000, 2001, 2007; Pauwels, 2012; Ucen, 2007). One of the distinctions most relevant for this chapter is between agrarian populism on the one hand, and political populism on the other. Although both kinds of populism focus on the distinction between the 'good people' and the 'bad elites', agrarian populists defend farmers' and peasants' interests against the urban elite, while political populists fight for the man in the street, opposing the (party) political establishment. Hence, in some respects agrarian populism is a form of radical ruralism, which usually manifests itself in times of modernisation and can be recognised on the basis of the distinct composition of its support base, which primarily consists of farmers and/or peasants (Canovan, 1981). Political populism focuses less on sociological processes and therefore draws upon a sociologically more diverse electorate. The distinction, however, does not imply that agrarian and political populism are coherent or homogenous phenomena, and movements and parties can display features of both kinds of populism. Moreover, different forms of agrarian and political populism exist, as will be demonstrated below.

Political populism

In the literature, three forms of political populism are frequently discussed: (1) national populism; (2) conservative or neo-liberal populism; and (3) social populism (e.g. Mudde, 2007; Pauwels, 2012).[3] In Western Europe, the largest group of political populists adheres to right-wing ideologies, such as conservatism, nationalism and neo-liberalism. These parties are right-wing in their 'rejection of individual and social equality and of political projects that seek to achieve it ... their opposition to the social integration of marginalised groups; and ... in their appeal to xenophobia' (Betz, 1994: p. 4; see also Bobbio, 1996). Although right-wing populist parties share certain ideological features, considerable differences in outlook can also be found amongst these parties. Most importantly, a distinction can and should be made between national populist parties on the one hand, and conservative and neo-liberal populist parties on the other (e.g. Betz, 1993, 1994; Carter, 2005; Kitschelt, 1995). The first group of parties adheres to nativism (a combination of nationalism and xenophobia) and focuses first and foremost on cultural, ethnic and religious inequalities, while the second group of parties espouses a conservative or neo-liberal ideology and mainly emphasises economic inequality.

National populist parties adhere to nativism. In their view the nation-state should be inhabited exclusively by natives: non-native elements, people as well as ideas, are seen as a fundamental threat to the ideal homogenous nation-state in which territory and nation coincide (Mudde, 2007: p. 19). National populist parties are known for their call to close borders to non-Western immigrants and rigid assimilation programmes. Since the 1990s they have especially targeted Muslims, claiming that their rising numbers and their unwillingness to integrate are incompatible with European Judeo-Christian values. Given its perceived 'totalitarian' nature, Islam represents a fundamental threat to individual freedom and liberal democracy (Betz, 2007: p. 47). Examples of national populist parties include the Austrian FPÖ, the Hungarian Jobbik, and the Italian Lega Nord (LN).

[3] Canovan (1981) makes a distinction between 'populist dictatorship', 'populist democracy', 'reactionary populism' and 'politicians' populism'. However, this distinction mainly focuses on the different forms that populism can take in autocracies and democracies and is therefore not used in this chapter.

Conservative and neo-liberal populist parties differ from national populist parties in two respects. First, nationalism and xenophobia are not central to their ideology. Although the parties may hold nationalist and xenophobic beliefs, they pay considerably less attention to immigration and integration issues than national populist parties (Mudde, 2007: p. 30). Moreover, the policy proposals promoted by conservative and neo-liberal parties to solve immigration and integration problems are often less radical than those of national populist parties. Hence, while national populist parties oppose multiculturalism, conservative and neo-liberal populist parties merely remain sceptical of it (Decker, 2004: pp. 219-220). The core of the ideology of conservative and neo-liberal populist parties is formed by economic liberalism. These parties advocate anti-egalitarian measures and aim to reduce government and state intervention, defending the 'ordinary people' against an allegedly 'corrupt elite' (Pauwels, 2010: p. 1009). Examples of conservative or neo-liberal populist parties include Lijst Dedecker (LDD) in Belgium, FI in Italy, and Prawo i Sprawiedliwość (PiS) in Poland.

A third group of political populists adheres to left-wing ideologies, such as communism and democratic socialism. These populists share with other left-wing parties their anti-capitalism and 'their identification of economic equality as the basis of existing political and social arrangements, and their espousal of collective economic and social rights as their principal agenda' (March and Mudde, 2005: p. 25).[4] At the same time, they devote less attention to class politics and doctrinal purity than their non-populist counterparts. Although social populist parties originate from different backgrounds, with some parties having their roots in Maoism or Marxism and others in mainstream social democracy, they have largely identical ideas and political styles. They appeal to the average worker or to industrious men and women, who they pit against the corrupt, economic and political elite. Bankers, big companies and insurers in particular are all believed to exploit the common people. Key points in campaigns and programmes of social populist parties are 'anti-elitism, empowerment, inclusiveness, morality and welfarism', which combine into a campaign that is primarily geared towards establishing and protecting the economic, financial and social rights of the lower and middle classes (March, 2007: p. 65). It should be noted, however, that it is not uncommon for social populism to be combined with a cross-class appeal and with nationalism and thus to take on both left-wing and right-wing characteristics. Examples of social populist parties include the Dutch SP, the German Die Linke and the Slovenian Smer.

In recent years, however, ideological differences between the three types of parties have become less clear-cut. In an attempt to become more acceptable to mainstream parties and voters, national populist parties, for instance, increasingly resort to liberal arguments to defend their opposition to immigration. The national populist critique of Islam, for example, stems from the observation that Islamic values are at odds

[4] In fact, a number of scholars have argued that socialism is inherently populist because the concept of the 'proletariat' largely overlaps with the concept of the 'people' (March, 2007: p. 65).

with liberal democratic values, such as the autonomy of the individual, democracy, emancipation of homosexuals and women, equality of men and women, freedom of expression, and separation of church and state (Akkerman, 2005; Betz, 2007; Betz and Meret, 2009; Zuquete, 2009). However, the liberalism that is invoked by national populist parties is driven by fear of Islam (Akkerman, 2005) and therefore is not liberal in the classic meaning of the term. The distinction between national and social populist parties has also become blurred. Many national populist parties have adopted economic programmes which are centred on egalitarian populism and welfare chauvinism in recent years (e.g. Derks, 2006; Mudde, 2007), while a number of social populist parties, especially in Central and Eastern Europe, have incorporated nationalist positions into their programmes.

Agrarian populism

When it comes to agrarian populism, Canovan (1981; see also Mudde, 2000, 2001) makes a distinction between farmers' populism, peasants' populism and a populism of intellectuals. The first form of agrarian populism, farmers' populism, or farmers' radicalism, emerged in the late nineteenth century and refers to movements with primary support amongst (commodity) farmers dissatisfied with the increasing power of big business, including banks and the transport sector. These movements called for government protection of farmers' interests, usually through the constitution of a people's government. Examples of farmers' populism include the German and United States' agrarian movements of the 1890s and the Canadian Co-operative Commonwealth Federation and Social Credit, both founded in the 1930s.

The second form of agrarian populism, peasants' populism, comprises a variety of grassroots movements that campaign for 'land and liberty'. It opposes bureaucracy, capitalism and the state and promotes democracy, cooperatives, family property and small businesses. The movements were particularly successful between the two world wars in Central and Eastern Europe. The Balgarski Zemedelski Naroden Sayuz (BZNS), for example, dominated Bulgarian politics between 1900 and 1923, despite being set up as a movement, not a party. Also in Czechoslovakia, Poland and Romania, peasants' populism gained a considerable following in the early twentieth century.

A third form of agrarian populism is represented by the Narodnichestvo of nineteenth-century Russia. Known for their curiosity towards, and idealisation of, the peasant lifestyle, the Narodniki sought to replace both Marxist and Western models of economic development with an agrarian form of socialism. Given their desire to lead the peasant population, Canovan (1981) describes this form of agrarian populism as revolutionary intellectual populism. In more recent years, aspects of revolutionary intellectual populism have been detected in the discourse of leaders and thinkers in the Third World such as Frantz Fanon of Martinique and Julius Nyerere of Tanzania.

All three populisms emerged in the late nineteenth or early twentieth centuries, and are in one way or another a response to the industrialisation and/or modernisation processes that took place during this period. Although clearly distinct, the commonalities between them make it possible to group the movements together (Canovan, 1981: p. 133). The shared ideological core of agrarian populism consists of the notion that:

> ... the peasants were biologically and morally the healthiest stratum of society and that they were destined to create a society more balanced and more just than the existing system which was dominated by the urban bourgeoisie and a corrupt bureaucracy dependent upon its favors.
>
> (Dziewoniwski, 1996: p. 171 as cited in Mudde, 2001: p. 39)

The following section will demonstrate that agrarian populism is far less prevalent in contemporary Europe than it was in the late nineteenth and early twentieth centuries.

8.4 Empirical manifestations of populism

Populist parties have become important players in European party systems. Over the past decades, they have entered national parliaments in fourteen countries in Western Europe and twelve countries in Central and Eastern Europe: in Austria, Belgium, Denmark, Finland, France, Germany, Greece, Ireland, Italy, the Netherlands, Norway, Sweden, Switzerland and the UK, as well as in Bulgaria, the Czech Republic, Croatia, Estonia, Hungary, Latvia, Lithuania, Poland, Romania, Serbia, Slovakia and Slovenia. The actual number of populist parties that have successfully contested elections has been exponentially greater, with at least 29 Western European parties that gained parliamentary representation between 1980 and 2013 being described as populist in the literature and 40 Central and Eastern European parties gaining parliamentary seats since 1990 being labelled populist by populism scholars. Of these 69 parties, the majority can be classified as national populist (sixteen in Western Europe and sixteen in Central and Eastern Europe), while the remaining parties are either conservative/neo-liberal populist (eight in Western Europe and nine in Central and Eastern Europe) or social populist (four in Western Europe and nine in Central and Eastern Europe) (Table 8.1 and 8.2).[5]

[5] Five parties have been categorised as 'other' types of populists, since their core ideologies cannot be classified as conservative or neo-liberal, nationalist or socialist. The classification has been made on the basis of the existing literature on populism (Albertazzi and McDonnell, 2007; March, 2008; Mudde, 2007) as well as country studies for Western Europe (Havlik and Pinkova, 2012; March, 2008; Mudde, 2007; Ucen, 2007) and country studies for Central and Eastern Europe.

Table 8.1. Populist parties in Western Europe 1980-2013.

Country	Party	Ideology
Austria	Bündnis Zukunft Österreich (BZÖ)	Nationalist
	Freiheitliches Partei Österreichs (FPÖ)	Nationalist
Belgium	Lijst Dedecker (LDD)	Cons/Neo-lib
	Front National Belge (FNb)[1]	Nationalist
	Vlaams Blok (VB)/Vlaams Belang (VB)	Nationalist
Denmark	Dansk Folkeparti (DF)	Nationalist
	Fremskridtspartiet (FRP)[1]	Cons/Neo-lib
Finland	Suomen maaseudun puolue (SMP)/Perussuomalaiset (PS)	Nationalist
France	Front National (FN)	Nationalist
Germany	Die Linke	Socialist
Greece	Laikós Orthódoxos Synagermós (LAOS)[1]	Nationalist
	Panellinio Sosialistiko Kinima (PASOK)	Socialist
Ireland	Sinn Féin (SF)	Socialist
Italy	Alleanza Nazionale (AN)/Movimento Sociale Italiano (MSI)	Nationalist
	Forza Italia (FI)/Popolo della Libertà (PdL)	Cons/Neo-lib
	Lega Nord (LN)	Nationalist
Netherlands	CentrumDemocraten (CD)[1]	Nationalist
	Centrum Partij (CP)[1]	Nationalist
	Leefbaar Nederland (LN)[1]	Other
	Lijst Pim Fortuyn (LPF)[1]	Cons/Neo-lib
	Partij voor de Vrijheid (PVV)	Nationalist
	Socialistische Partij (SP)	Socialist
Norway	Fremskrittspartiet (FRP)	Cons/Neo-lib
Sweden	Ny Demokrati (ND)[1]	Cons/Neo-lib
	Sverige Demokraterne (SD)	Nationalist
Switzerland	Schweizer Autopartei (AP)/Freiheits-Partei der Schweiz (FPS)[1]	Cons/Neo-lib
	Schweizerische Volkspartei (SVP)	Nationalist
United Kingdom	British National Party (BNP)[2]	Nationalist
	United Kingdom Independence Party (UKIP)[2]	Cons/Neo-lib

[1] Party no longer represented in national parliament.
[2] Party only represented in European parliament.

In Western Europe, populist parties such as the Alleanza Nazionale (AN), DF, FI, FPÖ, the Fremskrittspartiet (FRP), LN, the Schweizerische Volkspartei (SVP), SP and VB have been represented in national parliaments for almost twenty years. A number of these parties, including the FRP and the SVP, but also the Finnish Perussuomalaiset

Table 8.2. Populist parties in Central and Eastern Europe 1990-2013.

Country	Party	Ideology
Bulgaria	Ataka	Nationalist
	Bulgarska sotsialisticheska partiya (BSP)[1]	Socialist
	Grazhdani za evropeysko razvitie na Balgariya (GERB)	Cons/Neo-lib
	Nacionalno dviženie za stabilnost I väzhod (NDSV)[1]	Cons/Neo-lib
Czech Republic	Komunistická strana Čech a Moravy (KSČM)	Socialist
	Sdružení pro republiku – Republikánská strana Československa (SPR-RSČ)[1]	Nationalist
	Věci veřejné (VV)[1]	Cons/Neo-lib
Croatia	Hrvatska demokratska zajednica (HDZ)	Nationalist
	Hrvatska stranka prava (HSP)	Nationalist
Estonia	Eesti Rahvusliku Sõltumatuse Partei (ERSP)[1]	Nationalist
	Erakond Res Publica (RP)[1]	Cons/Neo-lib
Hungary	Független Kisgazda, Földmunkás és Polgári Párt (FKgP)[1]	Agrarian/socialist
	Fidesz – Magyar Polgári Szövetség (Fidesz)	Cons/Neo-lib
	Jobbik Magyarországért Mozgalom (Jobbik)	Nationalist
	Lehet Más a Politika (LMP)	Other
	Magyar Igazság és Élet Pártja (MIÉP)	Nationalist
Latvia	Jaunais laiks (JL)[1]	Cons/Neo-lib
	Tautas Kustība Latvijai (TKL)[1]	Nationalist
Lithuania	Darbo Partija (DP)	Other
	Liberalų Demokratų Partija (LDP) /Tvarka ir teisingumas (TT)	Cons/Neo-lib
	Valstieciu ir Naujosios Demokratijos Partiju Sajunga (VNDPS)	Other
Poland	Liga Polskich Rodzin (LPR)[1]	Nationalist
	Prawo I Sprawiedliwość (PiS)	Cons/Neo-lib
	Ruch Palikota (RP)	Other
	Samoobrona Rzeczypospolitej Polski[1]	Agrarian/socialist
Romania	Partidul Democrației Sociale in România (PDSR)	Socialist
	Partidul România Mare (PRM)[1]	Nationalist
	Partidul Socialist al Muncii (PSM)[1]	Socialist
	Partidul Unității Naționale a Românilor (PUNR)[1]	Nationalist
Serbia	Srpska Radikalna Stranka (SRS)[1]	Nationalist
	Socijalistička partija Srbije (SPS)	Nationalist
Slovakia	Aliancia Nového Občana (ANO)/Strana Slobodné Slovo – Nory Mojsejovej (SSS)[1]	Other
	Hnutie za demokratické Slovensko (HZDS)[1]	Nationalist
	Komunistická strana Slovenska (KSS)	Socialist

Table 8.2 Continued.

Country	Party	Ideology
Slovakia	Obyčajní Ľudia a nezávislé osobnosti (OL`aNO)	Cons/Neo-lib
	Slovenska Nacionalna Stranka (SNS)[1]	Nationalist
	Smer	Socialist
	Strana občianskeho porozumenia (SOP)[1]	Other
	Združenie robotníkov Slovenska (ZRS)[1]	Socialist
Slovenia	Slovenská národná strana (SNS)[1]	Nationalist

[1] Party no longer represented in national parliament.

(PS) and Italian Popolo della Libertà (PdL), are currently supported by more than one-fifth of the electorate and have thus outpolled many mainstream parties. Moreover, the AN, FI/PdL and LN, the Bündnis Zukunft Österreich (BZÖ) and FPÖ, LPF and SVP have all assumed office in recent years, while the DF and the Partij voor de Vrijheid (PVV) have acted as government support parties, thus enabling them to influence policymaking directly or indirectly. All in all, a number of populist parties have established themselves as permanent fixtures of Western European party systems (Akkerman and De Lange, 2012; De Lange, 2012a).

In Central and Eastern Europe the picture is both more complex and more volatile. It is more complex, because it is challenging to classify parties according to existing typologies. First of all, many parties adhering to one of the three great ideologies (conservatism, liberalism or socialism) have gone through brief periods of populism (e.g. the Bulgarska sotsialisticheska partiya (BSP) in Bulgaria or the Partidul Democrației Sociale in România (PDSR) in Romania), while others have become increasingly populist over time (e.g. Fidesz in Hungary or the PiS in Poland; see Chapter 9).[6] Second, many parties combine nationalism with socialism, making it impossible to qualify them as either national populists or social populists. In other words, many populist parties have emerged in Central and Eastern Europe since 1989, but it is difficult to classify and list them.[7] A description of the Central and Eastern European population of populist parties is further hindered by the fact that Central and Eastern European party systems remain highly uninstitutionalised. Thus, many populist parties may have achieved an electoral breakthrough, but did not managed to persist. In fact, many populist parties listed in Table 8.2 are defunct or at least no longer represented in parliament. The list of

[6] Some scholars have therefore proposed a distinction in Central and Eastern Europe between hard and soft populists (e.g. Lang, 2005).

[7] To facilitate classification, Ucen (2007) developed an elaborate classification scheme that distinguished between centrist or new populism, national populism, social populism, the radical left and the radical right.

populist parties in Central and Eastern Europe is therefore a tentative one that should be improved upon when more detailed country studies become available.

Nevertheless, a number of populist parties in Central and Eastern Europe can be identified. Among the most institutionalised and well-known populist parties in this region are the conservative/neo-liberal populist Fidesz in Hungary and PiS in Poland, the national populist Ataka in Bulgaria, Hrvatska demokratska zajednica (HDZ) and Hrvatska stranka prava (HSP) in Croatia, Jobbik in Hungary and the social populist Smer in Slovakia. The electoral successes of these parties have been noteworthy, with some almost or actually gaining a majority of the votes. As a consequence, Central and Eastern European populist parties have governed as frequently as their Western European counterparts (De Lange, 2012b).

Explaining the electoral success of populist parties is a challenging and complex task. On the basis of previous research it has been established that comprehensive explanations should include both demand-side factors (factors related to whether a breeding ground for populist voting exists) and supply-side factors (factors that describe the populist parties themselves and the environment in which they operate). Especially the interaction between demand and supply can account for the emergence and persistence of populist parties (Mudde, 2007).[8] Although it is beyond the scope of this chapter to empirically investigate every explanation that has been discussed in the literature on populism (for an overview of relevant studies see Mudde, 2007; Van der Brug and Fennema, 2007), we discuss and empirically examine two types of individual-level demand-side variables that have been found to play a role in explaining support for populist parties: socio-demographic background characteristics and political attitudes and preferences. A number of scholars have demonstrated that certain groups in society, such as men, the lower educated, and the young and the old, are more inclined to vote for populist parties than others. It has also been shown that certain attitudes and preferences, for example Euroscepticism, political cynicism and distrust, and opposition to ethnic minorities, immigration and multiculturalism, make citizens more likely to support populist parties (e.g. Arzheimer and Carter, 2006; Betz, 1994; Lubbers *et al.*, 2002; Van der Brug *et al.*, 2005).

To examine these factors more closely, we compare the socio-demographic background characteristics, attitudes and preferences of citizens who support populist parties with the background characteristics and attitudes and preferences of citizens who do not

[8] Although these and other findings presented in this section are based primarily on research that examines the support for nationalist populist parties, it is expected that the findings also relate to the support of agrarian, conservative/neo-liberal and social populist parties. With the exception of attitudes that are related to the core ideology of these parties (e.g. opposition to ethnic minorities, immigration and multiculturalism in the case of national populist parties, or attitudes towards the welfare state in the case of social populist parties), the explanations discussed are general enough to explain why certain citizens feel attracted to anti-establishment parties that defend the interests of the 'ordinary people'.

support these parties. Using the European Election Studies (EES) of 2009 (Van Egmond *et al.*, 2010), which consists of representative samples drawn from 27 European countries and includes information about 28 populist parties in both Western and Central and Eastern Europe,[9] we map the differences between populist and non-populist voters on four background characteristic (age, education, gender and income) and three attitudes and preferences (Europe, immigration and satisfaction with democracy).[10]

Figure 8.1 demonstrates that the electorates of non-populist and populist parties are in some respects quite similar in age. Both non-populists and populists have an electorate that consists of 15% young voters (aged between 18 and 30 years) and 21% old voters (aged between 66 and 99 years). They differ, however, when it comes to middle-aged voters. Of the citizens supporting populist parties, 33% are aged between 31 and 50 years and 31% are aged between 51 and 65 years. For citizens supporting non-populist parties, the percentages are 35 and 29% respectively. In other words, populist parties fare relatively well among the baby-boomer generation. These findings contradict

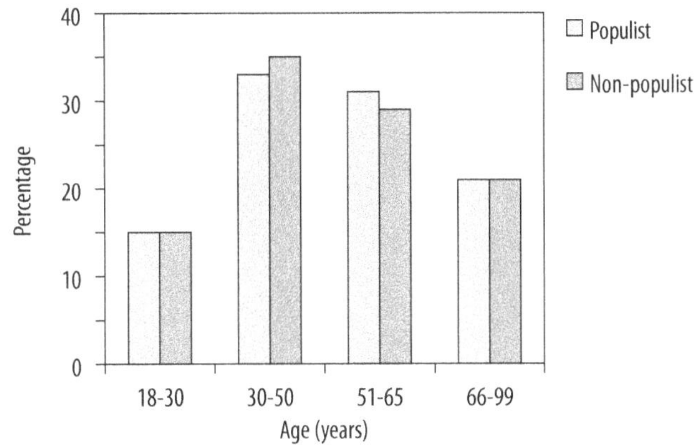

Figure 8.1. Differences in age between electorates of non-populist and populist parties.

[9] In each country included in the EES, the sample size was about 1000 respondents. The populist parties included in the dataset are: AN, BNP, BZÖ, DF, Die Linke, Fidesz, FN, FNb, FPÖ, Jobbik, LDD, LN, LPF, MIÉP, PdL, PiS, PRM, PS, PVV, Samoobrona, SD, SF, SNS (Slovak and Slovenian), SP, UKIP, and VB. The dataset also includes the German Republikaner, a national populist party that has never gained federal parliamentary representation, but has been represented at the Länder (state) level.

[10] Education was measured as the age in years at which respondents completed or ended their full-time education and was recoded into three categories (0-16 = low levels of education; 17-20 = middle levels of education; and 21-100 = high levels of education). Social class was measured on a 5-point scale as the social class to which respondents think they belong and recoded into a 3-point scale. The attitude towards immigration was measured on a 5-point scale as to the level of agreement with the statement: 'Immigration should be decreased significantly', and was recoded into a 3-point scale. Satisfaction with democracy was measured on a 4-point scale, ranging from 'very satisfied' to 'very unsatisfied' and was recoded into two categories. Attitude towards the EU was measured on an 11-point scale, ranging from 'Unification has already gone too far' to 'Unification should be pushed further', and was recoded into a 3-point scale.

previous research, which has shown that both the oldest and youngest voters are most likely to support populist parties (e.g. Arzheimer and Carter, 2006).

Figure 8.2 demonstrates that support for populist parties is greater among men than among women. It shows that 53% of the citizens supporting populist parties are male and 47% are female, while 44% of the citizens supporting non-populist parties are male and 56% are female. This finding is in line with previous research, which has established that men are more likely to vote for populist parties than women (e.g. Arzheimer and Carter, 2006; Betz, 1994; Lubbers *et al.*, 2002).

Figure 8.3 also confirms previous research. It highlights that the percentage of citizens with lower levels of education is higher among those supporting populist parties than among those who do not support populist parties (25 versus 21%). Conversely, the percentage of citizens with higher levels of education is considerably lower amongst those that support populist parties as compared to those who do not support populists

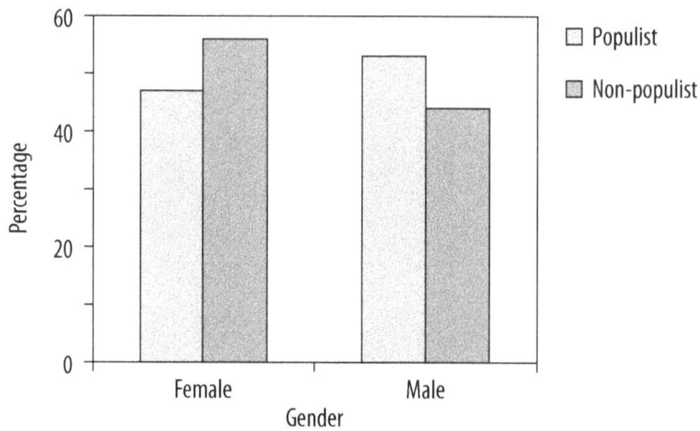

Figure 8.2. Differences in gender between electorates of non-populist and populist parties.

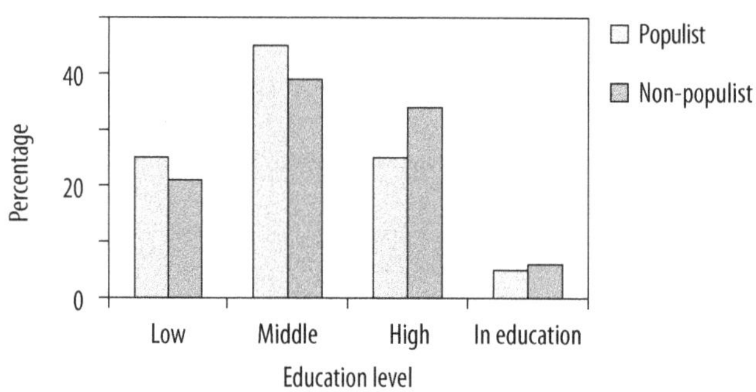

Figure 8.3. Differences in educational levels between electorates of non-populist and populist parties.

(25 versus 34%). Previous research has pointed to the crucial role that education plays in explaining support for populist parties (e.g. Arzheimer and Carter, 2006; Lubbers *et al.*, 2002). The effect of education could be due to various causal mechanisms, such as the capacity of the less well educated to deal with complexity and information overload, the competition experienced by the less well educated in the housing and the labour markets (e.g. Kriesi *et al.*, 2008), or the exposure of the more highly educated to liberal values in the education system (e.g. Weakliem, 2002).

Figure 8.4 reveals that the electorate of populist parties consist of 47% of citizens who identify as lower class, while the electorate of non-populist parties consists of 39% of citizens who identify as such. Among citizens from the middle classes, the figures are 45 versus 50% and for the upper classes, 8 versus 11%. Again, these findings corroborate what has been found in previous studies. People from lower social classes are more inclined to vote for populist parties than people from higher social classes (e.g. Kitschelt, 1995; Lubbers *et al.*, 2002).

It can thus be concluded that populist parties draw disproportionate support from less well educated men who belong to the lower social classes. However, a number of scholars have argued that socio-demographic variables are only of minor importance in explaining the support for populist parties (e.g. Ivarsflaten, 2008; Kitschelt, 1995; Van der Brug, 2003; Van der Brug *et al.*, 2000). According to these authors, the attitudes of citizens matter more, and they point towards attitudes with regard to the European integration process, immigration and integration, and the state of democracy in particular.

Figure 8.5 indicates that those who support populist parties are more often Eurosceptics than those who do not support populist parties. Among those who vote for populist parties, 45% believe that the European unification process has gone too far, while only

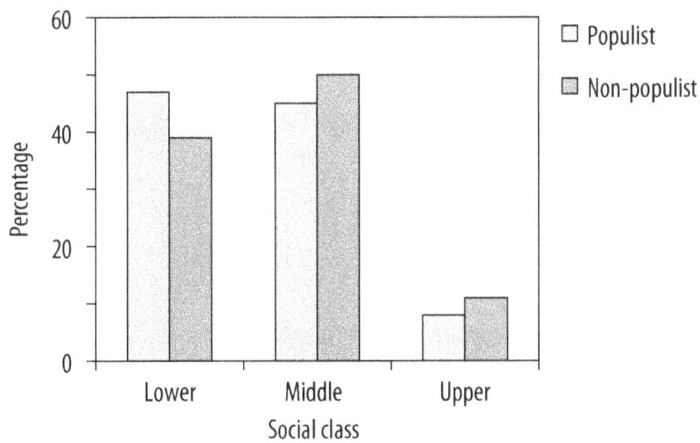

Figure 8.4. Differences in social class between electorates of non-populist and populist parties.

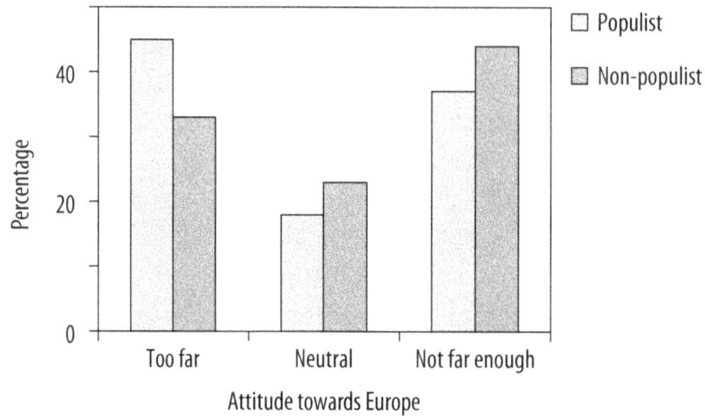

Figure 8.5. Differences in attitudes towards Europe between electorates of non-populist and populist parties.

37% think that it has not gone far enough. Among the citizens who do not support populist parties, the percentages are 33 and 44%, respectively. Our findings complement those of previous studies, which have indicated that Euroscepticism is a good predictor of support for populist parties (e.g. Arzheimer, 2009; Ivarsflaten, 2005; Van der Brug *et al.*, 2005; Werts *et al.*, 2012).

Figure 8.6 shows that among citizens who support populist parties, 71% agree with the claim that immigration should be decreased significantly, whereas this is 58% among those who support non-populist parties. Conversely, 13% of citizens who support populist parties disagree with the claim, compared to 26% of citizens who

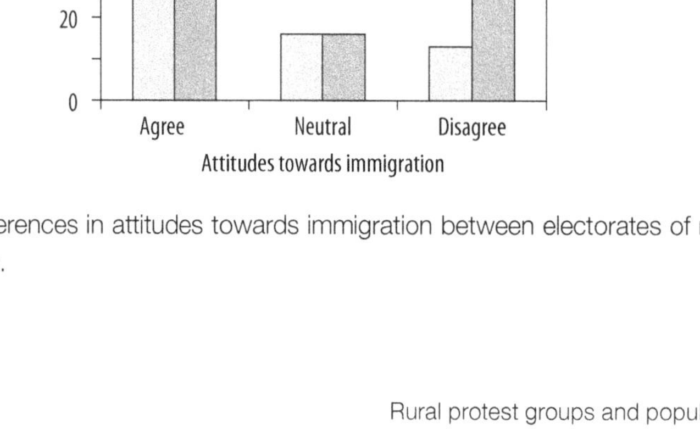

Figure 8.6. Differences in attitudes towards immigration between electorates of non-populist and populist parties.

do not support these parties. However, it can be expected that immigration issues do not have the same impact on the support for left-wing and right-wing populist parties. After all, many scholars have demonstrated that citizens who want to curb the influx of immigrants are inclined to vote for national populist parties because they agree with the ideas of these parties (e.g. Ivarsflaten, 2008; Van der Brug *et al.*, 2000, 2005). We therefore also look at attitudes towards immigration among supporters of left-wing populist parties and contrast them to the attitudes of supporters of non-populist parties (Figure 8.7). Indeed, the supporters of left-wing populist parties are less likely to agree with the statement that immigration should be decreased significantly than the supporters of non-populist parties (39 versus 58%). Again, the pattern reverses when we examine the percentages that disagree with the statement: 46% of those who support left-wing populist parties disagree with it, while 25% of those who do not support populist parties disagree with it.

Figure 8.8 demonstrates that citizens who support populist parties are more likely to be dissatisfied with democracy (56% not satisfied versus 44% satisfied) than citizens who do not support these parties (44% not satisfied versus 56% satisfied). Our findings are thus in line with previous studies which have demonstrated that dissatisfaction with the political system has a large impact on support for populist parties (e.g. Bergh, 2004; Betz, 1994; Bélanger and Aarts, 2006; Norris, 2005; Schumacher and Rooduijn, 2013; Swyngedouw, 2001).

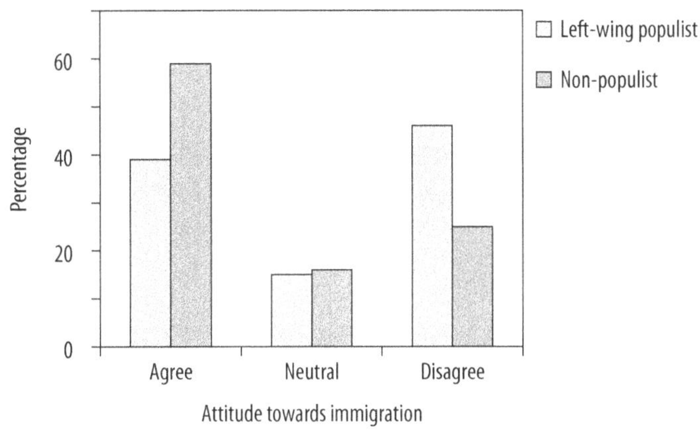

Figure 8.7. Differences in attitudes towards immigration between electorates of non-populist and left-wing populist parties.

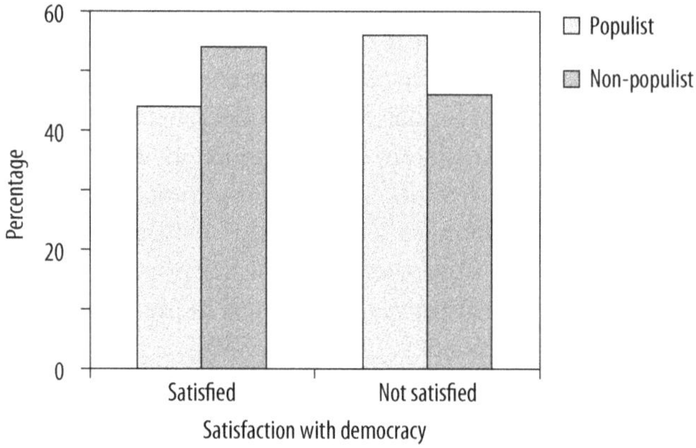

Figure 8.8. Differences in satisfaction with democracy between electorates of non-populist and populist parties.

8.5 Agrarian populism, populist voting and rural support

In the literature, agrarian populism is discussed as a clearly defined and existing type of populism (e.g. Canovan, 1981; Mudde, 2001; Taggart, 2004). However, in contemporary Europe few examples of agrarian populism can be found. In Table 8.1 and 8.2 only two agrarian populist parties have been identified: Független Kisgazda, Földmunkás és Polgári Párt (FKgP) in Hungary and Samoobrona Rzeczypospolitej Polski (Samoobrona) in Poland (Lang, 2005; Mudde, 2001).

The Independent Smallholders, Agrarian Workers and Civic Party, or FKgP, was founded in 1908 and highly successful during the interwar period. Like many agrarian populist parties in Central and Eastern Europe, it was dissolved after the communist regime was installed and was refounded shortly before the fall of communism. In the first free elections in Hungary in 1990 the FKgP won 11.7% of the vote (Mudde, 2000). The support for the party was mainly agrarian, despite the industrialisation and urbanisation that took place during communist rule, while the distribution pattern of support across the Hungarian regions in 1990 matched that of the pre-communist period (Körösényi, 1992). The party managed to maintain its support in 1994 (8.8% of the vote) and 1998 (13.2%) and even governed with the centre-right from 1990 to 1994 and again in 1998, but collapsed in 2002 (0.8%) as a consequence of internal strife (Batory and Sitter, 2004: p. 528). Agrarian issues took centre stage in the populist programme of the FKgP. The party claimed to defend the material interests of farmers, as well as those of the countryside more broadly speaking. For this reason, it was initially in favour of Hungary's accession to the EU, which it expected to benefit

Hungarian farmers through financial transfers related to the Common Agricultural Policy. However, when the consequences of EU integration became clearer, the FKgP became more critical, opposing, for example, landownership by EU citizens (Batory, 2002: pp. 532-533).

In Poland, Samoobrona or Self-Defence was founded in 1992, both as a farmers' trade union and a political party (see also Chapter 9).[11] Led by the charismatic Andrzej Lepper, the party did not manage to enter parliament in 1993 (2.8% of the vote) or 1997 (0.1%), nor did it fare particularly well in the presidential elections of 1995 (1.3%) and 2000 (3.0%). Its breakthrough came in 2001, when it won 53 out of 460 seats in the Sejm. In 2005 it won an additional three seats, making the party an attractive coalition partner to the governing PiS. After initially supporting the government from the opposition benches, Samoobrona assumed government responsibility in 2006 together with the national populist Liga Polskich Rodzin (LPR). This decision proved disastrous for both parties, leading to their electoral demise in the 2007 elections (De Lange and Guerra, 2010). The populist programme on which Samoobrona campaigned in the 1990s and 2000s consisted of three key elements: 'a sense of crisis stemming from the betrayal of ordinary people by a corrupt establishment, contempt for the rules of representative democracy (and the 'self-proclaimed new elites' who are its elected leaders), and the juxtaposition of the healthy heartland (symbolised by Pope John Paul II's moral authority) with schemes of foreign economic domination' (Jasiewicz, 2008: p. 14). The party opposed the economic inequalities that emerged in Poland after the government embraced a free market philosophy and can therefore be qualified as left-wing. It also defended the interests of farmers affected by the economic transition of the 1990s, most of which were located in the northeastern and western regions of the country (Krok-Paszkowska, 2003). Samoobrona was hostile to Poland's accession to the EU, arguing that the Common Agricultural Policy and other EU policies would destroy the Polish agricultural sector, as well as the industrial sector and the trade sector. Consequently, the party believed EU membership would benefit the cultural, economic and political elite at the expense of ordinary people (Jasiewicz, 2008: p. 11).

However, both the FKgP and Samoobrona abandoned their agrarian populism in favour of a left-leaning populism in the 2000s. According to Ucen (2007: p. 53), the collectivisation of farming and land in previous decades implied that the farming population increasingly voted for interest groups and post-communist parties. To curb this trend, their stances became increasingly indefinable. Moreover, they adopted a belligerent form of political populism, which alienated both actors in national politics and their former voters. In Samoobrona's 2005 and 2007 campaigns, for example, agricultural and rural issues did not play a major role (Jasiewicz, 2008: p. 14). As a

[11] Samoobrona is not the only agrarian party to take part in Polish politics after *die Wende*. However, a number of scholars have argued that the other possible candidate, the Polskie Stronnictwo Ludowe (PSL), is not populist (e.g. Mudde, 2001).

consequence of this development, both parties became less attractive in electoral terms and less relevant in political terms, and eventually disappeared from parliament.

Agrarian populist parties are currently not represented in Western European parliaments. Although agrarian populist parties had some success in the 1950s, 1960s and 1970s (e.g. the Boerenpartij (BP) led by the farmer Hendrik Koekoek in the Netherlands (Chapter 13), or Poujadism in France), electoral support for these parties dwindled with the onset of post-industrial society. Interestingly, a number of agrarian populist parties have managed to remain relevant and representative by morphing into national populist parties. This is most notably the case in Finland and Switzerland, where parties that defended the interests of farmers in the 1970s and 1980s now campaign against corrupt politicians, Europe and immigration. In Finland, the Suomen maaseudun puolue (SMP) was founded by Veikko Vennamo, a former member of parliament for the Agrarian League or Maalaisliitto, in 1959 (Chapter 10). The party entered the Eduskunta in 1970, gaining almost 10% of the parliamentary seats and was initially known for its agrarianism, its defence of the 'forgotten people' and its virulent attacks on the political establishment. However, in the 1990s the party disintegrated after Finland became post-industrial and prominent SMP members, including the member of parliament Raimo Vistbacka and party chair and secretary Timo Soini founded the PS. Contrary to the SMP, the PS is known for its anti-elitism, anti-consensus, anti-intellectualism, anti-multiculturalism, and its Euroscepticism, making it a national populist party (Arter, 2012).[12]

Although agrarian populism is far less prominent in Europe than political populism, most national populists do have a clearly defined stance on agrarian issues. Generally speaking, these parties have a nationalist and protectionist economic programme, which aims to defend national businesses, including agrarian enterprises. It also includes an emphasis on national preference, which usually results in a call to consume national products and to be self-sufficient when it comes to growing and processing produce. Moreover, the agricultural sector is perceived as a central employer and investor, making it a key sector for economic growth and national survival (Mudde, 2007: p. 127).[13] This view is easily illustrated by looking at the programmes of parties in both Central and Eastern Europe and Western Europe. The German Republikaner, for example, argue that '[a]griculture is an essential and elementary component of our national economy. It should secure our nourishment and keep us from political

[12] A similar story can be told about the Swiss SVP. Although initially only an agrarian party and not a populist one, it converted to national populism in the 1990s under the leadership of Christoph Blocher. With regard to the change, McGann and Kitschelt (2005: p. 153) observe that 'taking a more militantly anti-internationalist and authoritarian position did not alienate the party's existing traditionalist constituency [of farmers and small business owners], but did broaden its appeal to new groups, particularly blue-collar workers'.

[13] An agrarian component is also present in national populist parties' opposition to European integration. Most parties are highly critical of the Common Agricultural Policy and feel that the EU favours big corporations that operate internationally over local farmers (Mudde, 2007).

dependence and blackmail'.[14] In a similar fashion, the French FN claims that '[t]he relocation of companies' production processes to the national territory in the context of the French politics of reindustrialisation will permit production to take place in the close vicinity of consumers and distributers' and should be '[a]ssociated with the strong valorisation of 'buying French'' (for the FN, see Chapter 11).[15] The Belgian VB also underlines the importance of the agricultural sector, stating that '[t]he Vlaams Belang considers agriculture to be a strategic sector, the knowledge and skills of which should be protected. If we fail to do so, our production of food will become entirely dependent on foreign lands'.[16] In Central and Eastern Europe, national populist parties also campaign on agricultural issues. According to the LPR in Poland, for example, '[A] nation that fails to nourish itself will never be truly free (is destined to be enslaved)'.[17] In fact, it could be argued that these issues are more prominent in this region, because the agricultural sector is large and employs more people than in Western Europe. However, Central and Eastern European parties, such as Ataka, the LPR, and Magyar Igazság és Élet Pártja (MIEP), do highlight issues that are region specific, most notably the land privatisation that has taken place after the fall of communism (Mudde, 2007: p. 128).

In addition to adhering to agricultural protectionism, many national populist parties are also more broadly concerned with the appearance of the countryside and the disappearance of a rural way of life, as these issues often tie into their 'identitarian' programme. The Dutch PVV, for example, merges such concerns with manifest populism in the following statement: 'wind turbines do not run on wind, but run on subsidies. Nevertheless, the government continues to place these monstrosities in our beautiful Dutch countryside ... Everywhere they are zooming. They ruin the landscape, kill the birds and cost lots of money. The PVV says: give up this useless left-wing hobby'.[18]

Interestingly, a number of social populist parties share some of the views of national populist parties, even though they often take more moderate stances on agrarian and rural issues. The Dutch SP, for example, argues that '[t]he landscape and nature are too easily sacrificed to commercial short-term interests. The farmers that manage our scenery are ruined as a result of lethal international competition. If we want to let the peasantry do its job ... protect the countryside ... the coming government should make different choices in this area'.[19]

---

[14] Republikaner, as cited in Mudde, 2007: p. 127.

[15] Front National programme: http://tinyurl.com/nvvv3wa.

[16] Vlaams Belang programme: http://tinyurl.com/8xthtrs

[17] LPR programme, as cited in Mudde, 2007: p. 127.

[18] PVV programme: http://tinyurl.com/bu5h8z5.

[19] SP programme, 2010: http://tinyurl.com/qz9a604

Does the concern with agricultural and rural issues also translate into support from farmers and people living in the countryside? To answer this question we can analyse the data from the EES 2009 once again, looking in particular at the area/community in which the respondents live and the sector in which they are employed.[20] Figure 8.9 shows that, of the citizens supporting populist parties, 35% live in rural territories and villages and 34% in mid-sized and small-sized towns. The percentages of citizens supporting non-populist parties that live in these community types are slightly lower (32 and 28% respectively). Conversely, the percentages of citizens supporting non-populist parties living in cities and large towns are higher than the percentages of citizens supporting populist parties living in these community types (29 versus 20%). In other words, populist parties fare relatively well in the countryside and less well in urbanised parts of Western European countries.

When it comes to occupational sectors, a less distinct pattern emerges. Figure 8.10 demonstrates that the group of citizens working in the agricultural sector and supporting populist parties is only marginally larger than the group of citizens in that sector that do not support these parties (9 versus 6%). Thus, agrarians are a bit overrepresented in the electorates of populist parties. Although the differences are small, they are in line with previous research, which has shown that some national populist parties benefit from strong support from people working in the agricultural sector or in small enterprises. In the late 1990s, for example, 36.4% of farmers and small business owners supported the SVP in Switzerland, while another 32.7% cast their vote for the Austrian FPÖ (McGann and Kitschelt, 2005: p. 155). Moreover, in terms of attitudes towards agricultural issues, national populist parties might attract voters who have an affinity with agrarian populism. In Poland, for example, supports of the LPR are almost as likely

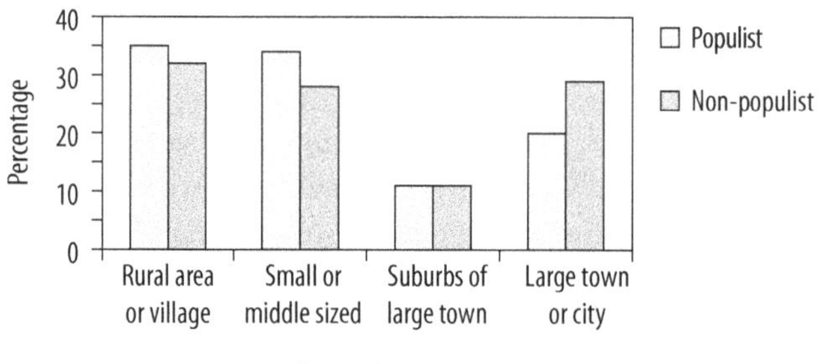

Figure 8.9. Differences in community type between electorates of non-populist and populist parties.

[20] We analysed the variables 'rural-urban area' (q115) and 'sector of employment' (q111). Since these variables have only been collected in Western European countries, we are unable to assess differences within communities and occupational sectors in Central and Eastern Europe.

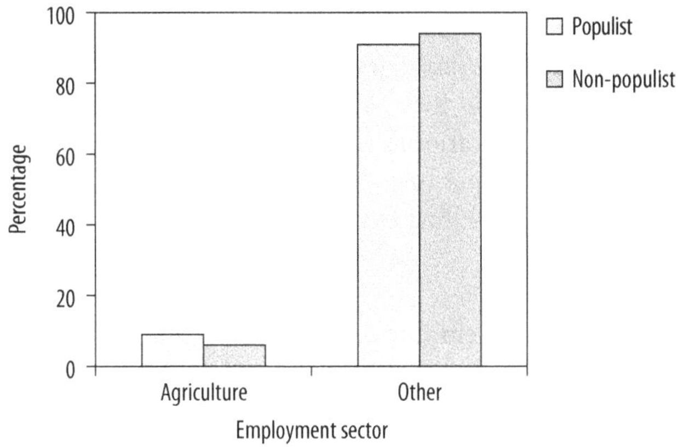

Figure 8.10. Differences in employment sector between electorates of non-populist and populist parties.

to be in favour of agricultural state subsidies as supporters of Samoobrona (7.6 versus 8.0, on a scale of 0 to 10) (Jasiewicz, 2008: p. 18).

8.6 Conclusion

Europe has witnessed a slow but steady upsurge of populist parties since the 1980s. The number of populists has risen significantly in both Central and Eastern Europe and Western Europe. Parties such as the BZÖ and FPÖ in Austria, Die Linke in Germany, the DF in Denmark, the FN in France, the FRP in Norway, the AN, LN and PdL in Italy, the PVV and SP in the Netherlands, the SVP in Switzerland, and the VB in Belgium have become institutionalised and *salonfähig*, and thereby define parts of Western European party systems. In Central and Eastern Europe, populist parties such as Ataka in Bulgaria, Fidesz and Jobbik in Hungary and the PiS in Poland have also become key players in politics. On the basis of their core ideologies, these parties can be classified as either agrarian populist, conservative or neo-liberal populist, national populist or social populist.

Although conservative/neo-liberal, national and social populist parties have increased their presence in European governments and parliament, the role played by agrarian populist parties has actually diminished over the past decades. While agrarian populism was a major political force – especially in Central and Eastern Europe – before the Second World War, it is currently a marginal phenomenon. The few agrarian populist parties that have been represented in post-war Europe have either disappeared from parliament and/or have transformed themselves into either national or social populist parties. However, despite the marginalisation of agrarian populist parties, a weak link between the agrarian and the rural on the one hand, and the populist on the other remains. First of all, national populist parties, and to a lesser extent social populist parties, defend the interests of both the agrarian sector and the countryside, arguing that they are vital parts of the national economy and identity. Populist parties thus seem to have a clearly defined stance on agrarian questions. Second, farmers and people living in the countryside are more likely to support populist parties than citizens living in urban areas.

This chapter constitutes a first attempt to unravel the relationship between the agrarian, the rural and the populist. To assess whether there is indeed a relationship between the agenda of national and social populist parties and the support of agrarians and people living in rural areas and small villages, case studies of specific populist parties and their voters are required. Future research could, for example, analyse the programmes of populist parties and speeches of populist leaders more systematically and might also examine the attitudes of their supporters in detail through qualitative interviews and/or quantitative surveys. The following chapters offer such studies, examining the Polish, Finnish, French and Dutch cases.

References

Akkerman, T. and De Lange, S.L., 2012. Radical right parties in office: incumbency records and the electoral cost of governing. Government and Opposition 47(4): 574-596.

Akkerman, T., 2005. Anti-immigration parties and the defence of liberal values: the exceptional case of the List Pim Fortuyn. Journal of Political Ideologies 10(3): 337-354.

Albertazzi, D. and McDonnell, D., 2008. Introduction: the sceptre and the spectre. In: Albertazzi, D. and McDonnell, D., (eds.) Twenty-first century populism. Palgrave Macmillan, New York, NY, USA, pp. 1-11.

Arter, D., 2012. The breakthrough of another West European populist radical right party? The case of the True Finns. Government and Opposition 45(4): 484-504.

Arzheimer, K. and Carter, E., 2006. Political opportunity structures and right-wing extremist party success. European Journal of Political Research 45(3): 419-443.

Arzheimer, K., 2009. Contextual factors and the extreme right vote in Western Europe, 1980-2002. American Journal of Political Science 53(2): 259-275.

Barr, R.R., 2009. Populists, outsiders and anti-establishment politics. Party Politics 15(1): 29-48.

Batory, A. and Sitter, N., 2004. Cleavages, competition and coalition-building: agrarian parties and the European question in Western and East Central Europe. European Journal of Political Research 43: 523-546.

Batory, A., 2002. Attitudes to Europe: ideology, strategy and the issue of European Union membership in Hungarian party politics. Party Politics 8(5): 525-539.

Bélanger, E. and Aarts, K., 2006. Explaining the rise of the LPF: Issues, discontent, and the 2002 Dutch election. Acta Politica 41(1): 4-20.

Bergh, J., 2004. Protest voting in Austria, Denmark, and Norway. Scandinavian Political Studies 27(4): 367-389.

Betz, H.-G. and Meret, S., 2009. Revisiting lepanto: the political mobilization against Islam in contemporary Western Europe. Patterns of Prejudice 43(3-4): 313-334.

Betz, H.-G., 1993. The two faces of radical right-wing populism in Western Europe. The Review of Politics 55: 663-686.

Betz, H.-G., 1994. Radical right-wing populism in Western Europe. Macmillan, Basingstoke, UK, 288 pp.

Betz, H.-G., 2007. Against the 'green totalitarianism': anti-Islamic nativism in contemporary radical right-wing populism in Western Europe. In Schori Liang, Ch. (ed.) Europe for the Europeans: the foreign and security policy of the populist radical right. Ashgate Publishing, Farnhem, UK, pp. 33-54.

Bobbio, N., 1996. Left and right: the significance of a political distinction. University of Chicago Press, Chicago, IL, USA, 148 pp.

Bos, L., Van der Brug, W. and De Vreese, C. H., 2011. How the media shape perceptions of right-wing populist leaders. Political Communication 28(2): 182-206.

Bruneau, I., 2015. The Confédération Paysanne and the political field: a conflicted history. Chapter 5. In: Strijker, D., Voerman, G. and Terluin, I.J. (eds.) Rural protest groups and populist political parties. Wageningen Academic Publishers, Wageningen, the Netherlands, pp 101-125.

Canovan, M., 1981. Populism. Harcourt Brace Jovanovich, New York, NY, USA, 351 pp.

Canovan, M., 1999. Trust the people! Populism and the two faces of democracy. Political Studies 47(1): 2-16.

Canovan, M., 2002. Taking politics to the people: Populism as the ideology of democracy. In: Mény, Y. and Surel, Y. (eds.) Democracies and the populist challenge. Palgrave, New York, NY, USA, pp. 25-44.

Carter, E., 2005. The extreme right in Western Europe: success or failure? Manchester University Press, Manchester, UK, 271 pp.

De Lange, S.L. and Guerra, S., 2010. The league of Polish families between East and West, past and present. Communist and Post-Communist Studies 42(4): 527-549.

De Lange, S.L., 2012a. New alliances: why mainstream parties govern with radical right-wing populist parties. Political Studies 60(4): 899-918.

De Lange, S.L., 2012b. Radical right-wing populist parties in office: a cross-national comparison. In: Backes, U. and Moreau, P. (eds.) The extreme right in Europe: current trends and perspectives. Vandenhoeck and Ruprecht, Göttingen, Germany, pp. 171-194.

Decker, F., 2004. Der neue Rechtspopulismus. Verlag Leske+Budrich, Wiesbaden, Germany, 320 pp.

Derks, A., 2006. Populism and the ambivalence of egalitarianism. How do the underprivileged reconcile a right wing party preference with their socio-economic attitudes? World Political Science Review 2(3): 175-200.

Di Tella, T.S., 1997. Populism into the twenty-first century. Government and Opposition 32(2): 187-200.

Dziewanowski, M.K., 1996. Polish populism. In: Held, J. (ed.) Populism in Eastern Europe. Racism, nationalism, and society. Boulder: East European Monographs, pp. 163-196.

Freeden, M., 1998. Is nationalism a distinct ideology?. Political Studies 46(4): 748-765.

Germani, G., 1978. Authoritarianism, fascism, and national populism. Transaction Books, New Brunswick, NJ, USA, 292 pp.

Havlik, V. and Pinkova, A., 2012. Party populism in East-Central Europe. In Havlik, V. and Pinkova, A., (eds.) Populist political parties in East-Central Europe. Muni Press, Brno, Chezch Republic, pp. 285-299.

Hawkins, K.A., 2009. Is Chávez populist? Measuring populist discourse in comparative perspective. Comparative Political Studies 42(8): 1040-1067.

Hawkins, K.A., 2010. Venezuela's chavismo and populism in comparative perspective. Cambridge University Press, Cambridge, UK, 304 pp.

Hawkins, K.A., Riding, S. and Mudde, C., 2012. Measuring populist attitudes. Available at: http://tinyurl.com/peedtov.

Ionescu, G. and Gellner, E., 1969. Introduction. In: Ionescu, G. and Gellner, E. (eds.) Populism: its meaning and national characteristics. Weidenfeld and Nicolson, London, UK, pp. 1-5.

Ivarsflaten, E., 2005. The vulnerable populist right parties: no economic realignment fuelling their electoral success. European Journal of Political Research 44(3): 465-492.

Ivarsflaten, E., 2008. What unites right-wing populists in Western Europe? Comparative Political Studies 41(1): 3-23.

Jagers, J. and Walgrave, S., 2007. Populism as political communication style: an empirical study of political parties' discourse in Belgium. European Journal of Political Research 46(3): 319-345.

Jagers, J., 2006. De stem van het volk! Populisme als concept getest bij Vlaamse politieke partijen. Dissertation, Universiteit Antwerpen, Antwerpen, Belgium, 391 pp.

Jasiewicz, K., 2008. The new populism in Poland the usual suspects? Problems of Post-Communism 55(3): 7-25.

Kitschelt, H.P., 1995. The radical right in Western Europe: a comparative analysis. University of Michigan Press, Ann Arbor, MI, USA, 352 pp.

Körösényi, A., 1992. Revival of the past or a new beginning? The nature of postcommunist politics. In: Bozóki, A., Körösényi, A. and Schöpflin, G. (eds.) Post-communist transition: emerging pluralism in Hungary. Pinter, London, UK, pp. 111-132.

Kriesi, H., Grande, E., Lachat, R., Dolezal, M. and Bornschier, S., 2008. West European politics in the age of globalization. Cambridge University Press, Cambridge, UK, 448 pp.

Krok-Paszkowska, A., 2003. Samoobrona, the Polish self-defence movement. In: Kopecky, P. and Mudde, C. (eds.) Uncivil society? Contentious politics in Post-Communist Europe. Routledge, London, UK, pp. 114-133.

Laclau, E., 1977. Politics and ideology in Marxist theory: capitalism, fascism, populism, New Left Books, London, UK, 203 pp.

Laclau, E., 2005. On populist reason. Verso, London, UK, 276 pp.

Lang, K.-O., 2005. Populism in Central and Eastern Europe: a threat to democracy or just political folklore? Slovak Foreign Policy Affairs (41):6-17.

Lubbers, M., Gijsberts, M. and Scheepers, P., 2002. Extreme right-wing voting in Western Europe. European Journal of Political Research 41(3): 345-378.

March, L. and Mudde, C., 2005. What's left of the radical left? The European radical left since 1989: decline and mutation. Comparative European Politics 3(1): 23-49.

March, L., 2007. From vanguard of the proletariat to vox populi: left-populism as a 'shadow' of contemporary socialism. SAIS Review 27(1): 63-77.

March, L., 2008. Contemporary far left parties in Europe: from Marxism to the mainstream? Friedrich Ebert Stiftung, Berlin/Bonn, Germany, 24 pp.
Mazzoleni, G., 2003. The media and the growth of neo-populism in contemporary democracies. In: Mazzoleni, G., Stewart, J. and Horsfield, B. (eds.) The media and neo-populism. Praeger, London, UK, pp. 1-20.
McGann, A.J. and Kitschelt, H., 2005. The radical right in the Alps: evolution of support for the Swiss SVP and Austrian FPÖ. Party Politics 11(2): 147-171.
Mény, Y. and Surel, Y., 2002. The constitutive ambiguity of populism. In: Mény, Y. and Surel, Y. (eds.) Democracies and the populist challenge. Palgrave, New York, NY, USA, pp. 1-21.
Mudde, C., 2000. Populism in Eastern Europe; 1-2. East European Perspectives 2(5): 5-6.
Mudde, C., 2001. In the name of the peasantry, the proletariat, and the people: populisms in Eastern Europe. East European Politics and Societies 14(2): 33-53.
Mudde, C., 2004. The populist Zeitgeist. Government and Opposition 39(3): 541-563.
Mudde, C., 2007. Populist radical right parties in Europe. Cambridge University Press, Cambridge, UK, 385 pp.
Norris, P., 2005. Radical right: voters and parties in the electoral market. Cambridge University Press, New York, NY, USA, 349 pp.
Panizza, F., 2005. Introduction: Populism and the mirror of democracy. In: Panizza, F. (ed.) Populism and the mirror of democracy. Verso, London, UK, pp. 1-31.
Pasquino, G., 2008. Populism and democracy. In: Albertazzi, D. and McDonnell, D. (eds.) Twenty-first century populism. Palgrave Macmillan, New York, NY, USA, pp. 15-29.
Pauwels, T., 2010. Explaining the success of neo-liberal populist parties: the case of Lijst Dedecker in Belgium. Political Studies 58(5): 1009-1029.
Pauwels, T., 2012. The populist voter: explaining electoral support for populist parties in the Netherlands, Belgium and Germany. Dissertation, Université Libre de Bruxelles, Brussels, Belgium, 259 pp.
Rooduijn, M. and Pauwels, T., 2011. Measuring populism: comparing two methods of content analysis. West European Politics 34(6): 1272-1283.
Rooduijn, M., De Lange, S.L. and Van der Brug, W., 2014. A populist zeitgeist? Programmatic contagion by populist parties in Western Europe. Party Politics 20(4): 563-575.
Schumacher, G. and Rooduijn, M., 2013. Sympathy for the 'devil'? Voting for populists in the 2006 and 2010 Dutch General Elections. Electoral Studies 32(1): 124-133.
Swyngedouw, M., 2001. The subjective cognitive and affective map of extreme right voters: using open-ended questions in exit polls. Electoral Studies 20(2): 217-241.
Taggart, P., 1995. New populist parties in Western Europe. West European Politics 18(1): 34-51.
Taggart, P., 2000. Populism. Concepts in the social sciences. Open University Press, Buckingham, UK, 140 pp.
Taguieff, P.-A., 1995. Political science confronts populism: from a conceptual mirage to a real problem. Telos 103: 9-43.
Ucen, P., 2007. Parties, populism, and anti-establishment politics in East Central Europe. SAIS Review 27(1): 49-62.
Van der Brug, W. and Fennema, M., 2007. What causes people to vote for a radical-right party? A review of recent work. International Journal of Public Opinion Research 19(4): 474-487.

Van der Brug, W., 2003. How the LPF fuelled discontent: empirical tests of explanations of LPF support. Acta Politica 38(1): 89-106.

Van der Brug, W., Fennema, M. and Tillie, J., 2000. Anti-immigrant parties in Europe: ideological or protest vote? European Journal of Political Research 37(1): 77-102.

Van der Brug, W., Fennema, M. and Tillie, J., 2005. Why some anti-immigrant parties fail and others succeed. Comparative Political Studies 38(5): 537-573.

Van Egmond, M., Sapir, E., Van der Brug, W., Franklin, M. and Hobolt, S.B., 2010. 2009 European election voter study – codebook. University of Amsterdam, Amsterdam, the Netherlands.

Weakliem, D.L., 2002. The effects of education on political opinions: an international study. International Journal of Public Opinion Research 14(2): 141-157.

Werts, H., Scheepers, P. and Lubbers, M., 2013. Euro-scepticism and radical right-wing voting in Europe, 2002-2008: social cleavages, socio-political attitudes and contextual characteristics determining voting for the radical right. European Union Politics 14(2): 183-205.

Weyland, K., 2001. Clarifying a contested concept: Populism in the study of Latin American politics. Comparative Politics 34(1): 1-22.

Zuquete, J.P., 2009. The European extreme right and Islam: new directions? Journal of Political Ideologies 13(3): 321-344.

9 The Polish Self-Defence party: from agrarian protest to the politics of populism, 1991-2007

B. Stanley

School of Law, Politics and Sociology, The University of Sussex, Brighton, United Kingdom; University of Social Sciences and Humanities, Warsaw, Poland; b.stanley@sussex.ac.uk

Abstract

This chapter explains the significance for post-communist Polish party politics of the Self-Defence trade union protest movement and political party. The first section contextualises the emergence of the protest movement and trade union in the early years of post-communist transition, describing its transformation into a wider movement of populist direct action against the politics of liberal-democratic transition and, subsequently, into a political party. The second section focuses on Self-Defence as a political party, detailing the substance of its political appeal and explaining its position in the Polish party system. The third section analyses the nature of the party's support. The fourth and fifth sections describe respectively the breakthrough of the party to the parliamentary sphere and the role it played as a party of opposition and as a coalition partner. The chapter concludes that although Self-Defence's political potential was limited and it was ultimately eclipsed by more capable political actors, it played an important part in reorienting the basic line of competition in the Polish party system.

9.1 Introduction

This chapter explains the significance of the Self-Defence trade union, protest movement and political party to post-communist Polish party politics. The first section contextualises the emergence of the protest movement and trade union in the early years of post-communist transition and describes its transformation into a wider movement of populist direct action against the politics of liberal-democratic transition, and subsequently into a political party. The second section focuses on Self-Defence as a political party, detailing the substance of its political appeal and explaining its position in the Polish party system. The third section analyses the nature of the party's support. The fourth section describes the breakthrough of the party to the parliamentary sphere,

D. Strijker, G. Voerman and I.J. Terluin (eds.) Rural protest groups and populist political parties
DOI 10.3920/978-90-8686-807-0_9, © Wageningen Academic Publishers 2015

and the fifth section discusses the role it played as an opposition party and coalition partner. The paper concludes that although Self-Defence's political potential was limited and it was ultimately eclipsed by more capable political actors, it played an important part in reorienting the basic line of competition in the Polish party system.

9.2 From movement to party: the emergence of Self-Defence, 1991-1993

While Self-Defence was most prominent in Polish politics during the years 2001-2007, its emergence as a political party spanned the first decade of transition. Its roots lay in a protest movement founded soon after the end of communism by Andrzej Lepper, a farmer from the northern Polish town of Darłowo and the unquestioned leader of the movement and party until his suicide in August 2011. While the 'shock therapy' of Poland's economic reform after 1990 had a major impact on many social groups, farmers were hit particularly hard from the outset. State agricultural collectives were liquidated, leading to the emergence of regional pockets of agrarian unemployment, poverty and social exclusion. The removal of domestic and export subsidies, freeing of prices, depressed levels of demand and high interest rates left many farmers struggling to adapt to very different circumstances (Morgan, 1992: p. 154). Many of those who responded by taking on credit found themselves increasingly indebted as they struggled to meet repayments on their loans and faced having their machinery repossessed by bailiffs.

Together with a number of indebted farmers, Lepper organised protests in the Darłowo region during the summer and autumn of 1991, setting up a Regional Committee for the Self-Defence of Farmers (Wojewódzki Komitet Obrony Rolników). Farmers from neighbouring regions joined the protests, and the scope of the agricultural protest movement rapidly expanded into a nationwide organisation, with Lepper elected to lead a Committee for the Defence of Farmers (Komitet Obrony Rolników), set up in Warsaw. During this period, Lepper came to public attention as a result of protests outside the Polish parliament (Sejm) and the signing of an agreement with the government to block bailiffs from taking action against indebted farmers. In the parliamentary election of October 1991, Lepper's Regional Committee put up a list of candidates in the Koszalin-Słupsk constituency under the name of Farmers' Self-Defence (Samoobrona Rolników). However, this venture was not a success, with the list gaining only 1.35% of the vote (Państwowa Komisja Wyborcza, 1991: p. 468).

The organisation took more formal shape during 1992. In January, Lepper and his followers founded the trade union Agrarian Self-Defence (Związek Zawodowy Rolnictwa 'Samoobrona'). At the founding meeting, Lepper argued that existing agrarian trade unions – the most prominent being the agrarian branch of Solidarity

– were not able to ensure the government delivered on its earlier promises (Chłopska Droga, 1992). However, the foundation of Agrarian Self-Defence did not spur on the government to realise its promises, and in April 1992 Lepper led union members in a two-week occupation of the Ministry of Agriculture (Pilot, 1992). While the government refused to enter into dialogue with the union, President Lech Wałęsa invited their representatives to his residence and promised to address their concerns. This episode – and particularly President Wałęsa's intervention – brought the movement to national prominence, and local structures of Agrarian Self-Defence emerged across Poland. As Piskorski (2010: p. 54) notes, it also marked the first point at which the Self-Defence movement reached out to other social and professional groups.

Subsequently, the Self-Defence movement took an increasingly radical turn. Lepper announced that unless his demands were met, the union would create a paramilitary organisation to counteract the 'lawlessness' of state institutions (Naszkowska, 1992). These demands were clearly unrealisable and the threatened 'Peasant Battalions' never properly materialised, but Self-Defence was rapidly gaining a reputation for radicalism. While occupations and road blockades remained the dominant forms of protest, Self-Defence also engaged in more aggressive forms of direct action. During the 1993 general election campaign, a Self-Defence protest in the town of Praszka culminated in activists removing the physically disabled mayor from his offices in a wheelbarrow and carting him around the town square (Jachowicz and Pączek-Haładyj, 1993). In another incident in 1994 – after which Lepper received a prison sentence for incitement – Self-Defence activists occupied a farm in Kobylnica that was in the process of being put into administration. The administrator was attacked, his head shaven, and he was beaten with birch sticks. In a 1999 interview, Lepper asserted, in reference to this incident, that 'birching was always something that happened in Poland. It is a Polish tradition and a warning for everyone ... a warning to bailiffs and banks that no more farm enclosures will be auctioned off' (Kurski, 1999). Along with blockades, manure tipping and scythe-wielding, these methods of protest evoked a popular idiom of summary justice that appealed to the constituencies Lepper courted.

Undeterred by failure in 1991, the Self-Defence movement registered a political party in June 1992, initially under the name of Self-Defence Alliance (Przymierze Samoobrona). With Lepper as leader, the party gathered together activists from Agrarian Self-Defence, the Union of Metallurgists (Związek Zawodowy Metalowców) and several former members of the Green Party (Partia Zielonych) (Gładkiewicz, 2004: p. 147). With the aid of advisors, the party drafted an economic and social programme to 'halt the breakdown of the economy and the state', which went beyond the specific demands of the party's agrarian constituency to address broader issues such as the liquidation of state enterprises, the closure of mines and cuts to social welfare provision (Gładkiewicz, 2004: p. 148; Piskorski, 2010: p. 62). At the end of 1992, activists set up a Committee for the Self-Defence of the Nation (Komitet Samoobrony Narodu) which aimed to create a broad social movement in support of the political party. A number of unions joined the movement, and other political groups opposed to the direction of reforms approached Lepper with proposals for cooperation. Lepper also appealed to the unemployed, the homeless and other excluded social groups to create structures of their own within the Self-Defence movement (Piskorski, 2010: pp. 62-63).

By the spring of 1993, the three main pillars of the Self-Defence movement were in place: the union, the political party and the social movement. The turn toward more formal arenas of political and social participation brought with it a change in the movement's tactics. Seeking to project itself as more electable, Self-Defence scaled down its use of direct action and engaged in more conventional methods of agitation such as written polemics, petitions and appeals to society and the authorities (Gładkiewicz, 2004: pp. 147-148).

9.3 The supply side: party objectives, ideology and place within the party system

Of all parties in the post-communist era, Self-Defence was most clearly a product of the transition process and not the inheritor of any coherent historical tradition. While Lepper's unpredictability and political opportunism led him at times to take inconsistent positions, the general thrust of the party's ideological appeal remained quite consistent over time. The essence of the party's appeal can be best understood with reference to its socioeconomic programme from 2003, which provided a comprehensive diagnosis of the ills of transition and a statement of proposed remedies. It embraced a call for the nationwide satisfaction of positive rights: 'the right to work, for appropriate remuneration, the right to have housing needs satisfied, the right to health and healthcare, the right to education and equal life chances' (Samoobrona Rzeczypospolitej Polskiej, 2003: p. 8).

Self-Defence argued that successive governments had failed to ensure the realisation of these rights, blaming not the hidden hand of the market but a conscious economic conspiracy of domestic elites and foreign capital against the interests of the Polish people, which could be summarised as:

- the erroneous monetary policy of the Polish Central Bank and fiscal policies of the state;
- the rapacious and criminal privatisation of banks and entire branches of Polish industry;
- the uncontrolled importation of goods (Samoobrona Rzeczypospolitej Polskiej, 2003: p. 8).

Self-Defence argued that the combination of these policies led to a situation in which foreign investors and competitors benefited from the openness of the Polish economy, while domestic entrepreneurs and agricultural producers were at the mercy of banks dominated by foreign capital. As a result, Poland had effectively ceded economic sovereignty to foreign interests, rendering impossible 'planning for the future of the state and the national economy in accordance with the needs of society as a whole' (Samoobrona Rzeczypospolitej Polskiej, 2003: p. 8).

Self-Defence's overriding aim was to regain this lost economic sovereignty, with the exercise of strict controls on the operation of banks dominated by foreign capital, the reorientation of monetary policy away from the narrow pursuit of low inflation to a growth and employment remit, the consumption of 60% of Poland's foreign currency reserves in the creation of cheap credit and the emission of government bonds to finance a budget deficit. By increasing the money supply in this fashion, domestic economic entities could finance their ventures and the state would have funds for a wide and ambitious programme of spending on public works and social security.

While Self-Defence's programmatic priorities were mostly economic in nature, it was not averse to tapping the power of religious rhetoric and clerical authority in support of its programmatic priorities, and it advocated more extensive state patronage of the arts and media to protect against 'a flood of cheap cultural models' (Samoobrona Rzeczypospolitej Polskiej, 2003: p. 8). The party's position on European integration could best be characterised as an unenthusiastic, suspicious and grudging acceptance of the unavoidability of accession, interspersed with spasms of hostility. While it appeared by 2003 to have come to acknowledge the weight of opinion among Poles that 'Poland's joining the European Union can be a watershed moment and a great chance for Polish society' (Samoobrona Rzeczypospolitej Polskiej, 2003: p. 57), it was very critical of the conditions that had been negotiated for accession, raising fears of the decline of Polish agriculture in the face of foreign competition, loss of sovereignty over monetary policy and the possibility of German claims for restitution of land and property.

For much of the first two decades of transition, the left-right divide in Polish party politics was largely understood in terms of the 'regime divide' between the broadly social-democratic successors to the communist party and its satellites on the one hand, and the conservative and nationalist parties with roots in the Solidarity movement on the other. As a party without clear roots in either the communist regime or the anti-communist opposition, Self-Defence did not occupy a distinct position with respect to this divide. Instead, it stood out by virtue of its populist appeal. By linking the negative experiences of the agricultural sector to those of a far wider stratum of transition losers, it articulated the core tropes of populist ideology in textbook fashion. Rather than identify with 'quasi-popular' conceptions of the nation, the appeal of Self-Defence was pitched squarely at the ontological divide between elites and people. '[T]hey call us boors. From the beginning of modern times Poles have been divided into gentlemen and boors' (Lepper, 2002a: p. 8).

Self-Defence's challenge to the elite was, averred Lepper, simply another iteration of an age-old pattern:

> [t]he authorities in Poland can be called 'them'. They rule, they make laws, they give, they take, they permit – or not – others to live. The greatest success of Self-Defence is that when talking about us, Poles do not say 'them', but 'us'
>
> In the last elections … [w]e took away from them the certainty of their domination. Achieving the support of more than one-tenth of the electorate, we served a reminder of what is ours. And that is what really builds the greatest anger among 'them' – the political elite. Post-Solidarity and postcommunist – as they call each other …
>
> (Lepper, 2002a: p. 9)

Portraying all elites as indistinguishable, the party set itself against all other political parties. Self-Defence identified the Round Table settlement in 1989 – which paved the way for transition – as a moment of elite agreement and 'an arrangement made by representatives of the former communist party and part of the anti-communist opposition' (Samoobrona Rzeczypospolitej Polskiej, 2004). Those parties with roots in this 'deal' were scorned as emissaries of foreign capital, abusing the fruits of hard-won democracy (Samoobrona Rzeczypospolitej Polskiej, 2005). Against this elite, Self-Defence declared itself 'the voice of all social groups which, as a result of the reforms of the 1990s, have found themselves on the edge of poverty and despair' (Samoobrona Rzeczypospolitej Polskiej, 2004).

The invocation of authenticity and ordinariness is a key aspect of populism's appeal to the people, as it allows populists to lay claim to genuine representativeness. Lepper expressed this sentiment in his insistence that '[o]ur country should be ruled by the people and the representatives of their majority' (Lepper, 2002a: p. 196). Here, 'representation' refers not to the elite's enlightened stewardship of the general good but to the reflective expression of the popular will, made conceptual flesh in the form of one who originates from the people and can speak as one of them:

> I am the voice of the poor, deprived and humiliated … Self-Defence and Andrzej Lepper never were, are not, and never will be 'them'. We are 'us'.
>
> (Lepper, 2002a: p. 11)

Making a virtue of its lack of historical ties and involvement in the implementation of transition, the party emphasised its newness while attacking other parties – in an oft-repeated phrase – as 'having had their turn' ('oni już byli'). In its consistent combination of the classic elements of populist discourse, Self-Defence was as close as Polish politics has yet come to a 'pure' populist party.

9.4 The demand side: Self-Defence's sources of support

By broadening the scope of its appeal and inveighing against the failures of all transition elites to date, Self-Defence sought to attract support from sections of society who had – or could be persuaded to believe they had – lost out as a result of the transition process. In some cases there was a clear material basis for distinguishing between the inhabitants of a 'Poland of transformation' and a 'Poland outside transformation' (Rychard, 2006: p. 139). Low-skilled workers, those living in impoverished regions and those with low levels of educational attainment faced significant structural barriers to material prosperity, and those whose real income levels declined were material losers of transition by definition. Yet material location was neither an infallible structural predictor of 'transition loserdom' nor the only criterion by which individuals might assess their transition-era status. Miners retained many of the material privileges they enjoyed during communism, but the fact that they now had to fight to do so and that they lived with the constant spectre of rationalisation and restructuring, struck at a less immediately tangible but no less meaningful legacy of communism: their sense of social prestige and respect. The cohort of retirees was not among the social groups most adversely affected by transition, but was widely assumed to have suffered disproportionate hardships, and was regularly appealed to on this basis.

To characterise the nature of Self-Defence's support, the following analysis focuses on the 2005 election, when the party was electorally at its strongest. Some conclusions can be drawn about the nature of Self-Defence's support from an analysis of aggregate-level data, although it is necessary to be mindful of the ecological fallacy. Figure 9.1 displays a correlation matrix of levels of support for Self-Defence at county level (*powiat*) in 2005 and 2007, and a number of regional-level indicators relevant to economic divergence. As the correlation coefficients between these indicators suggest, there were significant patterns of regional economic divergence in Poland. Levels of agricultural land use (agrarea05) were associated with lower salaries, lower levels of business activity, higher levels of unemployment and higher levels of welfare dependency; correspondingly, levels of urbanisation (urbpop) were associated with higher salaries, higher levels of business activity, lower levels of unemployment and lower levels of welfare dependency.

As the correlation of SO05 and SO07 indicates, the regional distribution of support for Self-Defence did not change significantly between 2005 and 2007, despite the

	SO05	SO07	avsal	agrarea05	unempwk	econent	neconent	benefit	fixedass	postwork	prework
SO07	82										
avsal	-39	-34									
agrarea05	55	49	-32								
unempwk	45	46	-39	25							
econent	-49	-43	42	-50	-39						
neconent	-46	-41	30	-44	-29	86					
benefit	56	58	-44	47	76	-51	-43				
fixedass	-41	-38	65	-39	-41	48	37	-46			
postwork	6	1	11	-4	-9	-2	-18	-13	2		
prework	37	35	-40	49	27	-46	-26	50	-42	-58	
urbpop	-56	-49	43	-68	-27	60	50	-46	52	10	-66

avsal	– average salary
agrarea05	– agricultural area in 2005
unempwk	– unemployed as percentage of working age people
econent	– number of economic entities in the region
neconent	– number of new economic entities in the region
benefit	– percentage of population in receipt of benefits
fixedass	– value of fixed assets in enterprises
postwork	– percentage of population older than working age
prework	– percentage of population younger than working age
urbpop	– percentage of population living in urban areas

Figure 9.1. Correlation matrix for support for Self-Defence (SO) and regional indicators. Source: author's dataset, compiled from data provided by the Polish State Electoral Commission (http://tinyurl.com/q4ydvsq) and the Polish Central Statistical Office (Główny Urząd Statystyczny, 2012). The correlation matrix was generated using the R package 'corrplot' (Wei, 2011). For clarity, correlation coefficients are given as percentages.

sizeable drop in support for the party. Three positive correlations stand out. Firstly, an increase in the proportion of agricultural land in a county was associated with an increase in levels of support for Self-Defence. Secondly, support for Self-Defence was higher in counties with higher levels of working-age unemployment. Thirdly, the higher the proportion of the population of a county dependent on state benefits, the higher the support for Self-Defence. A number of negative correlations can be identified. The strongest of these was the negative relationship between levels of voting for Self-Defence and the proportion of the population of a county living in urban areas. There was also a significant relationship between support for Self-Defence and the economic dynamism of a county: the greater the number of extant and new economic entities in a county, the lower the level of support for Self-Defence. Lower average salaries also had a negative association with levels of support for Self-Defence. In all cases, these relationships persisted in 2007, although with slightly lower coefficients.

Scholars analysing aggregate patterns of voting behaviour in Poland since the end of communism have often commented on the influence of regions and their historical legacies on present-day choices. Zarycki and Nowak (2000: pp. 346-349) identify four regions of particular significance, three corresponding to the areas into which Poland was partitioned between Austria-Hungary, Prussia and Russia during the nineteenth and early twentieth centuries, and one to the territories acquired by Poland after the Second World War. The Austrian partition left a legacy of Polish culture and national sentiment, but also poverty and poor farming structures. By comparison, in the Prussian partition, the state repressed Polish national identity but left a legacy of agrarian reform. The authoritarian and repressive Russian sector bequeathed a large peasant population with a 'less focused' sense of national identity than their compatriots in the former Austrian partition.

Figure 9.2 plots the regional distribution of support for Self-Defence in 2005. The highest level of support was achieved in the coastal county from which the party originated, with high levels of support in some of the surrounding counties. Support for the party was generally higher in central and eastern Poland, with particularly high levels of support in areas of central and eastern Poland that were formerly under the Russian partition. Support was slightly lower in the former Prussian partition and the northern lands acquired from Germany after the Second World War. Three distinct areas saw lower than average levels of support for Self-Defence: the major cities; the industrial regions of Upper Silesia (mining) and Szczecin and Gdańsk (shipbuilding); and the majority of the former Austrian partition. These distributions support the notion that Self-Defence's support was predominantly, but not exclusively, concentrated in rural areas and away from the major industrial centres, and that they did not mobilise strongly conservative constituencies.

Individual-level analysis bears out these findings to a certain extent.[1] Respondents with primary and secondary levels of education were more likely to have a greater affinity for Self-Defence than those with higher education. The same pattern obtained for

[1] For reasons of space, the summary of the statistical analysis from which these results are derived has been omitted. These data are available from the author on request.

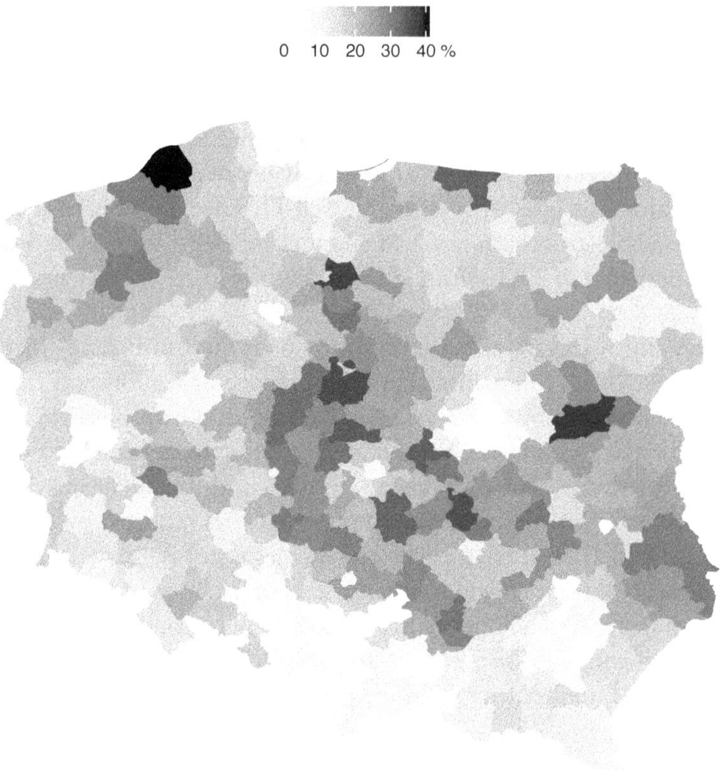

Figure 9.2. Regional distribution of support for Self-Defence (SO), 2005. Source: author's rendition of data provided by the Polish State Electoral Commission (Państwowa Komisja Wyborcza, 2012). The map was generated using the R package 'ggplot' (Wickham, 2010).

those living in villages or in small towns compared with city-dwellers. People with low levels of education who were living in villages were particularly likely to feel a stronger affinity for Self-Defence. Compared with white-collar workers, both blue-collar and agricultural workers were more likely to have a more positive view of Self-Defence, while the managerial class was more negatively disposed toward them. Region had some significance: compared with those living in the Russian partition, those in the Austrian partition were less likely to have a positive view of Self-Defence. Surprisingly, though, atheists and those with an ambiguous attitude towards religious faith and observance felt less affinity for the party than the devout.

Ideological attitudes had less influence on respondents' attitudes. Befitting the party's indistinct left-right profile, respondents' left-right placement had no effect on their preferences. Attitudes to decommunisation (i.e. the removal of former communists, typically above a certain level of seniority, from positions of authority in post-communist states) also had little effect on attitudes to the party, supporting the notion that Self-Defence stood outside the left-right politics of the regime divide. The remaining results suggest that the politics of transition did not have a significant impact on voters' attitudes to a party intent on challenging those ideas. Positive attitudes to European integration, privatisation and regressive taxation had a negative but insubstantial relationship with affinity for Self-Defence.

9.5 From the margins to the mainstream: electoral failure and success, 1993-2001

Dissatisfaction with transition increased in most sections of society during the first decade of transition. The old, those living in small towns and villages, those of lower educational attainment, those in the lower income quartile, the unemployed, retired, and those receiving invalidity benefits were more likely to state that post-1989 reforms had negatively affected them (Czapiński, 2006: p. 184). However, there are a number of reasons why parties such as Self-Defence did not initially succeed in mobilising a significant proportion of these social groups.

The dominance of the regime divide at the elite level also initially drove voting choices, as it resonated with some of the 'coherent and active social identities' of groups of voters (Grabowska, 2004: p. 358) and provided a clear heuristic for voters disoriented by the remarkable proliferation of political parties in the first few years of transition. As Self-Defence had no clear roots on either side of this divide, it initially struggled to attract support. Parties which appealed to transition losers were also hampered by the sceptical attitude of these social groups to democracy itself (Paczynska, 2005: pp. 598-600). Parties attempting to mobilise these groups faced the problem of not only how to persuade significant proportions of these groups that it was worth voting for them, but also how to persuade them that it was worth voting at all. Self-Defence's electoral record in the 1990s was testament to these difficulties. In the 1993 parliamentary elections it gained a mere 2.8% of the vote, and in the presidential election two years later, Lepper achieved less than half of that vote share, gaining 1.3%. In the 1997 parliamentary elections it reached a nadir of 0.1% (Rose and Munro, 2009: pp. 204-205).

As long-term survey trends show, in the mid-1990s Poles were on the whole more positive than in previous years about the direction of change and less negative about the political and economic situation (CBOS, n.d.). The general upswing in the public mood during the mid-1990s posed a problem for a party of protest. However, if Self-Defence suffered from an upturn in the public mood, they might be expected to benefit from a downturn. From 1998 onwards the movement returned to public consciousness through its involvement in a renewal of agrarian protest actions. These actions took place against the backdrop of a clear decline in attitudes to transition reforms and an increasingly negative appraisal of the prevailing political and economic situation. The 'second wave' of economic reforms implemented by the 1997-2001 post-Solidarity coalition were received badly by the general public, and the severity of the decline in the public's attitude towards the government reflected the increasingly radical mood. Between 1998 and the parliamentary elections of 2001, Self-Defence was involved in the organisation of numerous protests that went beyond the agricultural sector to incorporate a wide range of discontented groups on both sides of the regime divide.

At first, it seemed that Self-Defence had not benefited from its renewed public profile as a result of this wave of activity. In the presidential election of 2000, Lepper gained 3.1% of the vote, which at first glance did not seem to be a particularly impressive return on Self-Defence's activity during this period. However, this result must be placed in due context. This election saw a resounding first-round victory for the incumbent, Aleksander Kwaśniewski, who was endorsed by the main opposition party and thus benefited to some extent from the anti-government protest vote. In such circumstances, it was quite an achievement for Lepper to attract over half a million voters. Nevertheless, as Figure 9.3 shows, Self-Defence's polling figures remained below the 5% threshold (indicated by the dotted line) after the presidential election and well into the coming year. This may help to account for the sense of surprise felt across the political and media establishment at Self-Defence's showing in the September 2001 parliamentary elections, where it gained 10.2% of the vote and became the third largest party in the Sejm, holding 53 of 460 seats.

While the circumstances preceding the 2001 parliamentary elections were conducive to the success of parties which based their electoral appeals on wholesale criticism of the preceding decade, the electoral triumph of Self-Defence cannot be attributed solely to its leading role in orchestrating protest during the late 1990s. Rather, it was a product of substantial supply-side and demand-side volatility in the Polish party system and the particularly dramatic nature of party-system change in 2001 (Szczerbiak, 2002 for a detailed analysis of this 'unexpected political earthquake'). The 2001 elections were the first at which parties outside the mainstream could make a credible case that they represented a new quality in the party system, emphasising as they did the politics of transition rather than the politics of post-communism. Writing soon after these elections, Szczerbiak (2002: p. 41) suggested that the elections of 2001 'may herald a major re-shuffling of the party system'. These were prescient words, in that the elections did not themselves constitute the re-shuffling but rather cleared the ground for subsequent change. The period of transition-deepening undoubtedly made conditions more conducive to radicalism, but the eleventh-hour nature of Self-Defence's emergence makes something of a just-so story of explanations that invoke a 'populist zeitgeist'. While the 2001 elections could be classed as a 'populist breakthrough' in the sense that all parties that would later form the 'populist coalition' of 2006-2007 coalition

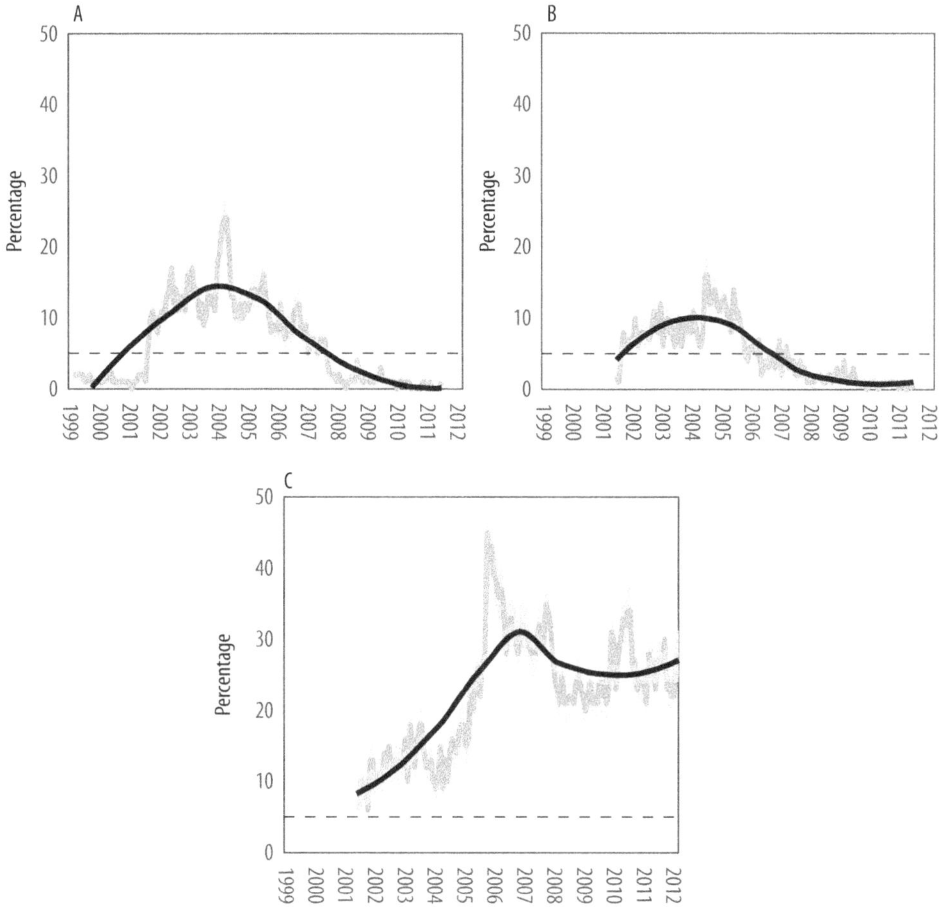

Figure 9.3. Support among the electorate for populist parties, 1999-2012. Source: author's own compilation of CBOS survey data (CBOS, 2000a,b, 2001a,b,c,d, 2002b, 2003a,b, 2004a,b, 2005, 2006a,b,c, 2007a,b, 2008a,b, 2009a,b, 2010a,b, 2011a,b, 2012). The graphs were plotted using the R package 'graphics' (R Development Core Team, 2012). Notes: The dark grey line is the actual level of support for the given party. Support is calculated on the basis of declared support and the weighted redistribution of those who declared an intention to vote in a hypothetical election but were unsure for which party they would vote. The black line is the smoothed mean of polling figures. The grey shaded areas represent 95% confidence intervals for both lines.

government first entered parliament at this time, prospects for their cooperation were at that point decidedly weak, and the 2001-2005 parliamentary term continued the process of party-system disintegration.

9.6 From opposition to short-lived power: Self-Defence in parliament, 2001-2007

The breakthrough of populists in 2001 sparked much soul-searching among liberal politicians and media outlets as to how parties with radical credentials could have gained nearly one-fifth of the vote. It is widely believed that Self-Defence made substantial use of public relations and media training to professionalise their image prior to the 2001 elections, although Lepper and other leaders subsequently downplayed the extent of this training and its contribution to the party's success (Piskorski, 2010: p. 195-196). However, any hope that Self-Defence would adjust to parliamentary procedure and etiquette swiftly evaporated. Aiming to benefit from the support of the third largest party, the governing coalition voted for Lepper as one of the deputy speakers for the new parliamentary term. He did not last long in this position, voted out after a series of actions unbecoming of his position, including involvement in a direct action protest.

Lepper responded to his ousting by going dramatically on the offensive with a speech directed against the entire political class of transition, railing against the elite 'in suit and tie, smelling of Dior and Chanel, [who have] been pampering each other these last twelve years'. Directing his invective at both post-communist and post-Solidarity, Lepper raged:

> ... you have brought this country to ruin ... and now you dream of courtesy? There won't be any in this hall, because the country is sinking, because the people don't even have enough for bread – and you want courtesy? You say I am primitive, that I use unparliamentary language, but I am simply repeating what people are saying. ... [P]eople don't want anything for free, they want to work hard and honestly, but they also want a dignified life in return for their labour, and you aren't able to provide them with this.
>
> (Andrzej Lepper, cited in Sejm Rzeczypospolitej Polskiej, 2005, term 4, session 6, day 2)

Lepper followed this tirade with accusations that five politicians from the liberal-conservative Civic Platform (Platforma Obywatelska, PO) and the social-democratic, post-communist Democratic Left Alliance (Sojusz Lewicy Demokratycznej, SLD) had taken bribes from gangsters, on the basis of 'evidence' that swiftly turned out to be concocted. The party's disruptive behaviour did not abate in 2002, with Lepper and other Self-Defence parliamentarians blocking access to the Sejm rostrum in order to prevent a vote on depriving Lepper of his parliamentary immunity, and Lepper bringing his own microphone into parliament to thwart the speaker's refusal to let him continue his tirades. Although parliamentary sanctions forced Lepper and other party deputies to moderate their more overtly obstructive behaviour in the face of parliamentary sanctions, the party retained its reputation for thumbing its nose at the proprieties of the salons of power. The leader's expensive tailoring and practised politeness did not quite disguise a penchant for crude language and cruder sentiment.

The behaviour of Self-Defence and the Catholic-nationalist League of Polish Families (Liga Polskich Rodzin, LPR), which also engaged in unconventional and obstructive methods of parliamentary behaviour, was deeply alarming for political elites, with President Kwaśniewski speaking darkly of parliamentary crisis (Lizut, 2002). The impact of populist opposition on the public mood was difficult to gauge. While Dudek (2007: p. 446) identified a decrease in the already low percentage of Poles who considered the Sejm a source of social authority, it is not easy to separate the specific effects of populist agitation from a more general decline in the prestige of the institution. Poles were divided on the legitimacy of these disruptive methods in a democratic system. Although 80% were of the opinion that deputies should respect the law, and only 15% regarded it as acceptable for deputies to use extra-legal methods if their cause was especially important, 39% concurred with the notion that it was the role of opposition deputies to 'single out abuses committed or mistakes made by the ruling coalition and use all available opportunities to criticise them', against the 47% who believed that it was the responsibility of the opposition to stand in support of the government in cases where particular parties disrupted ordinary parliamentary procedure. While 38% regarded the tactics used by Self-Defence and LPR as acceptable in democratic systems, 51% saw them as symptoms of anarchy and a threat to democracy (CBOS, 2002a).

These figures did not indicate that the behaviour of populists was an unambiguous threat to the Poles' faith in parliamentarianism. Instead, it was one of a number of elements contributing to a marked decline in public approval both of parliament and of the present administration (CBOS, n.d.). The apparent incapacity of the administration to deal decisively with the behaviour of radicals in parliament cannot but have had some impact on public attitudes towards its incapacity for strong and competent government. Yet this could not easily be separated from the impact of other problems the administration experienced; problems which themselves contributed to a more general radicalisation of political life. Millard (2006: p. 1011) and Brier (2009: p. 80) point to the themes of economic turbulence and corruption as salient issues during this period. European integration also provided a convenient point of contestation for radical parties.

Self-Defence gave its support to the formation of a government comprised of the SLD, its electoral coalition partner the social-democratic Labour Union (Unia Pracy, UP) and the agrarian Polish Peasant Party (Polskie Stronnictwo Ludowe, PSL). However, it quickly sensed the opportunity to reach out to SLD voters disaffected by the sight of a nominally left-wing government employing policies of economic retrenchment. Lepper was scathing about an economic strategy that remained within the liberal paradigm and failed to provide 'economic lustration', that is, a review of previous reforms 'which in every concrete case of incompetence, manipulation and abuse must result in accusation, trial and conviction'. According to Self-Defence, Poland could only be rescued from bloodshed by abandoning the restrictive monetary policy of liberal orthodoxy, devaluing the złoty to stimulate exports, and trusting the public sector to deliver growth through increased expenditure – naturally without making cuts to public services and benefits or meddling with labour laws. Lepper spoke with the freedom of one unencumbered by responsibilities, and his prognosis of a crisis similar in scale and scope to the contemporaneous Argentinian economic collapse was clearly unrealistic, but his evocation of a Poland in which 'shipbuilders strike for a piece of sausage in their soup after twelve hours of work outdoors in winter' chimed with the bleak social mood (Lepper, 2002b). While Self-Defence was not always opposed to the government's economic policy, it sought at all turns to bolster its credentials as defender of the interests of the 'little man' and scourge of 'big-business' privileges, mounting staunch opposition to restructuring and privatisation with the involvement of foreign capital, and pursuing its flagship project, the removal of Leszek Balcerowicz as head of the Polish Central Bank and its subordination to the Sejm (Sejm Rzeczypospolitej Polskiej, 2002).

The conclusion of European Union accession negotiations was one of the major tasks of the SLD-UP-PSL coalition, and the inevitable compromises this process entails afforded opposition parties numerous opportunities to attack. Self-Defence focused in particular on the issue of foreign ownership of farmland. Agriculture was a highly sensitive issue prior to accession, both because of the land issue and the adaptation of the agricultural sector to the Common Agricultural Policy. Responding to the final accession conditions negotiated at the Copenhagen summit of December 2002, Lepper berated the coalition government for yielding to the demands of the Member States and failing to secure terms better than those negotiated by other nations (Andrzej Lepper, cited in Sejm Rzeczypospolitej Polskiej, 2005, term 4, session 38, day 4) (Sejm stenographic transcript, 20/12/02). While agriculture was an area of special interest for Self-Defence in its own right, in highlighting the themes of subservience to the West, lack of respect for domestic producers, and the exposure of Poland to the threat of unfair foreign competition, it also served as a convenient encapsulation of their critique of transition capitalism in its entirety.

If Self-Defence led the attack on the coalition in terms of the economy and Europe, it was significantly less active on the issue of corruption. Although no administration had escaped the charge of underhand activity since the outset of transition, the 2001-2005 term saw more attention than ever before turned to what were alleged to be vast webs of illegality, with their roots in the beginnings of transition. The most notable of these scandals was the bribery affair dubbed 'Rywingate', which centred on the attempt in mid-2002 by film producer Lew Rywin to elicit a bribe from media house Agora – owner of the liberal daily *Gazeta Wyborcza* – in return for changes to a proposed media bill that would allow it to buy a satellite television station. The affair afforded Poles 'a window on the world of political networks, business links to politics, and general intrigue' (Millard, 2006: p. 1011). It might have been expected that Self-Defence would take a leading role in these investigations, given Lepper's fondness for conspiracy-theory explanations of the politics of transition. However, the chief beneficiaries of the Rywin commission were the post-Solidarity parties of the Civic Platform (Platforma Obywatelska, PO) and the conservative, statist Law and Justice (Prawo i Sprawiedliwość, PiS). The latter, in particular, began to exert more influence on public debate as the themes it emphasised – corruption and decommunisation – coalesced in the public consciousness. A controversial minority report was accepted by parliament as the definitive set of findings from the Rywin commission. This version of events supported PiS's broad narrative of the covert cooperation of reformist communists and liberal members of Solidarity, whose covert cooperation in a 'network' (*układ*) spanning the political, administrative and business sectors worked both to their enrichment and to the exclusion of their political opponents from the sphere of political legitimacy (Kaczyński, 2006). Self-Defence was ill-placed to compete with PiS for 'ownership' of the corruption issue, as recurring rumours about the role of the ex-communist secret services in the creation of Self-Defence undermined its credibility as an anti-corruption force.

The SLD-led government never recovered from the blow of Rywingate, and the last two years of the parliamentary term turned into an extended electoral campaign. During this period, Self-Defence struggled to find ways of broadening its potential support. As Figure 9.3 shows, support for the party fluctuated between approximately 12 and 18% for most of the 2001-2005 term, with the exception of a brief spike in popularity during 2004. The drastic fall in support for the SLD gave Self-Defence opportunities to appeal to disenchanted voters of the left. However, they entered the campaign for the dual presidential and parliamentary elections of September and October 2005 without any significant upward momentum, in contrast to PiS, whose support rose substantially over its period in office and who entered the elections with credible prospects of victory in both (Figure 9.3).

The results of these elections in 2005 confirmed Self-Defence's position as a minor but meaningful element of the party system. Lepper placed third in the first round of the presidential election, gaining 15.1% of the vote, a significant improvement on his previous figure of 3.1%. However, the party gained only 11.4% of votes for the Sejm, a mere 1.2% improvement over 2001. Given the lack of dynamism in the party's pre-election polling, there is no clear reason to argue that they should have gained substantially more than this. However, their failure to make significant gains was undoubtedly affected by the shifting dynamics of the electoral races, which turned the two front-runners and expected coalition partners, PiS and PO, against each other. With PiS recasting itself as the 'social' alternative to the liberal PO, Self-Defence faced competition for floating voters with left-wing economic sympathies.

After an acrimonious campaign, coalition talks between PiS – who gained both the presidency and the largest number of seats in parliament – and PO were swift to break down. Reluctant to call fresh elections and risk losing its slim advantage, PiS cast around for alternative solutions. With their poll ratings falling, Self-Defence were also conscious of the risks of new elections and enthusiastic about the prospect of power and prominence. PiS initially trod with caution, signing a 'stabilisation pact' with Self-Defence and LPR that committed the signatories to cooperate on a basket of initiatives dominated by PiS's flagship initiatives. This arrangement proved somewhat unstable and was rescinded the following month after a number of public pronouncements by Lepper and Roman Giertych appeared inconsistent with its provisions. However, after a failed attempt to dissolve parliament, PiS offered Self-Defence and LPR a formal coalition agreement in May 2006. Mainstream politicians and the liberal media were horrified at the formation of this 'exotic threesome' (Paradowska, 2006). No explicit *cordon sanitaire* had been erected around these parties during their time in opposition. However, it was widely assumed that a coalition partnership with them would break an unwritten rule, regardless of electoral mathematics.

The coalition agreement constituted a radical programme for fundamental reform of state institutions and sweeping changes to the administrative elite, buttressed by a significantly greater interventionist role for the state in the spheres of national identity and the economy, and a decisive break with the 'bended-knee' foreign policy of the past fifteen years. In keeping with its priorities, Self-Defence left a clear imprint on the economic provisions of the coalition agreement. Widening the scope of the Central Bank's remit to cover the pursuit of economic growth was a key element of the party's policy platform. The agricultural element of Self-Defence's agenda was also reflected in plans to renegotiate the production quotas with the EU and increase production of biofuels. The commitment to ensure that 'farmers have full information on what to produce, to whom it will be sold, what price they will receive and when they will get the money' (Prawo i Sprawiedliwość *et al.*, 2006: p. 9) came straight from Lepper's own pen. The concern of all three parties with the effects of 'heedless' privatisation was in evidence in the call for a new privatisation strategy which promised 'the unambiguous definition of the ownership perspective of subjects controlled by the State Treasury' and 'the decisive improvement of the transparency of privatisation procedures' (Prawo i Sprawiedliwość *et al.*, 2006: p. 3). Social policy also bore the imprint of the coalition's interests, promising the creation of a basket of guaranteed medical services and the 'yearly revaluation of pensions and invalidity benefits taking into consideration both inflation and the effects of economic growth' (Prawo i Sprawiedliwość *et al.*, 2006: p. 3).

The life of the coalition was short and turbulent due to the intensity of external opposition and the indiscipline and mutual distrust of the coalition partners. The coalition's attempts to implement its programme engendered fierce opposition from state institutions (in particular the Constitutional Tribunal), from liberal media organisations, and from politicians committed to defending the achievements of the Third Republic against those who aimed to institute a Fourth. Self-Defence did not play a direct role in the most significant policy controversies, as PiS had complete control over the key Foreign, Interior, Justice and Defence portfolios where the most far-reaching reforms were instituted.

Lepper's appointment as Minister of Agriculture was a non-negotiable condition of Self-Defence's participation in the coalition. Initially, assessments of his activities as minister were not entirely negative: he was credited, for example, with a significant role in the removal of a Russian embargo on Polish agricultural produce (Piskorski, 2010: p. 159-160). However, Self-Defence attracted much criticism for exploiting the extensive opportunities for patronage afforded by the numerous agencies in the purview of the ministry and in companies in which these agencies had holdings (Kokot, 2006; Rybak and Cieśla, 2006). Lepper left the majority of ministerial employees in their posts rather than conduct a purge, but this can be attributed to the party's lack of a cadre of competent alternatives (Piskorski, 2010: p. 160). Most Polish parties have been criticised for their use of patronage, so these failings were not sufficient in themselves to affect Lepper's standing unduly. What was more damaging for Self-Defence was the rapid deterioration in the capital of trust Lepper had built up among farmers.

This was partly a result of incompetence and lack of oversight of his ministry. For example, many farmers lost the right to obtain state help after a mini-drought ruined crops in the summer of 2006, as the ordinances Lepper signed attached conditions to the receipt of compensation that significantly narrowed eligibility (Naszkowska, 2006). While this could be explained away as the price of being an Agriculture Minister in a coalition with a liberal Finance Minister, failure to head off meat overproduction directly undermined his professional credibility. Lepper had prided himself on practical experience of the cycle of production and awareness of farmers' needs but drew accusations of indolence in the face of a predictable crisis (Naszkowska, 2007). This incident, in particular, provoked farmers' groups to turn on Lepper, accusing him of having 'entered the salon and stayed there' rather than concern himself with farmers' needs (Tylenda, 2007). Former associates of Lepper resolved to employ against him the very tactics he had taught them, although they admitted to lacking a leader with his charisma and talent (Łazarewicz, 2007). The other Self-Defence minister Anna Kalata, the Minister of Labour and Social Policy, also gained a reputation for incompetence and misuse of the trappings of office. A motion of no confidence in her stewardship of the ministry noted that of 46 projects discussed during that term by the parliamentary Commission on Social Policy, only 15 had originated from the government, raising the question of 'not only how it was possible to talk of deep reform, but also of ordinary, straightforward governance' (Sejm Rzeczypospolitej Polskiej, 2007).

The period of coalition government lasted from 5 May 2006 to 21 October 2007, during which there were five changes to its composition. The coalition was beset by tensions from the outset, and Self-Defence was at the centre of the most damaging ructions. Given PiS's numerical and strategic superiority, the minor parties swiftly became aware of their vulnerability. It became commonplace to refer to Self-Defence and LPR as the 'appetisers' ('przystawki') ripe to be 'consumed' by their larger partner.

The first crisis resulted from Lepper's refusal to support the government's budget unless his demands for substantial spending increases were met. With his government barely two months old, Kaczyński responded to Lepper's 'bellicosity' by expelling Self-Defence from the government, claiming that Lepper had failed to take the chance to prove that he could be a responsible politician (Milewicz, 2006). While Lepper and Self-Defence were soon granted a second chance, the departure of eight Self-Defence deputies after Lepper's first ousting, in protest at their leader's autocratic tendencies, encouraged PiS strategists to attempt to persuade others to defect. One of Lepper's close associates, Renata Beger, taped an attempt by PiS deputies to secure her defection, embarrassing PiS, but also exposing Self Defence's practice of compelling its deputies to sign promissory notes that were executable if they switched parties. Lepper had always dominated the party, but this episode indicated that this domination was founded as much in coercion as in genuine loyalty. Exposure of the power structure within the party took a more sinister turn in December 2006, when former Self-Defence councillor Aneta Krawczyk accused Lepper and senior party member Stanisław Łyżwiński of demanding sexual favours in return for a job in Łyżwiński's office. Although DNA tests refuted Krawczyk's assertions that either Łyżwiński or Lepper was the father of her child, the odium of the 'sex for jobs' scandal remained.

Lepper stayed at his post, but PiS was faced with a dilemma. The continued presence of such a party in the coalition compromised the credibility of the vaunted moral revolution. However, Self-Defence would be a destructive presence in opposition, particularly considering the opposition's desire for early elections. July 2007 saw the arrival of what – depending on your perspective – was either a *deus ex machina* or a conspiracy. The Central Anti-Corruption Bureau (Centralne Biuro Antykorupcyjne, CBA), a new agency set up by PiS, mounted a surveillance operation to provide evidence of corruption in the Ministry of Agriculture involving the reclassification of agricultural land. Lepper was tipped off that he was under surveillance and the trap failed. Since the CBA was widely considered to operate in a state of vigilance as to PiS's expected needs, if not quite at its behest, the rumours of an intrigue that was intended to remove Lepper from politics and absorb Self-Defence deputies into PiS had a certain plausibility, even if the affair remained far from explained. Lepper's second dismissal in the aftermath of this affair effectively ended the coalition, and after two months of stalemate parliament was dissolved.

Szczerbiak (2008: p. 27) characterised the early parliamentary elections of October 2007 as 'a plebiscite on a polarising and controversial government'. The real losers of this plebiscite were not PiS, which lost the election but gained nearly two million extra votes, but Self-Defence and LPR, both of whom were eliminated from parliament. Neither of these parties had any clear conception of how to broaden their support, and they ran lacklustre campaigns that the media largely overlooked in favour of the head-to-head battle between the two major parties, PiS and PO. Self-Defence, which had polled on or slightly above the electoral threshold for most of its time in office, only received 1.5% of the vote and has remained beneath the threshold ever since.

9.7 Conclusion

Self-Defence's fall from prominence was swift and, it seems, irrevocable. Even before Andrzej Lepper's still unexplained suicide in the summer of 2011, there were no reasons to expect that the party would be able to make a comeback, and since the death of the party's charismatic leader there are even fewer. It might be tempting to blame the party's controversial tenure in government for its subsequent plight. However, the fundamental problem that the party faced after its electoral failure in 2007 was not one of reputation but of relevance. From the outset of the coalition, PiS was much more proficient than its junior partners in creating a vivid innovative and plausible populist discourse. It was PiS, rather than Self-Defence, which made the concept of the post-communist 'network' the salient point of political reference for supporters of the coalition and defenders of the politics of transition alike. It was PiS which created an effective and memorable lexicon according to which 'real Poles' set themselves against the 'mendacious elites' and 'pseudo-intellectuals' of the liberal mainstream. In its position as senior coalition partner, PiS was far better placed to benefit from public perceptions of its ability to set the terms of political debate, against which Self-Defence appeared merely ineffectual.

PiS also offered voters who were disenchanted with the politics of transition a wider and more satisfying narrative. While Self-Defence focused on the economy, the 'economic anger' identified by Ost (2005: p. 9) is too narrow a specification of transition discontents. If economic anger was a product of the devaluation of certain types of economic assets in favour of others, then 'cultural anger' stemmed from the devaluation of 'authentic' Polish identities and values in the pursuit of imitative modernisation, while 'political anger' arose due to the exclusion of particular sociodemographic and attitudinal constituencies and their political representatives from the main currents of political action. These different facets of victimhood often overlapped, and by 2007 PiS was in a much better position than its coalition partners to convince voters that it, of all parties, could most effectively represent those who – in whichever sense, economic, cultural or political – considered themselves to be the losers of transition.

Self-Defence cannot be said to have achieved its aims as a political party or as a broader social movement. It did not succeed in bringing about even a partial change to the economic orthodoxy, and squandered the trust and electoral support of its core constituency by reaching for power it was not competent to wield. Like many populist parties, it was more successful as a party of opposition. Rather than legislative achievements, its legacy to Polish party politics consists in the role it played as a catalyst in the reorientation of the party system. While the parliamentary elections of 2011 demonstrated that new parties can still make breakthroughs, the fundamental line of division between the major political parties now runs not between historical identities but between the ideological poles of 'integration' and 'demarcation' and the electoral poles of a 'Poland in transition' and 'Poland outside transition'. It remains to be seen whether Polish political life will polarise further along these lines.

References

Brier, R., 2009. The roots of the 'Fourth Republic': solidarity's cultural legacy to Polish Politics. East European Politics & Societies 23: 63-85.

Centrum Badania Opinii Społecznej (CBOS), 2000a. Preferencje partyjne w grudniu. Komunikat z badań. CBOS, Warszawa, Polska.

Centrum Badania Opinii Społecznej (CBOS), 2000b. Preferencje partyjne w styczniu. Komunikat z badań. CBOS, Warszawa, Polska.

Centrum Badania Opinii Społecznej (CBOS), 2001a. Preferencje partyjne tydzień przed wyborami. Komunikat z badań. CBOS, Warszawa, Polska.

Centrum Badania Opinii Społecznej (CBOS), 2001b. Preferencje partyjne w grudniu. Komunikat z badań. CBOS, Warszawa, Polska.

Centrum Badania Opinii Społecznej (CBOS), 2001c. Preferencje partyjne w listopadzie. Komunikat z badań. CBOS, Warszawa, Polska.

Centrum Badania Opinii Społecznej (CBOS), 2001d. Preferencje partyjne w maju. Komunikat z badań. CBOS, Warszawa, Polska.

Centrum Badania Opinii Społecznej (CBOS), 2002a. Granice poselskiego protestu. CBOS, Warszawa, Polska.

Centrum Badania Opinii Społecznej (CBOS), 2002b. Preferencje partyjne w marcu. Komunikat z badań. CBOS, Warszawa, Polska.

Centrum Badania Opinii Społecznej (CBOS), 2003a. Preferencje partyjne w grudniu. Komunikat z badań. CBOS, Warszawa, Polska.

Centrum Badania Opinii Społecznej (CBOS), 2003b. Preferencje partyjne w styczniu. Komunikat z badań. CBOS, Warszawa, Polska.

Centrum Badania Opinii Społecznej (CBOS), 2004a. Preferencje partyjne w grudniu. Komunikat z badań. CBOS, Warszawa, Polska.

Centrum Badania Opinii Społecznej (CBOS), 2004b. Preferencje partyjne w marcu. Komunikat z badań. CBOS, Warszawa, Polska.

Centrum Badania Opinii Społecznej (CBOS), 2005. Preferencje partyjne na tydzień przed wyborami. Komunikat z badań. CBOS, Warszawa, Polska.

Centrum Badania Opinii Społecznej (CBOS), 2006a. Preferencje partyjne w grudniu. Komunikat z badań. CBOS, Warszawa, Polska.

Centrum Badania Opinii Społecznej (CBOS), 2006b. Preferencje partyjne w marcu. Komunikat z badań. CBOS, Warszawa, Polska.

Centrum Badania Opinii Społecznej (CBOS), 2006c. Preferencje partyjne w sierpniu. Komunikat z badań. CBOS, Warszawa, Polska.

Centrum Badania Opinii Społecznej (CBOS), 2007a. Preferencje partyjne na tydzień przed wyborami. Komunikat z badań. CBOS, Warszawa, Polska.

Centrum Badania Opinii Społecznej (CBOS), 2007b. Preferencje partyjne we wrześniu. Komunikat z badań. CBOS, Warszawa, Polska.

Centrum Badania Opinii Społecznej (CBOS), 2008a. Preferencje partyjne w grudniu. Komunikat z badań. CBOS, Warszawa, Polska.

Centrum Badania Opinii Społecznej (CBOS), 2008b. Preferencje partyjne w marcu. Komunikat z badań. CBOS, Warszawa, Polska.

Centrum Badania Opinii Społecznej (CBOS), 2009a. Preferencje partyjne w grudniu. Komunikat z badań. CBOS, Warszawa, Polska.

Centrum Badania Opinii Społecznej (CBOS), 2009b. Preferencje partyjne w lutym. Komunikat z badań. Centrum Badania Opinii Społecznej, Warszawa, Polska.

Centrum Badania Opinii Społecznej (CBOS), 2010a. Preferencje partyjne w grudniu. Komunikat z badań. CBOS, Warszawa, Polska.

Centrum Badania Opinii Społecznej (CBOS), 2010b. Preferencje partyjne w lutym. Komunikat z badań. CBOS, Warszawa, Polska.

Centrum Badania Opinii Społecznej (CBOS), 2011a. Preferencje partyjne w grudniu. Komunikat z badań. CBOS, Warszawa, Polska.

Centrum Badania Opinii Społecznej (CBOS), 2011b. Preferencje partyjne w sierpniu. Komunikat z badań. CBOS, Warszawa, Polska.

Centrum Badania Opinii Społecznej (CBOS), 2012. Preferencje partyjne w marcu. Komunikat z badań. CBOS, Warszawa, Polska.

Centrum Badania Opinii Społecznej CBOS (n.d.). Trends. Available at: http://www.cbos.pl/EN/trends/trends.php.

Chłopska Droga, 1992. Związek Zawodowy Rolników 'Samoobrona' – legalny.

Czapiński, J., 2006. Stosunek do przemian systemowych i ocena ich wpływu na życie badanych. In: Czapiński J and Panek T (eds) Diagnoza Społeczna 2005: Warunki i jakość życia Polaków. Wyższa Szkoła Finansów i Zarządzania w Warszawie, Warszawa, Polska, pp. 182-189.

Dudek, A., 2007. Historia Polityczna Polski 1989-2005. Wydawnictwo ARCANA, Kraków, Polska.

Gładkiewicz, W., 2004. Samoobrona Rzecypospolitej Polskiej. 3rd ed. In: Partie i koalicje polityczne III Rzeczypospolitej. Wydawnictwo Uniwersytetu Wrocławskiego, Wrocław, Polska, pp. 147-154.

Główny Urząd Statystyczny, 2012. Bank Danych Lokalnych. Available at: http://tinyurl.com/ksboys6.

Grabowska, M., 2004. Podział postkomunistyczny: społeczne podstawy polityki w Polsce po 1989 roku. Scholar, Warszawa, Polska.

Jachowicz, J. and Pączek-Haładyj, M., 1993. Praszka napięta. Gazeta Wyborcza October 8, 1993.

Kaczyński, J., 2006. The fall of post-communism: transformation in Central and Eastern Europe. Available at: http://tinyurl.com/kz32quj.

Kokot, M., 2006. Ludzie Leppera w państwowych spółkach. Gazeta Wyborcza November 9, 2006.

Kurski, J., 1999. Ładny plon mi wyrósł. Gazeta Wyborcza June 2, 1999.

Łazarewicz, C., 2007. Sławek ze świńskiej górki. Polityka 12: 2597.

Lepper A., 2002b. Strategia gospodarcza rządu. Website no longer available: http://www.samoobrona.org.pl/pages/21.Archiwum/01.Klub/05.Odezwy/index.php?document=08.strategia.html.

Lepper, A., 2002a. Lista Leppera. Wydawnictwo KAMEA, Warszawa, Polska.

Lizut, M., 2002. Sejm się chwieje. Gazeta Wyborcza October 19, 2002.

Milewicz, E., 2006. Premier odwołuje Leppera 'za warcholstwo'. Gazeta Wyborcza September 22, 2006.

Millard, F., 2006. Poland's politics and the travails of transition after 2001: the 2005 elections. Europe-Asia Studies 58: 1007-1031.

Morgan, W.B., 1992. Economic reform, the free market and agriculture in Poland. The Geographical Journal 158: 145.

Naszkowska, K., 1992. Bataliony chłopskie 'Samoobrony'. Gazeta Wyborcza April 30, 1992.

Naszkowska, K., 2006. Lepper najgorszym ministrem rolnictwa? Gazeta Wyborcza September 29, 2006.

Naszkowska, K., 2007. Dziś Lepper zażąda 300 mln zł z budżetu na skup świń. Gazeta Wyborcza January 30, 2007.

Ost, D., 2005. The defeat of solidarity: anger and politics in postcommunist Europe. Cornell University Press, Ithaca, NY, USA.

Paczynska, A., 2005. Inequality, political participation, and democratic deepening in Poland. East European Politics & Societies 19: 573-613.

Państwowa Komisja Wyborcza, 1991. Obwieszczenie Państwowej Komisji Wyborczej z dnia 31 października 1991 roku o wynikach głosowania i wynikach wyborów do Sejmu Rzeczypospolitej Polskiej przeprowadzonych dnia 27 października 1991 roku. Monitor Polski 41: 427-517.

Paradowska, J., 2006. Tercet egzotyczny. Polityka 19: 2553.

Pilot, D., 1992. Dobrowolnie stąd nie wyjdziemy. Gazeta Wyborcza April 27, 1992.

Piskorski, M., 2010. Samoobrona RP w polskim systemie partyjnym. Poznań, Polska.

Prawo i Sprawiedliwość, Samoobrona Rzeczypospolitej Polskiej and Narodowe Koło Parlamentarne, 2006. Koalicyjna deklaracja programowa 'Solidarne Państwo' z 27 kwietnia 2006 r., Załącznik nr.1, Cele i zadania rządu koalicyjnego w latach 2006-2009. Available at: http://tinyurl.com/ng6ezlv.

R Development Core Team, 2012. R: a language and environment for statistical computing. Vienna, Austria. Available at: http://www.R-project.org.

Rose, R. and Munro, N., 2009. Parties and elections in new European democracies. ECPR Press, Colchester, UK.

Rybak, A. and Cieśla, J., 2006. Jędruś Czarodziej. Polityka 32: 2566.

Rychard, A., 2006. Rozproszona Polska. Wstępna próba bilansu socjologicznego. In: Polska jedna czy wiele? Trio, Warszawa, Polska.

Samoobrona Rzeczypospolitej Polskiej, 2003. Program Społeczno-Gospodarczy Samoobrony. Samoobrona Rzeczypospolitej Polskiej,Warszawa, Polska.

Samoobrona Rzeczypospolitej Polskiej, 2004. Samoobrona RP. Kim jesteśmy i o co walczymy. Available at: http://www.samoobrona.org.pl/pages/02.Opis/.
Samoobrona Rzeczypospolitej Polskiej, 2005. Trzecia Droga Samoobrony RP. Website no longer available: http://www.samoobrona.org.pl/pages/01.Program/index.php?document=465.html.
Sejm Rzeczypospolitej Polskiej, 2002. Poselski projekt o zmianie ustawy o Narodowym Banku Polskim. Website no longer available:: http://tinyurl.com/ooecxsg.
Sejm Rzeczypospolitej Polskiej, 2005. Sprawozdanie stenograficzne z posiedzeń Sejmu RP IV kadencji. Available at: http://tinyurl.com/l9kmhn6.
Sejm Rzeczypospolitej Polskiej, 2007. Wniosek o wyrażenie wotum nieufności wobec Ministra Pracy i Polityki Społecznej Anny Kalaty. Available at: http://orka.sejm.gov.pl.
Szczerbiak, A., 2002. Poland's unexpected political earthquake: the September 2001 parliamentary election. Journal of Communist Studies and Transition Politics 18: 41-76.
Szczerbiak, A., 2008. The birth of a bi-polar party system or a referendum on a polarising government? The October 2007 Polish parliamentary election. Journal of Communist Studies and Transition Politics 24: 415-443.
Tylenda, M., 2007. Lepper musi odejść: rolnicy ruszą na Warszawę? Gazeta Wyborcza January 17, 2007.
Wei, T., 2011. Corrplot. CRAN, Vienna, Austria.
Wickham, H., 2010. Gsgplot2: elegant graphics for data analysis. Springer, New York, NY, USA.
Zarycki, T. and Nowak, A., 2000. Hidden dimensions: the stability and structure of regional political cleavages in Poland. Communist and Post-Communist Studies 33: 331-354.

10 Agrarian populism in Finland: continuity and change

A.C. Jungar

School of Social Sciences, Södertörn University, Stockholm, Sweden; ann-cathrine.jungar@sh.se

Abstract

Agrarian-based populism has an almost uninterrupted presence in Finnish politics. Its various manifestations are analysed in this chapter. The main focus is on party-based agrarian populism: the Finnish Rural Party (1959-1995) and its successor party, the True Finns (from 1995 onwards). The endurance of populism in Finland, apart from its organisation, is due to both continuity in relation to its main appeal – its anti-establishment position, the rhetorical construction of a united and threatened 'people', and the restoration of popular sovereignty – and transformations to the contextual references of these core populist ideas. The party that started as an agrarian populist party for rural smallholders is today a populist radical right-wing party, with nationalism, expressed in anti-immigration and anti-EU positions, becoming more salient and more radical for the True Finns. It is concluded that a process of radicalisation took place, which was key to their electoral breakthrough in 2011.

10.1 Introduction

Populism is a long-standing political current in Finland. The Finnish writer Ilmari Kianto revealed the presence of strong anti-elitist sentiments among the Finns in the novel *Pyhä viha*, (literally translated 'The sacred rage') in 1908:

> In the veins of the Finns flows a hatred of masters, a resentment against authority and the learned mind that they have absorbed with their mother's milk.[1]

[1] 'Suomalaisten suonissa virtaa jo äidinmaidossa imetty herraviha, epäluulo esivaltaa ja oppinutta mieltä kohtaan' (translated from Finnish by Heidi Granqvist).

DOI 10.3920/978-90-8686-807-0_10, © Wageningen Academic Publishers 2015

The 'hatred of the masters', for which there is a specific Finnish term, *herraviha*, has resonated both in radical socialist and agrarian political discourses. Anti-elitism and anti-establishment appeals have consequently been formulated in two overlapping and opposite ideological currents. Firstly, we can speak of the socialist rhetoric, in the form of class-based revolutionary appeals against the established conservative political parties and the capitalists. Immediately after Finland had gained independence from Russia in 1917, a civil war broke out between the 'whites' and 'reds', with the conservatives and the agrarians on the one side and the socialists on the other. The war was fought over class, the form of government and the young state's relationship with the nascent communist Soviet Union. The white side won the war, and as a result conservative and right-wing, even extreme-right movements and parties, dominated Finnish politics during the interwar period. Communism was forbidden in the 1930s and remained so until 1944 when a peace treaty was signed with the Soviet Union. The two former enemy camps, that is, the liberal-conservatives and the socialists/communists came together during the Second World War in the defence of Finnish independence and territorial integrity against Soviet aggression.

Secondly, while much has been written about these dramatic events and radical forces in Finnish interwar and postwar history, far less attention has been devoted to the anti-elite and anti-establishment appeals prevalent in the radical agrarian movements and the agrarian political parties that have existed since the early 1920s. Agrarian-based populism has, contrary to radical left-wing populism, survived and has had an almost uninterrupted presence in Finnish politics, although its electoral support has fluctuated. Agrarian populism, consequently, constitutes an important element in Finnish political culture. There is also a proper Finnish term for this political current, namely *vennamoisimi*, after Veikko Vennamo, the founder and long-standing leader (from 1959 to 1979) of the Finnish Rural Party. Finnish agrarian populism appeared shortly after the termination of the 'poujadist' movement in France, which was active between 1953 and 1958, named after its charismatic founder and leader Pierre Poujade. Although the Finnish version of populism has been far more successful, it is less well known internationally. The two populist movements were equally dependent on their charismatic party leaders, but the Finnish agrarian populists, contrary to the Poujadists, built a party organisation with national leverage, which ultimately contributed to the survival of this Finnish populist political party.

Finish populism is constituted by a stream of political appeals that mobilised Finnish voters at irregular intervals. By analysing how Finnish populists formulated fundamental political and social conflicts and how to resolve them, we will not only show how the ideology of populism expressed itself in a specific context, but also gain knowledge of a distinct line of political thought in Finland. The aim of this chapter is to analyse agrarian populism in its various manifestations in Finland, with the main focus on party-based agrarian populism; that is, the Finnish Rural Party (1959-1995) and its successor, the True Finns (1995-present) (Arter, 2012). The endurance of populism in Finland is not only due to the nature of its organisation, but also due to the continuity of its main ideals – the anti-establishment position, the rhetorical construction of a united and threatened 'people' and the restoration of popular sovereignty – and the transformation of the contextual references of these core populist ideas. The material consists of official party documentation, as well as speeches and writings by the party leadership. Before analysing the ideology and its populist appeal, we will describe the history and the electoral strength of Finnish agrarian populism.

10.2 Populist movements and parties

The roots of agrarian populism can be traced back to the late 1920s, the first decade of the independent Finnish republic. In 1809 Finland was lost by the kingdom of Sweden, of which it had been a part, to Russia, with Finland receiving the status of an autonomous Grand Duchy within the Russian Empire. In the aftermath of the revolution in 1917 Finland gained independence from Russia. Initially, rural restructuring, and later the economic depression, hit many small-scale landholders hard in Finland. This set in motion agrarian populist movements as early as the 1920s, among them the 'famine movements' in northern Ostrobothnia, situated on the west coast, which has been a populist electoral stronghold ever since (Maijala, 2005). These agrarian movements were formed at the same time as the populist agrarian movements in the United States and Canada, in the midst of an international economic downturn which particularly affected rural smallholders (Helander and Toivonen, 1971: p. 45). The populist appeals were similar on both sides of the Atlantic, targeting the political establishment for failing to address the hardships of the rural poor. Within these movements, solidarity and self-help associations were established, often within the realm of pious Christian congregations that provided financial assistance to rural smallholders whose properties were repossessed and sold at auction.

The first postwar populist political party in the Nordic region was established in Finland. In 1959, the Finnish Rural Smallholder's Party (Suomen pientalouspoikien puolue) was formed as a splinter party from the Agrarian Union (Maaseudun Liitto). In 1956, a lawyer, Veikko Vennamo, who had been a parliamentary representative of the Agrarian Union since 1945 and a deputy minister in the Ministry of Finance from 1954 to 1956, left the Union due to conflicts with the party leadership. The new party formed in 1959 was unsuccessful in its first parliamentary elections in 1962, but in 1966 the party took its first parliamentary mandate. Vennamo presented himself as the spokesperson for the rural smallholders, who he claimed were being neglected by the Agrarian Union.

The Finnish Rural Smallholder's Party was initially formed as an interest-based party for a specific segment of the rural population: farmers with the smallest properties (Sänkiaho, 1971). Veikko Vennamo, who was born on the shores of Ladoga in Karelia, the eastern part of Finland that was annexed by the Soviet Union in the Second World War, had made his reputation as the head of the office for the postwar resettlement of over 400,000 Finns from Karelia. Anti-establishment appeals flourished from the early days, with criticism of the established parties and the president, who were both depicted as corrupt and too subservient to Moscow in the eyes of the anti-communist party leader and his followers. In 1966, the Agrarian Union changed its name to the Centre Party (Keskusta) as a symbolic gesture hoping to broaden its appeal to non-rural voters. This was in a period of rapid depopulation of the countryside, as well as urbanisation and emigration to Sweden, and the party leader Veikko Vennamo immediately took the opportunity to rename his party the Finnish Rural Party (Suomen maaseudun puoluei) in 1967, proclaiming that the renamed party (abbreviated here as FRP) was now the only political party representing the threatened rural population.

The electoral breakthrough for the FRP came in the parliamentary elections of 1970. The party received 10.5% of the vote and the party leader was joined by 17 new representatives in the Finnish parliament, the Eduskunta (Figure 10.1). The rise was the result of its charismatic and popular leader Veikko Vennamo being able to mobilise the disaffected small farmers, including the many refugees from Karelia, as well as changes in the political opportunity structures that benefited the anti-establishment party (see below). The hard-working rural people were idealised and there were demands for

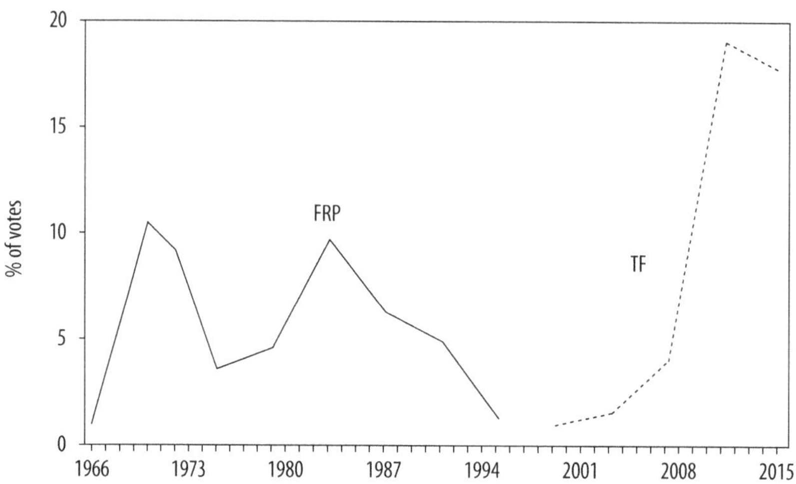

Figure 10.1. Electoral support in the parliamentary elections for the Finnish Rural Party (FRP) and the True Finns (TF), 1966-2015.

greater public commitment to the support of small farmers and the rapidly depopulating rural areas. The FRP defended the interests and values of small farmers and the rural population in the ongoing process of rapid urbanisation and modernisation, and it was fiercely anti-communist and in favour of traditional, primarily Christian, values. The party had its stronghold in the eastern and northwestern parts of Finland. While the largest agrarian party, the Centre Party, was in office and also exercised power through the long-term presidency of Urho Kekkonen (1956-1981), the FRP was able to present a credible and effective anti-establishment position.

Transformations in the opportunity structures explain the electoral success of the FRP in 1970. Firstly, in 1966 a broad surplus majority government was formed that included – for the first time since 1948 – the Finnish Communist Party (SKDL). This party had a substantial following in the countryside in northern Finland, but as a party of government, it was no longer perceived as a credible protest party and a substantial number of these voters turned to Vennamo's party (Sänkiaho, 1971: p. 35). Moreover, as Veikko Vennamo was one of the three candidates in the presidential elections of 1968, he received a lot of publicity for himself as well as the party. This was the first electoral campaign where the debates were televised and Vennamo was not only a gifted speaker on radio but also an excellent performer in the new media format. After the election of 1970, the party stabilised as a '5-10% party' until it failed in the 1995 election, leaving it with the support of 1.3% and only one MP (Figure 10.1).

The FRP was also the first Nordic populist party to take part in government. When the long-standing president, Kekkonen, resigned in 1981 and the Soviet Union no longer accused the FRP of being extreme right, the party was invited to take part in government between 1983 and 1990. As a matter of fact, fascist parties were forbidden in Finland in the peace treaty that was signed with the Soviet Union after the Second World War. Assuming governmental responsibility marked the beginning of the party's decline. Decreasing electoral support in the 1990s was also due to internal conflicts, and after the devastating electoral result in 1995, the financial situation forced the party leadership to declare the party bankrupt. However, the same party leadership immediately formed a new party, the True Finns.

It has been claimed that the True Finns represented the 'new radical right' in Finland (Pekonen, 1999), and presently the party can be conceived of as a populist radical-right party (Jungar and Jupskås, 2014). The party presents a milder version of agrarian populism, still speaking for the deprived ordinary people, pensioners, veterans, the sick and the rural population in general, all of whom have been under pressure since the weakening of the Finnish welfare state after the economic crisis in the 1990s and once again after the European economic downturn in 2008. Moreover, the True Finns are, contrary to their predecessor party, opposed to the European Union, and their criticism of the EU has gained momentum since the introduction of the euro in 1999.

The party has also strengthened its critique of immigration and multiculturalism, including the rights of the Swedish-speaking minority. The radicalisation of the True Finns came about after 2007, with its inclusion of members of nationalist-minded, anti-immigration and Islamophobic groups that were previously only found in civil society organisations or web communities (Koivulaakso *et al.*, 2012; Lydén, 2012).

The True Finns gained 4.1% of the vote in the 2007 general election, and almost quintupled their vote in the 2011 elections, obtaining 19.6%. In that year, the party took part in preliminary governmental negotiations, but abstained from participation in government as that would have meant accepting the EU bailouts to Greece and Portugal, which the party had vociferously campaigned against during the parliamentary elections. Even though the vote of the True Finns diminished to 17.7% in the 2015 parliamentary elections, party leader Timo Soini did not hesitate to claim the results as a 'second jytky' (great victory) following the one of 2011, and as a sign of party consolidation. It is a surprising achievement that the True Finns party successfully consolidated its support in the 2015 elections as none of the three key issues that explained its success in 2011 – anti-EU, anti-immigration, anti-establishment appeals – concerned the voters most in these elections (Jungar, 2015). Against the background of the deteriorating Finnish economy the electoral campaign revolved largely around traditional socio-economic issues, such as employment, the public debt and economic growth. Now the True Finns decided that the time for national office had come. They joined a centre-right cabinet and received four ministerial portfolios in a government with in total 14 portfolios. Timo Soino himself became Minister of Foreign Affairs.

10.3 Ideology and policies

In this chapter, populism is understood as a 'thin' ideology, that is, as a set of interconnected ideas about government, as well as 'the political' more generally (Canovan, 1999; Stanley, 2008). This thin ideology can and ultimately must combine with other ideologies to develop into an electorally attractive alternative. Historically, populism has been manifested in and married with various other 'fuller' ideologies. Populism can thus appear as left or right-wing, as well as take the form of agrarian populism.[2]

Finnish populism manifests more continuity than change in relation to the two main policy dimensions of party competition in Western Europe – the socioeconomic left and right versus authoritarianism and liberalism (Flanagan and Lee, 2003; Kriesi *et al.*, 2006). The FRP and its successor the True Finns combined a centre-left position on the socioeconomic dimension with, as a rule, a position on the authoritarian side of the liberal-authoritarian continuum (Jungar and Jupskås, 2014). While its socioeconomic

[2] See Jungar and Jupskås (2011) for an analysis of the position of the Nordic populist political parties in relation to the socioeconomic and the libertarian-authoritarian cleavage.

position has been stable over time, the authoritarian appeals have become more salient (Figure 10.2). The FRP was socially conservative and embraced a form of patriotic nationalism which, from the late 1980s, included anti-immigration appeals, while within the True Finns such appeals have taken centre stage alongside criticism of the EU. Thus, the True Finns have moved towards the authoritarian end on the value-cultural dimension and become radicalised in this respect. The socioeconomic left-right dimension includes party positions on state-market relationships, welfare, social justice and free enterprise. The party has supported moderate state regulation of the market

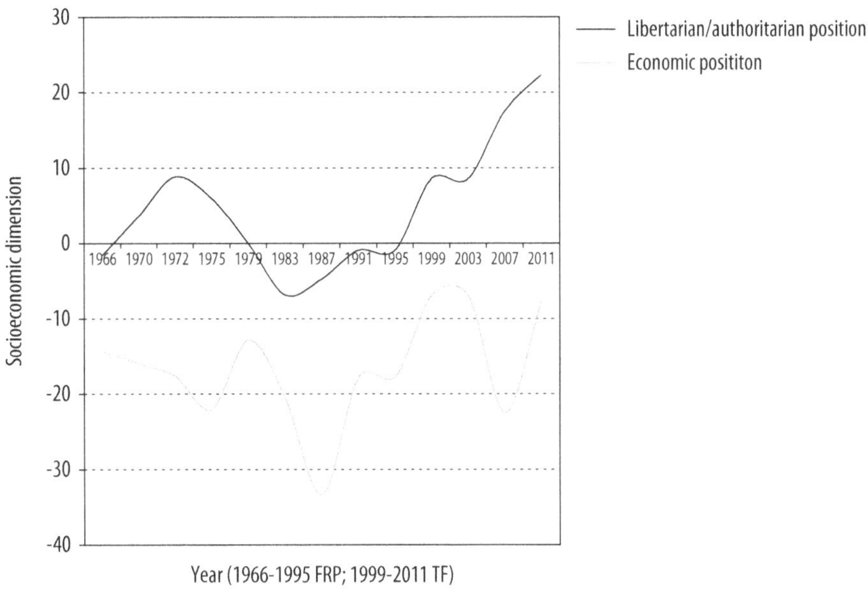

Figure 10.2. Finnish Rural Party (FRP) and True Finns (TF) policy change on the socioeconomic left-right and liberal-authoritarian dimensions between 1966-2011. Source: Comparative Manifesto Project. Notes: positive values mean right/libertarian positions and negative values mean left/libertarian, with zero being neutral. The economic left consists of the following variables: economic planning positive, protectionism positive, nationalisation positive, Marxist analysis positive, social justice positive, welfare state expansion positive, labour groups positive. The economic right is defined as the combination of: decentralisation positive, free enterprise positive, incentives positive, protectionism negative, economic orthodoxy positive, welfare state limitation positive, labour groups negative, middle class and professional groups positive. Libertarian is defined as military negative, national way of life negative, traditional morality negative, multiculturalism positive, underprivileged minority groups positive, anti-growth economy positive. Authoritarian is defined as military positive, national way of life positive, traditional morality positive, law and order positive, and multiculturalism negative.

and a tax-based redistributive welfare state, and is particularly favourable to small and medium-sized entrepreneurship.

The True Finns hold welfare chauvinist opinions as they claim that the 'proper' or the particularly 'deserving' people (pensioners, war veterans and the disabled) should be prioritised. The liberal-authoritarian dimension taps into issues related to democratic rights (minorities), morality (religion), authoritarianism (law and order) and post-materialism (cosmopolitanism, anti-growth). The salience of the liberal-authoritarian dimensions has increased within the True Finns, with nationalism resurfacing as a more pronounced resistance to immigration, multiculturalism and the rights of the Swedish-speaking minority, and strong support for Finnish culture and traditions. Policy issues relating to the liberal-authoritarian dimension are more salient than socioeconomic issues for the True Finns (Bakker *et al.*, 2010). The authoritarian turn of Finnish populism after 1995 is predominantly explained by these policy transformations, making the True Finns a case of new-right mobilisation and identifying the party as a member of the populist radical-right family (Jungar and Jupskås, 2014).

A more detailed study of the ideological transformations since the 1970s demonstrates that the agrarian legacy is still important (Figure 10.3). Agrarian policies concerning farming and forestry are still salient. The maintenance of and public support given to small-scale private farming was the key issue in the 1959 party programme, which stated that the rural areas must be supported and that both internal and external migration, from the countryside to Finnish cities and to Sweden respectively, had to be stopped (Suomen maaseudun puolue puolueohjelmat, 1959, 1967, 1973). The agricultural policies were criticised as they were perceived to discriminate against small-scale family farming in favour of larger agricultural producers. According to the FRP, foodstuffs would become cheaper and affordable through the maintenance of small-scale farming, as there would be a larger supply and a multitude of producers. The party proposed specific tax incentives, such as a tax-exempt minimum wage for farmers. Domestic agricultural production was to be protected against international competition, and Finland was to be self-sufficient with regard to the production of agricultural foodstuffs. In the 1980s, ecological considerations were introduced into the FRP's agricultural programmes (1981), which considered small-scale family farming particularly useful for the production of organic and clean foodstuffs. The party also took a firm stance against nuclear power and advocated natural domestic energy sources (trees, turf, etc.).

The rural policies of the True Finns are less targeted at farming and more concerned with securing both material and social infrastructure in the countryside, including roads, communications, education and welfare provision. Finnish populism presented itself as an interest party for small-scale farmers and the rural population. However, the countryside and its social values and traditions also served as the 'heartland' in the

Figure 10.3. Finnish Rural Party (FRP) and True Finns (TF) manifesto content for four policy categories: (1) agriculture and farmers; (2) morality; (3) EU; and (4) multiculturalism, 1970-2011. Source: Comparative Manifesto Project. Notes: the content of the dimensions are: 'Agriculture & farmers'= specific policies in favour of agriculture and farmers. Includes all types of agriculture and farming practices. Only statements that have agriculture as the key goal should be included in this category. 'Traditional morality' = favourable mentions of traditional and/or religious moral values. May include: prohibition, censorship and suppression of immorality and unseemly behaviour; maintenance and stability of the traditional family as a value; support for the role of religious institutions in state and society. 'EU negative' = negative references to the European Community/Union. May include: opposition to specific European policies which are preferred by European authorities; opposition to the net-contribution of the country to the EU budget. 'Multiculturalism negative' = the enforcement or encouragement of cultural integration. Appeals for cultural homogeneity in society.

populist rhetoric of both the FRP and the True Finns. The discursive use of so-called rural virtues are central to populist appeals in Finland, to which we now turn.

10.4 Populist appeals

Populism is often used as a derogatory and pejorative term in the context of accusing political opponents of opportunism; it is rarely used as a self-description by the political parties concerned. However, the Finnish populists are an exception to this norm, as both the FRP and the True Finns explicitly labelled themselves populists. The 1992 FRP party programme states that:

> The FRP is a populist [party] or a party that listens to the popular will, that is neither left- nor right-leaning. It emanates from the everyday thinking of people with the aim to empower the individual.
>
> (SMP, 1992)

In the 2011 electoral manifesto, the populist credo is formulated as part of the ideological core, as the True Finns proclaim that they are 'a patriotic and Christian social party... that supports a populist conception of democracy, which relies on the consent of the people' (Perussuomalaiset, 2011). The form of populism that is evoked in the party programmes is one of the sovereign people as competent democratic subjects with common sense and personal experiences. Popular sovereignty is juxtaposed with current democratic rule, characterised by non-responsive political representatives, bureaucrats and experts. The anti-establishment critique is expressed in the claim that popular sovereignty needs to be restored and individuals empowered as democratic citizens.

In this context, populism is defined as 'an ideology that considers society to be ultimately separated into two homogenous and antagonistic groups – 'the pure people' versus the 'corrupt elite', and which argues that politics should be an expression 'of the *volonté générale* (general will) of the people' (Mudde, 2007: p. 23). The definition includes four elements: (1) the existence of two homogenous groups – 'the people' and 'the elite'; (2) an antagonistic relationship between the elite and the people; (3) a positive valorisation of 'the people' alongside the denigration of 'the elite'; and (4) the idea of popular sovereignty.

Populism manifests itself in several analytical components, of which the empirical representations are contextually dependent. The above-mentioned elements – the anti-establishment position, the rhetorical construction of a unified and homogenous people, the elevation of the people and denigration of the elite, and popular sovereignty – are the points of departure for an analysis of how populism has 'adapted to the colours of its environment' (Taggart, 2000: p. 2). According to Paul Taggart, a key aspect of populism is its chameleon nature, meaning, as with all ideologies, populism has 'primary and secondary features, and that one of its primary features is that it takes on, as a matter of course, secondary features from its context' (Taggart, 2000: p. 2). However, populism does this to a large extent through narratives, myths and symbols that must resonate with the 'heartland', that is, by invoking an idealised community. Since the real-world expressions of populist ideas have been transformed and have changed over the five decades of parliamentary populism, this permits an investigation of the plasticity and adaptability of populist ideas in Finland.

The 'forgotten' people

Differing monolithic conceptions of the people are used within populist discourses, but the 'imagined people' are always conceived of as a homogenous group having common interests and a common will (Canovan, 1999, 2005). There are three major conceptions of the people in past and present populism, which are all based on a contrast with other groups. Firstly, the sovereign people are composed of those who live together in a political community and in relation to which the political elite is unrepresentative and unresponsive (the *demos*). Secondly, the people are identified as the ordinary and common people. This is the man – less frequently, the woman – in the street, the average, simple, honest and hard-working individual who also symbolises traditional virtues that are at odds with 'modern' and 'new' values, such as those embraced by leftist/liberal elites and the cultural establishment. The third populist construction of the people is that they are 'our' people, as opposed to some other people. The people are recognised on the basis of ethnicity, culture, language and history. From this point of view, the people are an *ethnos*, a unity whose homogeneity rests on shared characteristics or experiences. This ethnocultural community is constructed by processes of 'othering' on the basis of ethnic, linguistic, religious and cultural differences (Roma, immigrants, Islam) which are perceived to threaten the 'national' culture.

All three conceptualisations of the people are active in contemporary Finnish populism. The FRP was initially formed as an interest-based party 'for rural smallholders, free workers and small entrepreneurs' (Suomen pientalonpoikien puolueen ohjelma, 1959). The party addressed specific segments of the population, identified on the basis of their position in the labour market. However, in the late 1960s the FRP increasingly started to present itself as the party for a 'deserted' and an imagined homogenous rural population, whose values and habits were under threat. It was the party that acted on behalf 'of the forgotten people against the abuse of power' by the established political parties (Vennamo, 1989: p. 143). The 'ordinary' people were idealised and an image was created of a people under siege and in need of rescue. Populism presents itself as a cry for help by a 'forgotten people' (forgotten by politicians) or a people who have been lost along the path of socioeconomic or cultural transformation. Populism – and especially its leaders – presents itself as the people's friend in need. The present party leader of the True Finns, Timo Soini, has described Vennamo as 'the forgotten people's Batman, perceived by his supporters as utterly fair and honourable', and on another occasion as 'Moses, leading his forgotten people back to their lost land' (Soini, 2008). Timo Soini was politically educated and socialised within the FRP and was close to Veikko Vennamo, who adopted him as his political son. This type of leadership is reflected in the image that Soini has of his own present mission, that is, as the leader of the forgotten people, marginalised due to the free flows of capital, enterprise and people.

The term the 'forgotten people' (*unohdettu kansa*) is iconic in Finnish political vocabulary. Vennamo began using it during the 1968 presidential campaign. It is not unlikely that the concept – even though it is generally perceived of as an original Finnish invention – was borrowed from US political discourse. Vennamo was an avid reader of foreign newspapers, and his wife Sinikka, who formulated many of the party slogans with her husband, was also internationally oriented. After presenting the New Deal in 1932, President Theodore Roosevelt gave a radio speech from Albany in which he described those hard hit by the economic depression as the 'forgotten men'.[3] The term 'forgotten man' was also picked up by Richard Nixon in the late 1960s as a metaphor for the revolt of the white lower-middle class against the white pro-black upper-middle class. Nixon disliked the popular term 'the silent majority' and preferred to appeal to the 'forgotten American' (Carter, 1992: pp. 36-37).

The forgotten people were initially predominantly construed as a specific economic interest group, that is, the rural population affected by the rapid urbanisation and modernisation of Finland in the 1960s. The countryside was emptied and people were emigrating to Sweden as family farming became less and less profitable. Traditional norms and values grounded in agrarian society were being challenged by radical leftist ideologies and individualism. Associations with the 'lost Finland' had particular resonance among the Karelian refugee population, among whom there was widespread support for Veikko Vennamo. As a civil servant in the Ministry of Agriculture, Vennamo managed the settlement of the 400,000 Karelian refugees, which would serve him well in his later political career, which began in the Agrarian Union. The Agrarian Party, later the Centre Party, was the governing party in Finland starting in the 1950s, but even though the party's stance was farmer-friendly, there was widespread discontent among the farming population. The Communist Party became an appealing choice for family farmers in northern Finland, who were drawn to 'backwoods communism' (Alapuro, 1980; Allardt, 1970: pp. 45-63). Vennamo mobilised family farmers who were agitating for public subsidies for rural survival but were vehemently anti-communist and could not embrace radical leftist ideas either politically or in terms of values. At that time the FRP was described as a non-socialist leftist party (Helander and Toivonen, 1971: p. 65).

Vennamo made himself the spokesman for this threatened segment of the population, and increasingly described the rural population as forgotten, with their livelihood and values under threat, idealising its members as representatives of the authentic, hard-working and honest Finland (Ojanperä, 2000). The party leader made use of stereotypical virtues constructed and formulated in the Finnish nationalist movements and circles of the seventeenth and eighteenth centuries, as well as those of 'grace' and

[3] In the Albany speech, Roosevelt stated: 'These unhappy times call for the building of plans that rest upon the forgotten, the unorganized but the indispensable units of economic power, for plans like those of 1917 that build from the bottom up and not from the top down, that put their faith once more in the forgotten man at the bottom of the economic pyramid'.

'peaceful' resistance associated with pious religious movements. The Finnish people had developed a sense of justice through the hardships they had experienced under the Swedish and Russian empires, and in the political and literary nationalist narratives they were portrayed as honest, humble and hard-working (Ojanperä, 2000). They were described as being poor through no fault of their own, and on this basis as having a right to be bitter and dissatisfied. These appeals to nationalism and piety were combined with working-class rhetoric and demands for greater socioeconomic equality:

> The gentlemen got the roses and the people the nettles.
>
> (Vennamo and Kuitunen, 1973)

While the rural base prevails among the True Finns, in the contemporary rhetoric, the inhabitants of suburban environments are also being described as a new forgotten people. The success of the True Finns in opinion polls can be explained by this broadening of the voter base. The new party name 'The True Finns' (or the Finns) is associated with the 'ordinary people' or literally the 'basic Finn'. *Perussuomalaiset* literally means the 'basic Finns' – or the ordinary guy – the *jätkä* (hillbilly), who drinks beer, bets on the trots and likes cars. 'The true Finns do not pretend. We are honest, reliable', according to Soini in his blog (2011). To be a true Finn is thus not exclusively about embracing certain political attitudes, but also about having certain moral virtues.

Under the True Finns, the threatened people have also been redefined and reconstructed primarily as the Finnish-speaking people, who are contrasted with immigrants and Swedish-speaking Finns. The True Finns have become more aligned with other successful European right-wing populist parties, presenting immigration as not only a threat to jobs and social welfare, but also to Finnish culture and traditions. There is a tension within the True Finns party between welfare chauvinism and cultural nationalism: Finnish people should have first choice when it comes to both jobs and social benefits. Soini and other groups that came out of the old Rural Party tend to talk about immigration as a threat to Finnish jobs and welfare. However, others hold even more radical positions, their most extreme representative being Jussi Halla-Aho – a former MP and since 2014 a True Finns MEP – who has argued that immigration, especially Muslim immigration, is above all a threat to Finnish culture. He has been convicted of a hate crime after saying publicly that there is a connection between Islam and paedophilia.

Patriotic and nationalist sentiments are not entirely new within the True Finns, but within the Rural Party they took the form of criticism of foreign aid to underdeveloped countries (why should money be spent on poverty abroad when there is poverty in Finland?). From the late 1980s, the dangers and problems of immigration were voiced by representatives of the Rural Party, with Sweden and Norway taken as examples of the consequences of high levels of immigration. It was claimed that Finland should primarily help people from its own culture – from Karelia and Ingermanland – rather than Africans (Vennamo, 1989: pp. 316-317). Vennamo also touched upon the problems of the integration of Islam into Europe in relation to the *fatwah* on the British author Salman Rushdie in the late 1980s.

Finnish culture, defined in terms of the values of uprightness and honesty, Christian values and a Western cultural tradition, is juxtaposed with multiculturalism and Islam in particular. The Rural Party criticised public radio and television not only for their politically subservient journalists, but also for broadcasting improper and decadent programmes. In the run-up to the national elections of 2011, the True Finns presented a cultural policy platform based on a Finnish art canon and opposed the public funding of 'postmodern' artistic expression. 'Maassa maan tavalla' or 'when in Rome, do as the Romans do' is one of the True Finns' political slogans, which, for example, has also been used by Jutta Urpilainen, the former leader of the Social Democratic Party. Immigration should be restricted and those who choose to come to Finland should assimilate with Finnish culture.

'Finnishness' is given a different, stronger emphasis by the True Finns than was found in the Rural Party. This is obvious in the party's relationship to Swedish-speaking Finns and Finno-Swedishness. Finland is a bilingual country with Finnish and Swedish as official languages. Five percent of Finnish citizens have Swedish as their mother tongue and predominantly live in the southern and mid-western coastal areas of Finland. In Veikko Vennamo's rhetoric, the Swedish-speaking Finns were conceived of as a master race – a political, economic and cultural elite. In reality, the Swedish-speaking Finns constitute a socially heterogeneous group of both upper and lower middle class, overrepresented in government administration and business, as well as farmers and workers. Vennamo fomented an ambivalent relationship to Finno-Swedishness and understood that while theoretically there was a breeding ground for populism among the anti-communist family farmers in Österbotten, their linguistic identification overshadowed everything else. There was also smouldering resentment among some Rural Party supporters against Swedish-speaking regions of Finland that had resisted taking in Karelians when refugees were pouring in from the east during the Second World War. For example, Vennamo stated:

> The future of our people demands above all the building of homes for our own people and values that support a sense of popular community to secure our future. Only the Swedish speaking areas resisted the Karelians in the name of the holy Swedish earth. Then the proper interest of the country was forgotten.
>
> (Vennamo, 1989: p. 310)

For the True Finns, Finno-Swedishness is a foreign element in relation to the Finnish majority culture. According to the party, the Swedish-speaking minority is favoured at the expense of the Finnish-speaking population, with compulsory Swedish instruction at school, weighted study places for Swedish-speaking Finns in areas where service in Swedish is necessary, obligatory language requirements for civil servants, etc. The True Finns' demands for a national language and their desire for a monolingual Finland disguises a proposal for cultural assimilation behind an ambition to achieve equality or efficiency. The True Finns foment and cultivate a nostalgic reactionarism based on a revisionist history of Finland, considered as a land where Finnishness was once dominant and unchallenged and Swedishness represented Swedish colonial rule.

Anti-establishment appeals

Central to populist ideology is the negative valorisation of the elite and their positioning against the people. Anti-elitism and anti-establishment appeals constitute the historical core of Finnish populism. It is characterised by protest against and criticism of the established political parties, the government and the bureaucracy, but also of the economic, cultural and academic elites that have distanced themselves from the values and lives of ordinary people. The political elite is portrayed as both unrepresentative and unresponsive to the people. Continuity prevails through terms coined by Vennamo such as '*rötösherrat*' or the 'rotten gentlemen' – corrupt, wheeling and dealing politicians – and '*teoria-herrat*', the 'theoretical gentlemen', isolated in their academic ivory towers with no sense of what is happening on the ground. The terms again echo the True Finns' criticism of power cliques, political nepotism, corruption, the tyranny of experts and the all-powerful EU bureaucrats, who they claim have all turned a deaf ear to the opinions and values of ordinary citizens.

A recurrent theme in both old and contemporary populism is colluding and corrupt politicians. The Rural Party and Veikko Vennamo in particular (as a lawyer), have brought to court or conducted many investigations of illicit contacts between political parties and enterprises, and criticised the politicians' use of public money for private purposes. According to Vennamo, moral decay had infiltrated the political elite:

> The people are seriously worried and decent citizens upset about the rotten and immoral behaviour of public persons, which like a cancer has infiltrated our societal organism.
>
> (Vennamo, speech 16/06/1959, in Vennamo and Kuitunen, 1973)

The mission of the FRP was to make public the wrongdoing of public figures and institutions since 'a majority of the elected representatives are fortune seekers and aspiring for personal benefits rather than true and honest guardians of common interests' (Vennamo, speech 08/11/1962, in Vennamo and Kuitunen, 1973). Soini has followed in the footsteps of his predecessor and whenever possible talks about the corrupt practices of Finnish politicians, of which there have been quite a few cases over recent years, including, for example, the former prime minister, Matti Vanhanen (illegal party financing, gifts from companies, love affairs), the former foreign minister, Ilkka Kanerva (affairs with nightclub dancers) and party financing in general.

While Vennamo was party leader, the anti-establishment appeals related to misgovernment and corrupt politicians were the most salient, as is illustrated in Figure 10.4. Politicians should serve and represent the people as well as constitute examples of proper behaviour and virtues such as honesty and hard work. Vennamo assumed the role of a purifier, not in ideological terms but in terms of the quality of government and democracy (Lucardie, 2000). He explained the formation of the new agrarian party in terms of the Finnish people valorising and expecting pure hands and clean politics:

> The Finnish people have always understood the importance of cleanliness. The sauna is illuminating proof of this. Besides such corporal cleanliness, we now need to purify and elevate ourselves spiritually and societally.
>
> (Vennamo, 1959, in Vennamo and Kuitunen, 1973)

The anti-establishment appeals, formulated as a general critique of the political establishment, decreased drastically after Vennamo resigned as party leader and was replaced by his son Pekka Vennamo in 1989.

The political antagonisms are also expressed as a criticism of certain – often labelled 'politically correct' – hegemonic ideas, which the elites hold and embrace, while those who do not share these ideas are considered intolerant, conservative, non-informed and driven by passion rather than knowledge. The Finnish political culture of consensus that is reflected in surplus majority governments – whose failures have been increasingly scrutinised in the public arena – as well as a weak and absent political opposition, have fuelled the populist anti-establishment rhetoric. However, international relations constituted one policy area where consensus prevailed until the late 1980s.

In the past it was the Soviet Union but at present it is the EU that is the main external target of populist antagonisms. The main message is that Finnish political elites are too subservient and prone to sell out national sovereignty to secure their own power and the accompanying benefits. During the postwar era, all political parties with

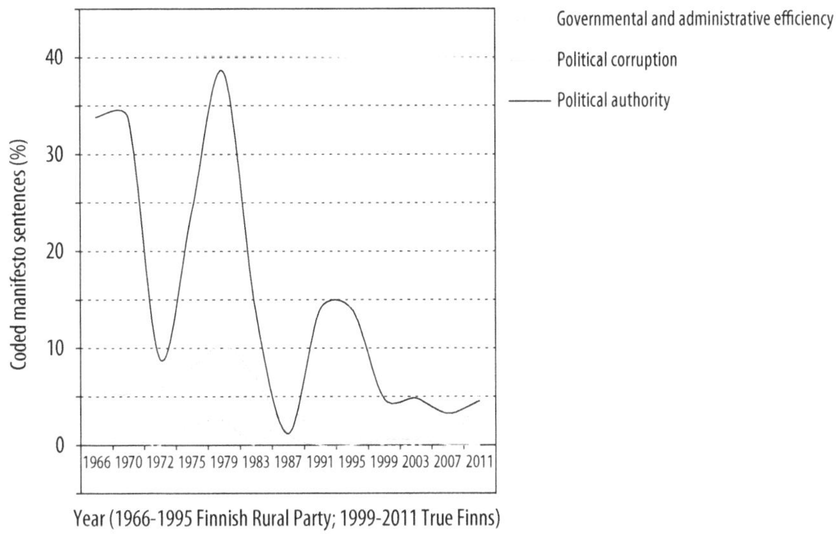

Figure 10.4. Saliency of statements on government performance, corruption and political authority, 1966-2011. Source: Comparative Manifesto Project. Notes: the content of the dimensions are: 'Governmental and administrative efficiency' = need for efficiency and economy in government and administration and/or the general appeal to make the process of government and administration cheaper and more efficient. May include: restructuring the civil service; cutting down on the civil service; improving bureaucratic procedures. 'Political corruption' = need to eliminate political corruption and associated abuses of political and/or bureaucratic power. 'Political authority' = references to the manifesto party's competence to govern and/or other party's lack of such competence. Also includes favourable mentions of the desirability of a strong and/or stable government in general.

ambitions to govern were expected to take the Paasikivi-Kekkonen[4] foreign policy line and uncritically protect friendly relations with the Soviet Union and accept the use of foreign policy to achieve domestic policy ends. The Rural Party, later in concert with the Constitutionalists, criticised President Kekkonen's autocratic stance and the uncritical support by other parties of his personal ambition to gain power. Vennamo's sobriquet for the Republic of Finland was 'Kekkoslovakia'. During the 1970s, Kekkonen's presidency was extended through an emergency powers act (1973) and in the next presidential election of 1978, he refused to debate with other presidential candidates. In his memoirs, written after his political career, Vennamo stated that '[a] Kekkonen dictatorship ruled our country. It distorted the entire society. A court of 'lickers' and

[4] Finland was a semi-presidential republic until 2000: Juho Kusti Paasikivi was the president of Finland between 1946-1956, and Urho Kekkonen between 1956-1981. The Paasikivi-Kekkonen line is a foreign policy doctrine of neutrality aimed at safeguarding Finland as an independent sovereign democracy and capitalist economy, while a neighbour to the Soviet Union.

'après-skiers' was established. A group of Kekkonen-imitating small dictators and liars saw the light of day' (Vennamo, 1989: p. 9).

In hindsight, the 1970s constitutes a dark chapter for democracy in Finland with regard to the presidential powers. At the time, the FRP proposed direct elections of the president (instead of being indirectly elected by a specific group, who in turn had been elected by a popular vote), the limitation of presidential power with regard to government formation and the restriction of the number of the presidential periods to two – which were all realised in the constitutional revisions in 1994 and 2000. The criticism of the presidential powers which intruded on those of the parliament reflects a democratic ethos and a populist belief in popular sovereignty. In a specific party programme on constitutional matters of 1974, the FRP considered 'that power should belong to the representatives elected by the people, the president elected by the people, to the Prime Minister elected by the parliament in accordance the rules of a balance of power. Inner circles and monopolies are not popular rule. The tsarist powers of the president have to be abolished' (Suomen maaseudun puolueen puolueohjelman perustuslakiemme eritysohjelma, 1974).

The ascendancy of the True Finns can partly be related to a similar climate in the Finnish public debate about the reform of the welfare state after the economic crisis of the 1990s and the almost complete support of the European project by all Finnish political parties. The euro was an effective target during Timo Soini's campaign prior to the 2009 elections for the European Parliament when the True Finns garnered almost 10% of the votes. Soini has successfully criticised the European bailout packages extended to Greece and Ireland in the wake of the financial crisis. Finnish populism's recipe for success has consisted of narratives about political conspiracies among colluding establishments dancing a tango out of time with the citizenry. The criticism of the EU by the True Finns is founded on their view that democracy is based on the nation-state. A supranational EU is considered impossible due to the mixture of peoples and demands for solidarity between them.

The competent people

'Yes, the people know' ('Kyllä kansa tietää') is another of Vennamo's cogent expressions still in use in Finnish political rhetoric. The quote captures populism's central emphasis on the people as both the democratic unit in which power should be vested and as the idealisation of the ordinary, honest and hard-working person, or the savvy and skills of 'the man (sometimes the woman) in the street'. Traces of different traditions can be found within Finnish populism in this elevation of the people. Firstly, there is the notion of the people conceived of as a nation, sharing common characteristics, influenced by German thinkers such as Herder and others. Partly within the old Rural Party, but above all within the True Finns, these types of ideas have prevailed in the form of the elevation of Fenno-Ugrian family solidarity and as Finnish cultural distinctiveness. Fenno-Ugrian solidarity is extended to neighbouring peoples who speaking languages belonging to the Fenno-Ugrian language groups, in particular Finnish and Estonian.

However, these ideas on the relationship between the individual and the people also find parallels in American populism. This is may be more contestable, but Vennamo's view of capitalism, as well as the idea of socialism representing the centralisation of power at the expense of the individual and the community, echoes American populism.

> My view is that capitalism based on the power of money and socialism based on force cannot succeed in the longer run. Why? Because they are cold ideas that neglect the individual and human life in general.
>
> (Vennamo, 1989: p. 301)

Hence, capitalism is found to commodify the people, while socialism is based on a system of thinking that undermines individual human talents, capabilities and ultimately liberty. Private property and freedom are important, but need to be cultivated within a community. In other words, earlier Finnish populism embraced a type of social individualism that was different from the collectivism embedded in socialism and the atomised individualism of capitalism and modern society. From this perspective, individual freedom encompasses a sense of duty and responsibility to the community. 'Every person and every life is valuable', Vennamo stated in many speeches. This expression is often re-used by Soini in his own writings and speeches.

Populism offers no full-bodied political ideology that explains social conditions or presents coherent visions of society. Rather, it elevates the people as a political subject, gives voice to its supporters' experiences and confirms their worldviews. In his 1987 master's thesis, entitled 'Populismi – politiikka ja poltinmerkki SMPn roolinmuutos' (Populism – politics and brand, the new role of the FRP), Soini wrote that 'populism tells the public what it wants to hear' (Soini, 1988: p. 9). Unlike other political ideologies, populism has no pretensions to enlightenment or the ambition to educate its voters: the democratic ethos is to reflect the people and in so doing become the voice of the people.

The democratic principle of equality begins with the idea that each person is the best defender of his or her own interests. As a consequence, each and every vote should have the same weight in the democratic process. Institutionally, this is reflected in the principle of one person, one vote. However, the inclusion of more groups in the *demos* has as a rule been accompanied by doubts concerning the actual political competence of the voters: Are they capable of rational political judgement? Do they have sufficient political knowledge? These fears have been reflected in different ways. According to liberals (and conservatives), popular rule had to be restricted by constitutionally guaranteed rights and power-sharing, among other things, whereas socialists maintained that true democracy had to be prepared by the take-over by an avant-garde, because the people were not aware of their own interests, that is, they still suffered from what was called a 'false consciousness'.

In *Democracy and its critics*, the well-known American political scientist Robert Dahl maintains that democracy is a system in which it is more probable that people will get what they want or what they believe they want than in other political systems (Dahl, 1989: p. 174). This categorical principle that no-one shall live under rules that they have not been able to influence, has often been accompanied by reservations and restrictions. Suffrage was introduced gradually because not everyone was considered sufficiently competent to vote. The revolutionary demand of universal suffrage was accompanied by fears of the consequences should the uneducated people or masses be given voting rights. The revolutionary demand for universal suffrage during the nineteenth century was accompanied by concerns about the consequences of the access to power of the un-illuminated people. Often – indirectly – populist voters are depicted as not being politically competent and having a poor knowledge of politics, as well as being guided more by passions and feelings than reason.

The populist idealisation of the citizens' political competence and explicit scepticism of knowledge and expertise in the policymaking process does not correspond to a view of democracy as a project of education and illumination. The purpose of politics is not to educate the voters or modify their political views. Timo Soini writes in his blog post: 'Your opinion cannot be wrong, since it is your opinion'. He continues, stating that in politics explanations are not needed 'since the sympathisers need no explanations and your opponents do not believe them'. To Soini, 'politics is a competition between opinions. There are different opinions and in the elections their respective support is measured'.

As such, populism breaks with democratic traditions in which the democratic procedure is seen not only as a legitimate procedure for formulating collectively binding decisions, but also as a process of the enlightenment and education of citizens. In short, democracy is an arena for critical reflection and re-evaluation. Democracy should be organised so that the citizens can develop and fully exercise their political citizenship and fully realise proper or common interests. From this perspective, democracy begins with the idea of enlightenment: with the educational effects of democracy being particularly important in contemporary deliberative and participatory models.

The sovereign people

> The True Finns support a populist democracy based on the support of the people and oppose an elitist or bureaucratic democracy. In a populist democracy the citizens elect as their representatives persons with similar values and opinions.

This brief explanation of democracy is given in the electoral manifesto of the True Finns in 2011. As such, this statement does not stand in opposition to other ideals of democracy. A convergence of opinions is a common indicator of the realisation of the popular will. In order to develop a broader picture of what characterises populist democracy one has to read further: 'In an elitist democracy, bureaucratic expertise is the source of power, rather than the opinions of the people'. In the 1970s, Vennamo spoke about the ideological foundations of the FRP and talked about how capacity and knowledge can be used for oppression:

> For us capacity and knowledge are instruments. They do not allow (those who possess them) to humiliate and oppress the individual, who possesses the spirit of humankind.
>
> (Vennamo and Kuitunen, 1973)

A distinct view of political representation has been formulated within Finnish populism. To represent is to be the mirror of the people, reflecting the voters in the political decision-making process in a manner that is as undistorted as possible. Both the FRP and True Finns advocate the increased use of direct democratic instruments such as referenda, citizen initiatives and recall. Every transfer of decision-making competences to the EU should be directly ratified by the citizens in referenda.

Populists are critical of institutions that restrict the expression of a popular political majority – that is, constitutionally granted minority rights, power-sharing and the delegation of competences to institutions that restrict the popular majority or which are difficult to hold politically accountable, such as the EU, expert bodies and constitutional courts. This is reflected above all in the True Finns' scepticism of the EU. For the True Finns, EU legislation infringes on the power of the national parliaments. Since democracy is based on the existence of a united *demos* (according to the True Finns) and since a European people will never exist, democracy on a supranational level is not realisable.

The majority principle is central to populist democracy. The opinion of the majority should be respected by the elected representatives and excessive consideration of minorities should be avoided. The latter should subordinate and adapt themselves to the majority and should not have more influence than is appropriate to their size. This is particularly reflected in the integration policies of the True Finns, who argue that immigrants and minorities should adapt to the culture of the majority. They also provide stark criticism of 'positive discrimination' in relation to minorities, such as measures intended to facilitate their integration into the labour market. In May 2011, the parliamentary group of the True Finns declared that they were against all types of positive and negative discrimination and that the state should not take measures to treat different groups – ethnic, linguistic, religious or political – differently. In this respect, they gave voice to individualism and, in addition, they called for discrimination against the majority to be taken as seriously as discrimination against minority groups (Perussuomalaisten eduskuntaryhmän julkilausuma, 25/05/2011).

The True Finns have also argued that minorities should not be given constitutionally guaranteed rights (or privileges). The focus on the majority principle emphasises their view of democracy as a competition between winners and losers, and if the constitution of the majorities and minorities change over time and according to policy issues, a strict use of the majority principle should not be a problem. However, if the majority and minority are always the same, the majority principle may result in the constant discrimination of majorities against minorities, which would be in conflict with the democratic principle that all citizens should be of equal concern, which in turn may necessitate specific measures to assist certain groups.

10.5 Conclusions

The seminal volume on populism edited by the social scientists Ghita Ionescu and Ernest Gellner (1969) describes populism in various parts of the world. However, there is no chapter on Scandinavia or the Nordic countries. One year after its publication, the FRP achieved a landslide victory which upset the Finnish political system. Ever since, the Nordic region has constituted fertile ground for populism, and this is still the case. Populism in Finland is the most long-lived, yet remains generally unknown internationally. Finnish populism is both a distinct element of Finnish political culture as well as an expression of general political ideas. The malleability or 'chameleonism' of populism lies in its ability to combine different ideas and adapt them to different contexts in time and space (Taggart, 2000). As a consequence, contemporary populism in Finland, as represented by the True Finns, combines different roots and traditions, reflected in the different factions within the present-day political party. Over the more than fifty years of its existence, the party has appealed to different groups of people and successfully transformed itself from an agrarian party into one with a more diverse appeal to voters.

Finnish populism has elaborated the central populist conceptualisations and skilfully adapted them to societal transformations. While it is perceived as a genuine Finnish political movement, since its inception it has taken inspiration from populist movements and parties elsewhere, as well as borrowed central concepts from abroad. The transnational dimension has been strengthened since the entrance of the True Finns into the European Parliament in 2009, where it forged contacts with like-minded political parties in Europe – starting with the Europe of Freedom and Democracy group within the European Parliament. After the European elections of May 2014, the True Finns, alongside the Danish People's Party, were welcomed into the Europe of Conservatives and Reformists. The question now is whether its populist appeals will once again adapt to these new circumstances.

References

Alapuro, R., 1980. Origins of agrarian socialism in Finland. In: Torsvik, P. (ed.) Mobilization, Center–Periphery Structures and Nation–building. Universitetsförlaget, Oslo, Norway, pp. 274-295.

Allardt, E., 1970. Types of protest and alienation. In: Allardt, E. and Rokkan, S. (eds.) Mass politics. The Free Press, New York, NY, USA.

Arter, D., 2012. Analysing 'successor parties': the case of the true Finns. West European Politics 35: 803-825.

Bakker, R., De Vries, C., Edwards, E., Hooghe, L., Jolly, S., Marks, G., Polk, J., Rovny, J., Steenbergen, M. and Vachudova, M.A., 2012. Measuring party positions in Europe: the chapel hill expert survey trend file, 1999-2010. Party Politics 00: 1-16.

Canovan, M., 1999. Trust the people! Populism and the two faces of democracy. Political Studies 47: 2-16.

Canovan, M., 2005. The people. Polity, Cambridge, England, 161 pp.

Carter, D.T., 1992. George Wallace, Richard Nixon and the transformation of American politics. The Thirteenth Charles Edmonson Historical Lectures. Baylor University, Waco, Texas, USA.

Dahl, R.A., 1989, Democracy and its critics. Yale University Press, New Haven, CT, USA.

Flanagan, S.C. and Lee, A.R., 2003. The new politics, culture wars, and the authoritarian-libertaria value change in advanced industrial democracies. Comparative Political Studies 36: 235.

Helander, V. and Toivonen, T., 1971. Populistisluonteinen pienviljelijäliikehdintä Suomessa ennen vennamolaisuutta. In: Helander, V. (ed.) Vennamolaisuus populistisena joukkoliikkeenä. Karisto Publishers, Hämeenlinna, Finland.

Ionescu, G. and Gellner, E. (eds.), 1969. Populism; its meanings and national characteristics. Weidenfeld and Nicholson, London, UK.

Jungar, A.C., 2015. A centre-right government takes form in Finland. Baltic Worlds. Available at: balticworlds.com/a-centre-right-government-takes-form-in-finland.

Jungar, A.C. and Jupskås, A., 2014. Populist radical right parties in the Nordic region: a new and distinct party family? Scandinavian Political Studies 37: 215.

Koivulaakso, D., Brunila, M. and Andersson L., 2012. Äärioikeisto Suomessa. Into, Helsinki, Finland.

Kriesi, H.P., Grande, E., Lachat, R., Dolezal, M., Bornschier, S. and Frey, T., 2006. Globalization and the transformation of the national political space: six European countries compared. European Journal of Political Research 45: 921-956.

Lucardie, P., 2000. Prophets, purifiers and prolocutors: towards a theory on the emergence of new parties. Party Politics 6: 175-185.

Lydén, M., 2012. Jag är inte rasist. Jag vill bara ha främlingsfientliga röster. Schildt and Söderströms, Finland.

Maijala, J., 2005. Maaseutuyhteisön kriisi 1930-luvun pula ja pakkohuutokaupat. Jyväskylän yliopisto, Jyväskylä, Finland.

Mudde, C., 2007. Populist radical right parties in Europe. Cambridge University Press, Cambridge, England.

Ojanperä, H., 2000. Kyllä kansa tietää Veikko Vennamon käsitys kansasta. Masters thesis. University of Jyväskylä, Jyväskylä, Finland.

Pekonen, K. (ed.), 1999. The new radical right in Finland. The Finnish political science association, Helsinki, Finland, 225 pp.

Sänkiaho, R., 1971. Populismi ja populistiset liikkeet. In: Helander, V. (ed.) Vennamolaisuus populistisena joukkoliikkeenä. Karisto Publishers, Hämeenlinna, Finland.

Soini, T., 1988. Populismi – politiikka ja poltinmerkki SMPn roolinmuutos. Masters Degree in Political science. Helsinki University, Helsinki, Finland.

Soini, T., 2008. Maisterijätkä. Tammi Publishers, Helsinki, Finland.

Soini, T., 2011. Blogg 2010-2011. Available at: www.timosoini.fi.

Stanley, B., 2008. The thin ideology of populism. Journal of Political Ideologies 13: 95-110.

Taggart, P., 2000. Populism. Open University Press, Buckingham, UK.

Vennamo, V. and Kuitunen E., 1973. Ajan valtimolla (juhlakirja). Suomen maaseudun puolue

Vennamo, V., 1989. Kekkosdiktatuurin vankina. Gummerus, Jyväskylä, Finland.

Official party documentation

Oikeutta kansalle, 1995. Perussuomalaisen puolieen erityisohjelma: Olen perussuomalainen. (party program)
Perussuomalainen, 2010-2011. (party journal)
Perussuomalaiset, 2007. Oikeudenmukaisuuden, hyvinvoinnin ja kansanvallan puolesta. Eduskuntavaaliohjelma 2007. (electoral manifesto)
Perussuomalaiset, 2011. Suomalaiselle sopivin Perussuomalaiset r.p:n eduskuntavaaliohjelma. Available at: www.perussuomalaiset.fi/getfile.php?file=1536. (electoral manifesto)
Perussuomalaisten eduskuntaryhmän julkilausuma syrjintää, rasismia ja väkivaltaa vastaan 25.5.2011. (statement from the True Finnish parliamentary group)
Perussuomalaisten maaseutuohjelma, 2001. (agricultural program)
Suomen maaseudun puolue, 1981. Maatalouden erityisohjelma: Maaseutu uuteen nousuun. (agricultural program)
Suomen maaseudun puolue – Finlands Landsbygdsparti, 1967. Puolueohjelma, Finland. (party program)
Suomen maaseudun puolueen puolueohjelma, 1973. Maatilatalouden erityisohjelma. (agricultural program)
Suomen maaseudun puolueen puolueohjelma, 1992. (party program)
Suomen maaseudun puolueen puolueohjelman, 1973. Metsien erityisohjelma. (program on forestry)
Suomen maaseudun puolueen puolueohjelman perustuslakiemme eritysohjelma, 1974. (special program on constitutional issues)
Suomen pientalonpoikien puolueen, 1959. Ohjelma. (party program)

11 The Front National and the new politics of the rural in France

G. Ivaldi[1*] *and J. Gombin*[2*]

[1]*URMIS-CNRS, Université de Nice Sophia Antipolis, Nice, France;* [2]*CURAPP-ESS, Université de Picardie-Jules Verne/CNRS, Amiens, France; gilles.ivaldi@unice.fr; joel.gombin@u-picardie.fr*

Abstract

The French Front National (FN) epitomises the post-industrial radical-right party family in Europe, manifesting a predominantly urban cultural and economic malaise. The party has, however, progressively formulated a more comprehensive rural agenda over the years, idealising agrarian values and specifically addressing farmers' interests. Since the early 2000s, the FN has grown stronger in rural areas. It has done so by challenging the status quo represented by the Common Agricultural Policy of the European Union and the traditional centre-right and Gaullist alignment of the farming world. More recently, the party has embraced a more leftist economic agenda, catering to intensifying demands for state intervention and public service provisions in rural territories. In 2012, the electoral success of the FN in the peripheries primarily reflected the territorialisation of the subjective and objective proletarianisation of the working and lower middle class.

11.1 Introduction

In the first round of the 2012 French presidential election, the Front National (FN) won 17.9% of the vote. Marine Le Pen improved her father's figures for 2002, when he made it through to the presidential second-round run off. Amid economic crisis, support for the FN increased among production and service workers and lower middle-class voters, who were the most severely hit by unemployment and economic difficulties. Reflecting the softening of the party's profile, Marine Le Pen assembled support from social groups who were traditionally averse to the FN, such as women (Mayer, 2013). Another notable feature of the electoral resuscitation of the FN in 2012 was its consolidation in rural areas. The radical right gained important footholds in the countryside and the distant outskirts of the large cities, indicating a move from original urban constituencies into rural France.

D. Strijker, G. Voerman and I.J. Terluin (eds.) **Rural protest groups and populist political parties**
DOI 10.3920/978-90-8686-807-0_11, © Wageningen Academic Publishers 2015

The 'ruralisation' of the FN calls for further analysis. Operationalising the 'rural' space is not unequivocal, and often faces the risk of excessively homogenising the diversity of local situations. The rural coalition of the radical right should not preclude the consideration that a variety of segmented socioeconomic interests and divergent political cultures coexist in rural France today. Over the past decades, important reconfigurations have occurred, both in the agricultural sector and in relation to the demography of rural areas, which call for a contextualisation of the FN vote.

This chapter examines the relationships between the Front National and the rural with regard to the policy positions taken by the party, but also in terms of the dynamics behind the electoral support for the radical right in France's countryside.

The first section examines the FN's extreme right origins as a mostly urban phenomenon with little roots in earlier rural protest movements. We argue that the party also failed to establish links with actors currently representing the interests of the rural world. We then turn to how the FN has progressively built a more comprehensive rural agenda, embedded in changes in the party's competitive positions. We focus on the recent Keynesian shift by the FN, where endorsing a new statist redistributive economic agenda became the party's response to the intensifying demand for social welfare and public service provisions in rural areas.

The second section examines the demand side of the ruralisation of the FN. Two important facets are discussed. The first concerns the increase in farmer support for the radical right. Farmers have become over-represented in the FN's electoral base since the early 2000s. Moreover, this support has a specific geography, revealing a challenge to the traditional centre-right and Gaullist alignment of the farming world in France. A second aspect concerns the 'peripheralisation' of the FN vote and the growth in support for the radical right in periurban areas – referring to outer satellite towns more distant from the large cities. The spatial distribution of the FN vote along the 'urban gradient' corroborates the geography of social and economic inequality. Territorial divisions based on socioeconomic positions reflect both the subjective and objective proletarianisation of the working class and the lower middle class.

11.2 Origins and party ideology

The historical roots of the Front National lie in the post-war French extreme right. The FN was formed in 1972 from a disparate coalition of ultra-nationalist groups, those with nostalgia for the Vichy regime, and former supporters of French colonial Algeria. The party was electorally irrelevant during the 1970s, making its first national breakthrough in the 1984 European elections running on a strong nationalist-populist platform fostering xenophobia and anti-Semitism.

Since its political take off, the FN has become a permanent feature of the French party system, both nationally and locally (Table 11.1). The party has established a distinctive

Table 11.1. Front National national electoral results since 1973.

Year	Election	% valid[1]
1973	Legislative	0.5
1974	Presidential	0.7
1978	Legislative	0.8
1979	European	1.3
1981	Presidential[2]	–
1981	Legislative	0.3
1984	European	11.0
1986	Legislative	9.6
1986	Regional	9.6
1988	Presidential	14.4
1988	Legislative	9.7
1989	European	11.7
1992	Regional	13.7
1993	Legislative	12.4
1994	European	10.5
1995	Presidential	15.0
1997	Legislative	14.9
1998	Regional	15.0
1999	European	5.7
2002	Presidential	16.9
2002	Presidential[3]	17.8
2002	Legislative	11.3
2004	Regional	14.7
2004	European	9.8
2007	Presidential	10.4
2007	Legislative	4.3
2009	European	6.3
2010	Regional	11.4
2012	Presidential	17.9
2012	Legislative	13.6
2014	European	24.9

[1] % of valid vote cast.
[2] The Front National did not run in the 1981 presidential election.
[3] Second-round run off.

appeal among French voters by politicising immigration and criminality issues. It advocates strong anti-immigrant and welfare chauvinist policies coupled with a hard stance on law-and-order and fierce rejection of the European Union. In 2011, Marine Le Pen took over party leadership after nearly four decades of unlimited rule by her father. She has modernised the party's strategies and programme to rebrand the old FN and present it in a more acceptable style. However, the FN has retained most of its core authoritarian, nativist and protectionist policies (Ivaldi, 2016).

A prototype of the new radical right

Most scholars agree on the location of the FN on the extreme right margin of French politics. For many years, the French FN has epitomised the prototype of the modern radical-right party family in Europe. It has been primarily considered as the political manifestation of the cultural and economic malaise caused by immigration, structural unemployment and criminality in industrialised urban areas. In the early stages, the FN was regarded as a prototypical case of Kitschelt's new radical right (1995: p. 91), mixing free market economics and culturally particularistic authoritarian appeals. According to Ignazi (2003), the French FN epitomises a third wave of post-industrial extreme right parties that are reacting to the process of cultural liberalisation. Betz (1994: p. 108) sees in the FN a variant of national-populism that combines ethno-pluralist policies with an economically protectionist agenda. In addition, Mudde (2007: p. 41) regards the French FN as representing the successful party family of the populist radical right whose key ideological features are nativism, authoritarianism and populism.

The main features of the FN differ from agrarian populism. According to Canovan (1981: p. 8), agrarian populism is a reaction against modernisation, recognisable in terms of the distinct composition of its support base, which primarily consists of farmers and/or peasants. More generally, agrarian parties are distinguished by their origins and their defence of the rural periphery, mobilising within the peasantry in opposition to the national, urban establishment. In contrast to other European versions of populism such as the True Finns in Finland (Chapter 10) or Self-Defence in Poland for example (Chapter 9), which have their roots in agrarian movements, the French FN finds no historical legacy in the rural world.[1]

The FN emerged and was consolidated as a predominantly urban phenomenon. Its politicisation of immigration and law-and-order issues concerned urban and suburban constituencies. This stands in sharp relief to the rural component that was found in earlier manifestations of right-wing populism in France, most notably the Poujadist

[1] Agrarian elements were found in the French extreme right during the interwar period. The most prominent organisation was the Green Shirts, founded by Henry Dorgerès (Ory, 1975; Paxton, 1997).

rebellion in the mid-1950s (Goguel, 1981: Chapter 7).[2] The Union for the Defence of Shopkeepers and Craftsmen (Union de Défense des Commerçants et Artisans, UDCA) was created by Pierre Poujade in 1953 as a right-wing movement of anti-tax revolt in response to France's economic modernisation and the rapid process of industrialisation. In the 1956 legislative elections, the Poujadist lists won 11.6% of the vote and 51 seats. The UDCA primarily assembled a petty-bourgeois constituency galvanising small-town rural shopkeepers and small farmers. They joined together to oppose big corporate businesses, large commercial interests, left-wing trade unions and 'corrupt politicians'. The Poujadist map was different from that of the Front National: in 1956, the Poujadist vote was concentrated in the rural departments of western France (Maine-et-Loire, Charente-Maritime, Indre-et-Loire, Deux-Sèvres, Aveyron, Gers) and the South, from Tarn-et-Garonne to Vaucluse. In contrast, during the 1980s, the support for the FN originated primarily in the industrialised and urban regions of the country, east of the Le Havre-Valence-Perpignan line (Perrineau, 1998).

Historically, agricultural protest and farmers' discontent have been channelled in the political arena by political forces other than the FN, which has never established formal relationships with farmers' interest groups or rural protest movements. In the 1990s, the party strove to form a number of flanking organisations. The latter aimed to expand its sectoral appeal to specifically target professional groups. In rural areas, this strategy was embodied by founding the National Association of Hunting, Fishing and the Environment (Cercle National Chasse Pêche Environnement, CNCPE) and the National Association of French Farmers (Cercle National des Agriculteurs de France, CNAF), but these were mostly empty shells.

French farmers have traditionally been represented by parties of the conservative rather than radical right.[3] This was due to strong links between farmers' professional unions (with high rates of membership) and the Gaullists (Rémy, 1972). Rural Catholicism has also been a strong deterrent to the FN vote among farmers (Mayer, 1995). Hervieu and Purseigle (2013) demonstrate that political preferences and attitudes among farmers continue to have a strong rightist orientation. Religious and moral beliefs are predominant among farmers. They also adhere to core right-wing values, in particular the ethics of 'hard work', while strongly opposing '*assistanat*' (welfare scroungers).

Farmers' interests have, however, become more politically fragmented over time. Socioeconomic segmentation and regional differentiation have also increased in the 'rural worlds' (Gombin and Mayance, 2010). On the left, grass-roots mobilisation in the

[2] Despite Le Pen's participation in the Poujade movement and the assimilation by the FN of the UDCA's legacy of fiscal populism.

[3] Nevertheless, there has been a tradition of 'rural communism' in Dordogne (Boswell, 1998). Zones of rural support for the French Communist Party (Parti Communiste Français, PCF) were found in the northern and western edges of the Massif Central, and along the Western Mediterranean coast (see also Mischi and Streith, 2004).

form of the Peasants' Confederation (Confédération Paysanne, CP) (see also Chapter 5; Bruneau, 2015) took the centre stage of the populist rural agenda in the late 1990s through direct action against globalisation, Americanisation and agro-business. The Confederation was led by José Bové, a long-time activist who began his political career in the anti-military, farmer and environmental protest movements of the 1970s. Actions by the Confédération Paysanne included the dismantling of McDonald's restaurants or, years later, the launching of anti-GMO campaigns through the ripping up of genetically modified crops in southern France. While the FN has shown convergence with the criticism of economic globalisation and the World Trade Organization (WTO) by the Confederation, the two organisations are nevertheless strongly antagonistic. The Confédération Paysanne's 'altermondialist' positions are deeply entrenched in the internationalist anti-globalisation movement. The latter has historically fostered links with both the ecologist and the anti-liberal sectors of the French left while strongly rejecting the FN.[4]

On the right, the rural-urban cleavage underpinned the formation of the Hunting, Fishing, Nature, Tradition party (Chasse, Pêche, Nature, Traditions, CPNT). This organisation grew from a small single-issue shooters' rights group in the 1980s to become a more distinctly agrarian party in the early 2000s. In the early stages, the CPNT contested European regulations on hunting and opposed pro-environment laws seen to be antagonistic to the collective interests of farmers. The party won its first votes in the 1989 European and 1992 regional elections. It then achieved its best performance in the 1999 European ballot, where it gained 6.8% of the total national vote cast and won six seats. In 2002, CPNT's leader Jean Saint-Josse ran a successful presidential bid, receiving 4.2% of the presidential vote. His conservative Eurosceptic and agrarian agenda incorporated a broader set of rural issues. These included tax incentives for farmers, the fight against 'desertification' (rural flight) and the need to sustain public provision of services in rural France. The party's strongholds were found in rural areas such as the Somme, Manche or the Aquitaine regions (Bussi, 2003).

In 2007, CPNT's presidential candidate, Frédéric Nihous, polled a mere 1.2% of the vote. While the FN received support from former CPNT voters across a number of regions in 2012,[5] the two parties remained independent from each other, both organisationally and strategically. Following Nihous's electoral setback in 2007, the CPNT took a centripetal trajectory. The party merged into the mainstream Union for a Popular Movement (Union pour un Mouvement Populaire, UMP) in 2012

[4] In the 2005 referendum, Bové campaigned with the Communists and a number of anti-globalisation groups against the European Constitution. He then ran independently in the 2007 presidential election, where he competed with both the PCF and the far left for the anti-liberal vote. More recently, the Confédération Paysanne has moved closer to the Greens' Europe Ecologie (EE) coalition.

[5] There was empirical evidence of regionalised vote transfers between 2007 and 2012. A significantly positive cantonal correlation was found between the 2007 Nihous vote and the 2012 Le Pen vote in Picardie, Nord-Pas-de-Calais, Aquitaine, Auvergne and Haute-Normandie.

after a short-lived attempt to build a coalition with the right-wing Eurosceptics of the Movement for France (Mouvement pour la France, MPF) under the Libertas banner in the 2009 European elections.[6]

Finally, the FN has been continuously excluded from collaborating with the main farmers unions, namely the dominant National Federation of Farmers' Unions (Fédération Nationale des Syndicats d'Exploitants Agricoles, FNSEA), the Rural Coordination (Coordination Rurale, CR) and the above-mentioned Peasants' Confederation. The FN has shown hostility in particular towards the more mainstream FNSEA, accusing the latter of representing the interests of big agro-businesses against small tenant farmers. In the 2012 presidential campaign, Marine Le Pen was criticised by the FNSEA for her strong opposition to Europe and her claim that she would dismantle the European Common Agricultural Policy (CAP). Ten years earlier, the FNSEA had publicly supported Jacques Chirac against Jean-Marie Le Pen in the second-round run off of the 2002 presidential election.

The accentuation of the FN's protectionist and Eurosceptic agenda has certainly led to greater ideological proximity to the strong anti-CAP positions taken by the predominantly right-wing Rural Coordination. Nevertheless, formal political rapprochements with the FN have been restricted to individual trajectories.[7] In 2012, the Coordination's president Bernard Lannes publicly refuted accusations of political collusion with the FN and reiterated his claim of independence from all political parties.

Issues and policies: a new 'ruralist' strategy of the Front National

The FN has developed a more comprehensive rural agenda over the years. This is first demonstrated by the gradual increase in salience of agriculture and countryside-related issues in the party's election platforms since the mid-1980s. Figure 11.1 shows that rural-based policy statements concerning agriculture and farming have been given more importance by the FN over time, while there has been a decline in the emphasis of such issues by parties in the mainstream right. A change in relative salience occurred in the mid-1990s, when the FN took the lead on agricultural issues amongst parties on the right. As will be discussed below, this shift in emphasis exemplifies some of the tactical 'catch-allist' choices made by the party as a means to expand its electoral appeal beyond its early urban strongholds.

As early as 1973, the FN pledged to fight rural flight and the impoverishment of farmers. The party argued that it would defend the interests of French farmers against foreign

[6] Locally, former cadres of CPNT such as Nathalie Huiart in the department of the Somme joined the FN through the newly formed Rassemblement Bleu Marine (RBM) in 2012.

[7] Organisational contiguity was increased by the merging in 1994 of the CR with the former more FN-inclined Fédération Française de l'Agriculture (FFA). FFA president Alexis Arette had briefly joined the FN (Bihr, 1998).

Figure 11.1. Change in the issue profile of the right and the Front National on agriculture and farming: 1986-2012. As % of quasi-sentences/paragraphs referring to specific policies in favour of agriculture and farmers in each legislative manifesto; right = unweighted average of Gaullists (RPR, UMP) and centre-right (UDF, MODEM) (source: Comparative Manifesto Project Database for France, authors' data for 2012 based on presidential manifestos).

landowners and investment banks, and that it would reinstate higher prices for farming commodities. In the 1980s, agriculture and farming concerns were incorporated into the new national-conservative agenda of the FN. The party embraced the paternalist and rural traditionalist ideology characteristic of French right-wing nationalism. The 'hard-working' farmers were thus romanticised and the peasantry idealised as the typical possessors of national identity and entrenchment. According to the 1984 party platform, 'the cultivator doesn't treat his land with contempt, unlike the nomad who leaves a field as soon as his flock has grazed it into a no-man's land' (Front National, 1984: p. 34).

The celebration of traditional agrarian virtues became ubiquitous to the FN's appeal. The 1986 programme referred to 'France as the land of our fathers, the land that they have cleared and defended for centuries; the country that they have shaped in its landscapes ... and that they have fertilised with their sweat and their blood' (Front National, 1985: p. 29). In 2012, Marine Le Pen did not deviate from this lyrical evocation of the French farmers: 'the French soul is intrinsically peasant', she emphasised during the presidential campaign. 'Our countryside is immensely rich, and this is where the best of France's civilisation is perpetuated'.[8]

[8] Speech in Châteauroux, 26/02/2012 (http://tinyurl.com/qc4m82n).

Changes have occurred, however, in the FN's treatment of farming and rural issues over time, reflecting more generally the party's economic policy trajectory. During the 1980s, agriculture and rural concerns formed part of the capitalist appeal of the FN to bourgeois voters. This was illustrated, for example, by the invocation of the core values of 'profit, property, economic inequality and honour' in the 1984 electoral manifesto (Front National, 1984: p. 27).[9] Anti-statist deregulation policies, support for small family businesses and the dissolution of bureaucracy were pervasive in the 1986 platform. The FN would call for the 'liberalisation and de-socialisation' of French farming while advocating free enterprise, the suppression of quotas, tax incentives for small family businesses and cuts in the contributions paid by farmers to national social security.

Interestingly, the 1986 programme introduced a first set of protectionist measures against imported agriculture commodities, while emphasising France's need for food self-sufficiency. This nationalistic tone heralded a more substantial policy shift during the mid-1990s. During that period, the FN toned down its rhetoric on free market capitalism to espouse a more protectionist agenda. This move was consistent with the party's platform against EU federalism and the loss of national sovereignty. The 1993 manifesto brought new social and anti-globalisation policy features into the party. With regard to farming in particular, the FN's programmatic priorities included measures rejecting the GATT (General Agreement on Tariffs and Trade)/WTO and the EU. While maintaining tax incentives and the pledge to put an end to a trade union monopoly in the agricultural sector, the FN advocated a mix of social policies. These included a specific allowance for young farmers, an increase in pensions, as well as a debt moratorium to protect small family farms, with a view to stopping rural flight.

In one sense, the FN's agricultural policies became more far-reaching. This was perceptible in the 2002 presidential manifesto, which claimed to represent the sectional interests of farmers against the threat of both the EU and economic globalisation. Most strikingly, the FN claimed that France should leave the EU (Front National, 2002: p. 70). The 2002 platform reiterated the party's positions against intensive farming, introducing anti-GMO policies and, in the specific context of the 'mad-cow' disease crisis, a call for more stringent sanitary safety measures. Reflecting on the emergence of new rural questions, the 2002 manifesto tapped into the issue of 'rurality' (*ruralité*) and reaffirmed the imperative to 'combat the desertification (rural flight) of rural areas' (Front National, 2002: p. 49). It maintained, nevertheless, a clear right-wing stance revolving around additional tax incentives and the creation of special economic zones in rural municipalities.

[9] In 1984, however, no specific farming policies were formulated by the FN.

The attention paid to agriculture and rural issues peaked in the 2012 elections. The FN displayed a significant policy shift towards the left on the economic dimension (Ivaldi, in press). The party embraced interventionist and protectionist policies, breaking further away from its capitalist preferences of the 1980s. Framed as a new set of social and economic issues, 'rurality' re-emerged in the presidential manifesto, together with the launch of the 'France ruralité' internet platform, Facebook page and Twitter feed. The party pushed an agenda of public service provision and higher government spending and health services in rural areas. The 2012 presidential platform claimed that 'public services should be available to everyone, everywhere' (Front National, 2012: p. 9) and pledged that the FN would 'develop public provisions for public services of proximity ... to combat the desertification of rural areas' (Front National, 2012: p. 13).[10] In Châteauroux, Marine Le Pen similarly called for 'the French constitution to guarantee equal access to public services regardless of income or place of residence' (Marine Le Pen, 2012, see note 43).

This was combined with the accentuation of the party's Eurosceptic and protectionist stances. In 2010 the FN called for France to leave the Eurozone. During the 2012 elections, the party advocated the renegotiation of both the 'shameful EU directives' and the 'treaties that hurt France', in order to 'eliminate free competition rules applicable to specific sectors such as transport or postal services'. According to the FN, postal services should be renationalised and price controls should be established over basic commodities such as food, petrol or energy. France should also leave the Common Agricultural Policy to implement a national system of agricultural subsidies – in the form of a new French Agricultural Policy (FAP). Measures should be taken to protect small farmers from larger agri-businesses, while a 'buy French act' would give priority to agricultural commodities produced in France. These themes dominated the FN's European election agenda in May 2014 (Ivaldi, in press). The party topped the French polls with 24.9% of the national vote, garnering 19.3 and 24.2% in the two main rural regions of the West and the Centre, respectively. In line with its protectionist stance, the FN strongly opposed the EU-US free trade agreement (Transatlantic Trade and Investment Partnership, TTIP). It was stigmatised as 'ultra-liberal savagery' that threatened the interests of French farmers, while allowing multinational corporations to bypass all existing safety, health and environmental standards.

Thus, in recent years, the endorsement of a neo-ruralist agenda by the FN has certainly further extended what Kitschelt (1995: p. 95) originally described as the party's 'catch-all' extremism. In 2012, Marine Le Pen appealed to the traditional petty-bourgeois and working class clienteles of the populist radical right. She also attempted and was able to identify her party with a larger coalition, which she repeatedly referred to as

[10] There were already signs of this policy shift in the 2007 election. The 2007 manifesto already included a pledge to increase government spending to expand public services (hospitals, crèches, retirement homes) in rural areas (Front National, 2007: p. 27).

the 'invisible constituency'. The latter included 'workers, farmers, students, pensioners, shopkeepers, civil servants and employees'. Their common interests would be threatened by 'the powerful', representing the dominant 'globalist' (*mondialiste*) ideology, that is, national politicians, the European Commission, financial markets, CAC 40 companies, the 'super wealthy', 'loutish businessmen' or 'large retail companies'.

Marine Le Pen has revisited the FN's traditional populist 'heartland' through redrawing the contours of the opposition that populism classically establishes between the 'pure people' and the 'corrupt elite' (Mudde, 2007: p. 23). In the context of economic crisis, these categories were constructed along socioeconomic rather than strictly political or cultural lines – for example, the hard-working 'little people' against big corporations, the neo-liberal EU and the global markets – ignoring, however, the objective divisions that exist among the so-called 'invisible constituency'. In 2012, this populism of the 'little man' helped the FN broaden its appeal to the lower sectors of the middle class that had been hit by the economic recession. The party reached out to 'forgotten rural France' pledging that it would tackle social segregation, high levels of unemployment and growing feelings of insecurity, both cultural and economic.[11]

11.3 The making of a new 'rural' constituency

Analysing a party's relationship to society requires that we look not only at its programmatic supply, but also at the demand side. In this section, we examine the relationship of the Front National to the rural and agricultural worlds by looking at its electoral socio-demographic make-up and geography, and their transformation over time.[12]

Farmers and the French radical right

Since the late 1980s, the FN has been characterised by its cross-sectional appeal in the French electorate, with an over-representation of petty-bourgeois and working-class voters (Mayer, 2002). Because the overall share of farmers in the population is very small and declining,[13] poll estimates of votes in the agricultural sector must be treated with care. However, the survey data available seems to indicate that farmers

[11] The FN's ruralist policies overlapped with its traditional law-and-order agenda, as it also called for an increase in the police force (*gendarmerie*) in rural areas.

[12] One problem is that FN voters tend to conceal their vote and are therefore continuously under-represented in national polls.

[13] In the last 50 years, the farming population has declined from about one third to less than 4%. In 2010, farmers accounted for 3.4% of the French population (see Ministère de l'Alimentation, de l'Agriculture et de la Pêche, Centre d'études et de prospective, Analyse n° 14, 'Les agriculteurs dans la société', April 2010, http://agriculture.gouv.fr/les-agriculteurs-dans-la-societe-francaise).

were continuously under-represented in the FN's electorate up to the late 1990s, while turning in greater numbers to the radical right after 1997.

Farmers were less inclined to support the radical right during the 1980s and the 1990s. In 1984, an estimated 8% of farmers voted for the FN in the European election, where the party made its first national breakthrough with 11%.[14] The electoral geography of the FN remained mostly industrial and urban (Perrineau, 1998). In the 1986 legislative elections, the FN polled about 7% of the farmers' vote, less than its percentage of the overall national figures. According to Boy and Mayer (1997), 7% of the farmers voted for Jean-Marie Le Pen in the 1988 presidential election – half the vote in the rest of the population – demonstrating the continuing resistance of farmers to the radical right.[15] Another clue that the 1988 support for Jean-Marie Le Pen was rather low among farmers and, more generally, in the rural world, is provided by Boussard and Chiche (1992). The results of their ecological analysis show that in 1988 the FN vote was lower in the 104 constituencies (*circonscriptions*) where the share of farmers was the highest, compared with the other 473 constituencies. Their study also demonstrates that constituencies with a higher proportion of farmers had the lowest share of the FN vote in the 1989 European elections.

While its sociology changed dramatically between 1988 and 1995, exhibiting in particular a proletarianisation of its support (Mayer, 2002; Mayer and Perrineau, 1996; Perrineau, 1996), FN support among farmers, and rural voters generally, remained more or less stable. In the 1993 legislatives, the FN candidates won 10% of the farmers' vote, compared with 12.4% nationally (Mayer, 2002). In 1995, 8% of farmers (and 0% of farm workers – sic) supported Jean-Marie Le Pen's presidential bid (Boy and Mayer, 1997), a proportion estimated at about 10% in Mayer (2007). According to Mayer (2002), the farmers' FN vote dropped to a mere 2% in the 1997 legislative elections – a figure which should be taken with caution, however, due to possible methodological bias.

The FN's breakthrough among farmers and rural voters occurred between 1997 and 2002. The evidence from cross-sectional surveys is corroborated by growing electoral returns in rural areas during the same period. Le Pen won his first notable success in the rural and agricultural worlds in the 2002 presidential election. On that occasion, the FN leader received levels of farmer support similar to levels gained from its core petty-bourgeois – businessmen and shopkeepers – and working-class constituencies, and among white-collar voters (Cautrès and Mayer, 2004). Mayer (2007) suggests that Le Pen might have received as much as 22% of the farmers' vote in the first round of the 2002 presidential election, giving him a decisive thrust to progress to the second-round

[14] For 1984 and 1986, we use pooled datasets of cross-sectional surveys conducted by the French polling institute BVA after the European and legislative elections.

[15] The same figure appears in Mayer (1992). The author argues that the vote for the FN leader was stronger among farmers most integrated into the farming *milieu* – which seems a bit counterintuitive.

run off. This first success was even more remarkable insofar as the FN had to confront direct competition from the ruralist CPNT candidate in rural areas.

While the core of radical-right support remained with the urban working class, the FN grew considerably stronger among rural voters, far beyond farmers, in the early 2000s. At the same time, the support for the party declined in urban areas. Let us recall here that the rural-urban division lost salience in France after the Second World War, mostly because of large-scale urbanisation, with France transforming from a predominantly agricultural and rural society into a largely urban industrial one during the Trente Glorieuses (1945-1970). Lévy (2003) argues, however, that the 2002 presidential election was a turning point in French politics: the cleavage between urban centres and their rural peripheries produced stronger effects than ever before in the Fifth Republic.

In 2007, Jean-Marie Le Pen's last presidential bid marked a critical electoral setback for the FN, which received only 10.4% of the vote. This reversal reflected important swings by FN voters to the centre-right UMP candidate Nicolas Sarkozy. The latter had embraced a strong authoritarian stance coupled with an agenda of stricter immigration control. Rural and farming voters showed, however, a significantly higher degree of loyalty to the FN's leader (Fourquet, 2008). Gombin and Mayance (2010) suggest that farmers were still more likely than other occupational groups to support the FN. Based on an ecological analysis of registered voters, they calculated that about 12% of registered farmers voted for Le Pen in 2007 – yielding an estimated 13.5% of the valid vote cast, compared with a 10.4% national average.

The FN made an impressive electoral comeback in the 2012 presidential election, winning 17.9% of the first-round vote. As will be discussed below, this electoral resuscitation was largely attributed to the growth in support for the FN in peripheral rural territories. The party also maintained its attractiveness to voters in the agricultural sector: the 2012 French election study shows that 21% of farmers voted for the radical right's candidate.[16] Based on ecological data, it can be estimated that 19% of registered farmers voted for Marine Le Pen – compared with 14.5% for other registered voters – representing 21.8 and 17.8% of the valid vote cast, respectively.

Overall, the FN's support among farmers – and more generally among rural voters – was very low in the early years, but it then grew more rapidly than average and today it is higher than in the rest of the electorate. This evolution is summarised in Figure 11.2, which displays the results from a multilevel growth model using aggregate election returns (Gombin, 2013b). This longitudinal approach allows the assessment of the evolution of FN support among farmers over time. As can be seen, the FN vote grew more rapidly in the cantons where farmers represent a larger share of the population.

[16] Source: French Election Study 2012 (TNS-SOFRES/CEE Sciences-Po Paris).

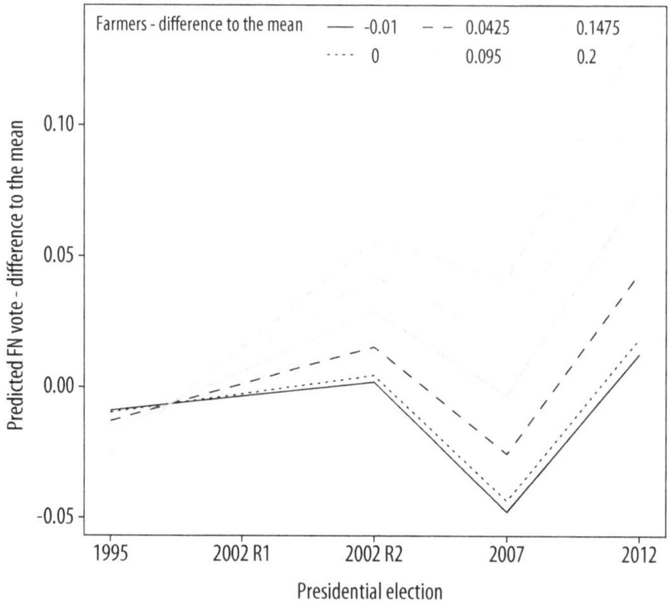

Figure 11.2. Predicted Front National (FN) presidential vote as a difference to the mean by share of farmers in the population, 1995-2012. The predicted vote for the FN expressed as a difference from the overall mean on the time span. Scores were estimated from a multilevel growth model, fitted on aggregate data at the cantonal and departmental levels. In 2002 there were two presidential elections (R1 and R2). The detailed specification of the model can be found in Gombin (2013b) (sources: Ministère de l'Intérieur; Ministère de l'Éducation nationale, INSEE).

For example, in 2012, Marine Le Pen won a 13 point surplus in cantons where shares of farmers were 20 points above the national average. Moreover, this evolution was largely independent of the party's performance among petty-bourgeois voters – craftsmen and shopkeepers. Among the latter, there was a marked decline in the propensity to support the radical right between 1995 and 2012, which was almost the exact inverse of the trend observed among farmers.

However, farmer support for the radical right is not uniform across the country. It has a very specific geography, which is also different from the more general political geography of the FN. This is illustrated in Figure 11.3, which presents an estimation of the 2012 FN vote among farmers across the country. Departments with the highest level of farmer support for the FN were located in the greater Paris area, as well as in the northern and western regions – from Nord to Landes, including Normandy, Brittany and the interior western part of the country. Strongholds of FN farmer votes were also found in Bouches-du-Rhône, Isère and Haute-Savoie.

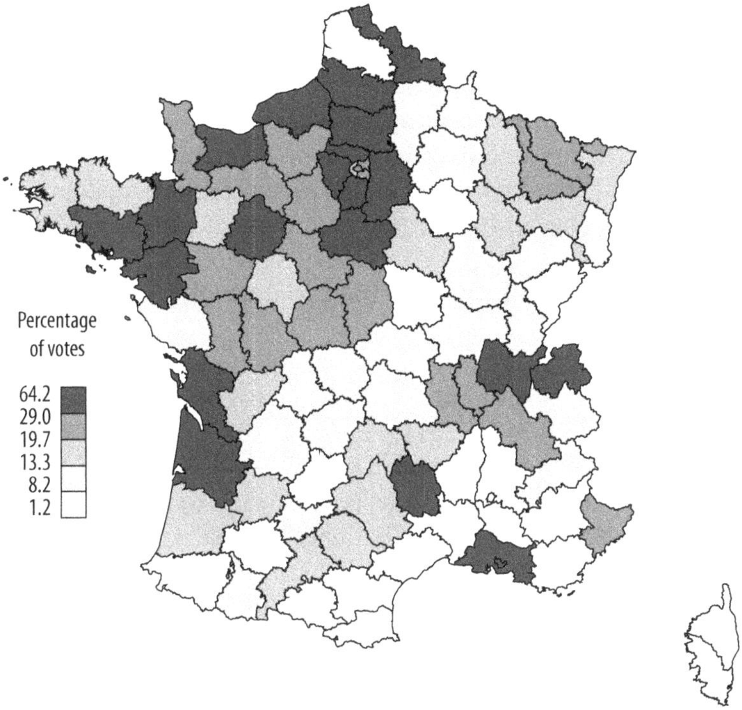

Figure 11.3. Estimated Marine Le Pen vote among farmers in the 2012 presidential election. Estimates are derived from a multilevel model fitted on aggregate data at the communal and departmental levels (sources: Ministère de l'Intérieur; INSEE).

The areas are characterised by rather large-scale, intensive farming, and they are also those in which farmers were the most likely to vote for Sarkozy and the UMP in 2012 (R=0.48). These regions have experienced economic difficulties, for example in the milk production sector, and have been among the most severely affected by the crisis in the agricultural sector. The CAP has historically proven an effective shield for farmers, helping them secure their incomes while national agricultural markets were opened to greater international competition. Consequently, the French farming world has proven to be one of the strongest supporters of the European integration process. However, over recent years, this protective effect may have become less effective for specific subgroups of farmers, crystallising discontent with the EU and fostering support for the FN's national-protectionist policies.

The FN's growing electoral success among farmers thus reflects the weakening of the links which originated in the pact forged between the agricultural sector and the Gaullist state in the late 1960s. This pact was well represented, for example, by the Pisani plan – an ambitious state-driven plan to help French agriculture modernise and become more competitive – and the creation of the CAP in the early 1960s. In contrast, smaller farm tenants in the south were more distant from this original arrangement and continued to vote more solidly for parties on the left. In this regard, the development of the FN vote among farmers represents both an alternative and a challenge to the traditional centre-right and Gaullist alignment of the farming world in France. It is also consistent with the conservative worldview which continues to dominate political attitudes and values among farmers (Hervieu and Purseigle, 2013: p. 225).

The peripheralisation of support for the radical right

The growth in farmer support for the radical right has been accompanied by a 'peripheralisation' of the FN vote. This phenomenon has been extensively documented by scholars in France (Bussi *et al.*, 2011; Fourquet, 2012; Girard, 2013; Lévy, 2003, 2007; Rivière, 2009). Originally an urban phenomenon, the geography of the FN gradually shifted away from cities, while support for the party augmented in the more remote urban peripheries. This process of 'peripheralisation' is illustrated in Figure 11.4 below, which presents the variation in support for the FN since 1995, according to the type of municipality. It shows in particular that the gap between larger urban cores and

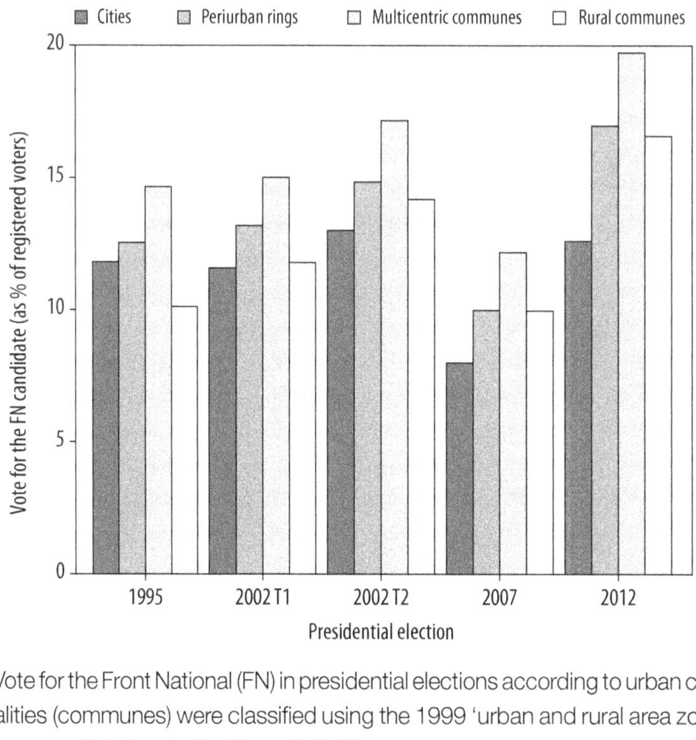

Figure 11.4. Vote for the Front National (FN) in presidential elections according to urban configurations. The municipalities (communes) were classified using the 1999 'urban and rural area zoning' devised by INSEE (sources: Ministère de l'Intérieur; INSEE).

their more remote private suburban housing peripheries has grown larger over time. In 2012, the FN vote was thus significantly stronger in rural communes than in the more urbanised areas. More importantly, electoral support for the radical right peaked in periurban towns and subareas in a situation of 'multicentricity', that is, adjacent to multiple centres of economic activity and employment.

A number of explanations have been put forward to account for this evolution. Based on a theory of 'social relegation', some authors see this evolution as the result of in-migrations and socio-demographic transformations. The metropolitanisation process (Davezies, 2008) in particular would push voters in low-income households further from urban centres towards the more distant outskirts of the metropolis, where they would feel socially excluded and subject to 'cultural insecurity' (Guilluy, 2010, 2011). In contrast, Lévy (2013) argues that the FN's over-representation in periurban areas could reflect residential choices by xenophobic voters who deliberately move out of urban zones to avoid direct contact with immigrants. The study by Cartier *et al.* (2008) indeed provides empirical evidence suggesting that the evolution in the geography of the FN vote has been correlative of the spatial redistribution of social groups that make up the core constituencies of the radical right, which did move away from deprived social housing neighbourhoods to more remote areas with individual private housing (Cartier *et al.*, 2008).

Many studies have pointed to the social and economic heterogeneity of periurban areas (Chalard, 2006; Girard, 2013; Rivière, 2008). Gombin and Rivière (2013) suggest, for example, that other factors should be taken into account; in particular, existing levels of economic inequality. As discussed above, understanding the growth in FN support in periurban areas also requires that we consider the 'supply side' of the equation. We have already emphasised how the FN has recently attempted to reach out to voters in the so-called 'invisible constituency' in urban peripheries and rural areas. The neo-rural repositioning of the FN testifies to the socioeconomic reconfiguration of the rural space and the territorialisation of class-based divisions. These have resulted in the politicisation of new issues concerning in-migrations, models of rural development and the transformation of the rural-urban divide.

This territorial divide is in fact nested in France's higher order class cleavage. Overall, there is a relationship between the geography of economic inequality and changes that have occurred in the spatial distribution of FN support. Access to urban centres has become more difficult for larger segments of the population, particularly in the lower class locations. Access to urban peripheries has become equally competitive. This has resulted in growing conflict between social groups, which differ not only in terms of their socioeconomic status but also with regard to their relationship to the territory they live in and how they use it. Territorial divisions based on socioeconomic positions reflect the proletarianisation of the working class and the lower middle class

(Gombin, 2013a) and the perception of this process by those concerned – a process which mainstream parties have so far failed to address. Rather than revealing a new rural-urban divide, changes in the FN's geographical contrast reflects the geographical clustering of voters from the middle and lower class locations in semi-rural areas, from which the FN is increasingly drawing electoral support. As suggested by Woods (2003: p. 310), new waves of rural protest can be found across developed countries, which illustrate a change from 'rural politics' to a 'politics of the rural' (see also Chapter 2). The latter mobilises a traditional array of farming issues, but also issues concerning housing, transport and the preservation of rural services.

11.4 Conclusion

While clearly rooted in urban post-industrial society, the FN has nevertheless embraced the classic agrarian and paternalist values of French right-wing nationalism. The party has adhered to a reactionary ideology that romanticises the peasantry and has striven to address the economic interests and social concerns of farmers. During the 1980s and the 1990s, it has progressively fashioned an agricultural policy agenda combining tax incentives, economic protectionism and welfare.

This new programmatic mix has enhanced the party's appeal to voters in the rural world. Over the years, the FN has expanded its support further from its original urban bastions. While under-represented in the FN electorate during the 1980s and the 1990s, farmers have become a more solid base for the radical right in France since the early 2000s. On the supply side, the FN's electoral success in the agricultural sector has been concomitant with a significant increase in salience of farming issues in the party's manifestos. On the demand side, the rise of the radical right among farmers has reflected their growing dissatisfaction with the agricultural status quo, represented by both the EU's CAP and mainstream farming policies.

More recently, rural themes have been connected with socioeconomic issues. In 2012, the FN embraced a more leftist and interventionist economic agenda. Its new 'politics of the rural' appealed to the 'forgotten rural France'. Its broad 'catch-all' strategy reached out to a more diverse rural constituency, comprising not only farmers but also voters in low-income households clustered in periurban areas.

Today, the FN is more rural and working class than it has ever been. While both the agricultural and industrial sectors continue to decline, the natural shrinking of its base of support could limit the FN's electoral progression. According to Kitschelt (1995: p. 23), radical-right parties relying too heavily on the narrow confines of the working class are bound to fail electorally. Moreover, working-class voters in France are more prone to abstain, and make therefore a more volatile support. Thus, the future of the French radical right will depend on its ability to access a larger share of the middle-class vote to establish a broader electoral coalition. Widespread economic insecurity and profound political discontent may further destabilise and polarise middle-class voters, paving the way for future electoral victories by the FN.

References

Betz, H.-G., 1994. Radical right-wing populism in Western Europe. Macmillan, Basingstoke, UK.

Bihr, A. 1998. Le spectre de l'extrême droite. Les Français dans le miroir du Front National. Editions de l'Atelier, Paris, France.

Boswell, L., 1998. Rural communism in France, 1920-1939. Cornell University Press, Ithaca, NY, USA.

Boussard, I. and Chiche, J., 1992. Les élections présidentielles et législatives de 1988, Européennes de 1989. Analyse écologique globale. In: Hervieu, B. (ed.) Les agriculteurs Français aux urnes. L'Harmattan, Paris, France.

Boy, D. and Mayer, N. (eds.), 1997. Que Reste-T-Il des variables Lourdes? L'électeur a ses raisons. Presses de Sciences Po, Paris, pp. 101-138.

Bruneau, I., 2015. The Confédération Paysanne and the political field: a conflicted history. Chapter 5. In: Strijker, D., Voerman, G. and Terluin, I.J. (eds.) Rural protest groups and populist political parties. Wageningen Academic Publishers, Wageningen, the Netherlands, pp 101-125.

Bussi, M., 2003. Le vote Saint-Josse: la protestation en campagne. In: Perrineau, P. and Ysmal, C. (eds.) Le vote de tous les refu. Presses de la F.N.S.P., Paris, France, pp. 311-338.

Bussi, M., Colange, C. and Rivière, J., 2011. Distance(s) à la ville et comportements électoraux. Quelques éclairages quantitatifs lors des derniers scrutins présidentiels. In: Mattéi, M.-F. and Pumain, D. (eds.) Données urbaines 6. Anthropos, Paris, France.

Canovan, M., 1981, Populism. Harcourt Brace Jovanovich, London, UK.

Cartier, M., Coutant, I., Masclet, O. and Siblot, Y., 2008. La France des 'petits-moyens': enquête sur la banlieue pavillonnaire. Textes à l'appui. Série Enquêtes de terrain. La Découverte, Paris, France, 320 pp.

Cautrès, B. and Mayer, N., 2004. Les métamorphoses du «vote de classe. In: Cautrès, B. and Mayer, M. (eds.) Le nouveau désordre électoral. Les leçons du 21 avril 2002. Presses de Sciences Po, Paris, France, pp. 143-159.

Chalard, L., 2006. Le vote d'extrême droite dans l'aire Métropolitaine Marseillaise. EspacesTemps.net November 10, 2006.

Davezies, L., 2008. La République et ses territoires: la circulation invisible des richesses. La république des idées, Seuil, Paris, France, 110 pp.

Fourquet, J., 2008. L'érosion électorale du lepénisme. In: Perrineau,P. (ed.) Le vote de rupture. Les élections présidentielles et législatives d'avril-juin 2007. Presses de Sciences Po, Paris, pp. 213-234.

Fourquet, J., 2012. Le sens des cartes. Analyse sur la géographie es votes à la présidentielle. Les Essais, Fondation Jean-Jaurès, Paris, France.

Front National, 1984. Droite et démocratie économique. Doctrine économique et sociale du Front National, 2nd edition.

Front National, 1985. Pour la France: programme du Front National. Editions Albatros, Paris, France.

Front National, 2002. Pour un avenir français: programme du Front National. Available at: http://tinyurl.com/pkrhkyg.

Front National, 2012. Projet présidentiel de Marine Le Pen. Available at: http://tinyurl.com/owymors.

Girard, V., 2013. Sur la politisation des classes populaires périurbaines. Politix 101(1): 183-215.

Goguel, F., 1981. Chroniques électorales: les scrutins politiques en France de 1945 à nos jours. 1. La Quatrième République. Presses de la Fondation Nationale des Sciences Politiques, Paris, France, 171 pp.

Gombin, J., 2013a. The Front National vote and its sectorial support. In: ECPR General Conference Sciences Po. September 4-7, 2013. ECPR, Bordeaux, France. Available at: http://tinyurl.com/myqcg27.

Gombin, J., 2013b. 'Nouveau' FN, vieille carte électorale? In: Congrès de l'Association Française de Science Politique. July 9-13, 2013. AFSP, Paris, France. Available at: http://tinyurl.com/l5yw2oy.

Gombin, J. and Mayance, P., 2010. Tous conservateurs ? Analyse écologique du vote de la population agricole lors de l'élection Présidentielle de 2007. In: Mayer, N., Muller, P., Hervieu, B. Rémy, J. and Purseigle, F. (eds.) Les mondes agricoles en politique. Presses de Sciences Po, Paris, France, pp. 193-216.

Gombin, J. and Rivière, J., 2013. Éléments quantitatifs sur la dimension spatiale des effets électoraux des inégalités sociales dans les mondes Périurbains Français (2007-2012). In: Congrès de l'Association Française de Sociologie. September 2-5, 2013. Nantes, France. Available at: http://tinyurl.com/knvhyqx.

Guilluy, C., 2010. Fractures Françaises. François Bourin Editeur, Paris, France, 206 pp.

Guilluy, C., 2011. Les métropoles et la France périphérique. Le Débat 166(4): 42-52.

Hervieu, B. and Purseigle, F., 2013. Sociologie des mondes agricoles. Armand Colin, Paris, France, 318 pp.

Ignazi, P., 2003. Extreme right parties in Western Europe. Oxford University Press, Oxford, UK, 27 pp.

Ivaldi, G., 2016. A new course for the French radical-right? The Front National and 'de-demonization'. In: Akkerman, T., De Lange, S. and Rooduijn, M. (eds.) Up to the mainstream? Radical Right-Wing Populist Parties in the New Millennium. Routledge, London, UK.

Ivaldi, G., in press. Towards the median economic crisis voter? The new leftist economic agenda of the Front National in France. In: French Politics, Special Issue.

Ivaldi, G., in press. The Front National and the politics of Euroscepticism in France. In: Wassenberg, B. and Moreau, P. (eds.) The 2014 European Elections: the rise of anti-Europeanism? Franz Steiner Verlag, Stuttgart, Germany.

Jungar, A.C., 2015. Agrarian populism in Finland: continuity and change. Chapter 10. In: Strijker, D., Voerman, G. and Terluin, I.J. (eds.) Rural protest groups and populist political parties. Wageningen Academic Publishers, Wageningen, the Netherlands, pp 217-241.

Kitschelt, H., 1995. Radical right in Western Europe. A comparative analysis. University of Michigan Press, Ann Arbor, MI, USA.

Lévy, J., 2003. Vote et gradient d'urbanité. L'autre surprise du 21 Avril. EspacesTemps.net june 5, 2003. Available at: http://tinyurl.com/mytq35w.

Lévy, J., 2007. Regarder, voir. Un discours informé par la cartographie. Les Annales de La Recherche Urbaine 102: 131-140.

Lévy, J., 2013. Liens faibles, choix forts: les urbains et l'urbanité. La vie des idées (May 29). Available at: http://tinyurl.com/o5s5q6w.

Mayer, N., 1992. Le vote des agriculteurs en 1988. Facteurs individuels, facteurs contextuels. In: Hervieu, B. (ed.) Les agriculteurs français aux urnes. Alternatives rurales. L'Harmattan, Paris, France, pp. 47-65.

Mayer, N., 1995. Les choix politiques ces agriculteurs. Cahiers Du Cevipof 12): 27-38.

Mayer, N., 2002. Ces Français qui votent Le Pen. Flammarion, Paris, France, 480 pp.

Mayer, N., 2007. Comment Nicolas Sarkozy a rétréci l'électorat Le Pen. Revue Française de Science Politique 57(3): 429-445.

Mayer, N., 2013. From Jean-Marie to Marine Le Pen: electoral change on the far right. Parliamentary Affairs 66(1): 160-178.

Mayer, N. and Perrineau, P., 1996. Le Front National à découvert. Presses de La Fondation Nationale Des Sciences Politiques, Paris, France.

Mischi, J. and Michel Streith, M., 2004. L'implantation du PCF. Bastions ruraux, bastions urbains. Études Rurales 171-172: 29-34.

Mudde, C., 2007. Populist radical right parties in Europe. Cambridge University Press, Cambridge, UK, 17 pp.

Ory, P., 1975. Le dorgérisme. Institution et discours d'une colère paysanne (1929-1939). Revue d'histoire moderne et contemporaine 22(2): 168-190.

Paxton, R.O., 1997. French peasant fascism: Henry Dorgères's greenshirts and the crises of French agriculture, 1929-1939. Oxford University Press, New York, NY, USA, 244 pp.

Perrineau, P., 1996. Les étapes d'une implantation électorale (1972-1988). In: Mayer, N. and Perrineau, P. (eds.) Le Front National à découvert. Presses de Sciences Po, Paris, France, pp. 37-62.

Perrineau, P., 1998. Le symptôme Le Pen: radiographie des électeurs du Front National. Fayard, Paris, France.

Rémy, P., 1972. Le gaullisme et les paysans. In: Tavernier, Y., Gervais, M. and Servolin, M. (eds.) L'univers politique des paysans dans la France contemporaine. Cahiers de la Fondation Nationale des Sciences Politiques. Armand Colin, Paris, France, pp. 255-272.

Rivière, J., 2008. Le vote pavillonnaire existe-t-Il? Comportements électoraux et positions sociales locales dans une commune rurale en cours de périurbanisation. Politix (83): 23-48.

Rivière, J., 2009. Le Pavillon et l'isoloir. Géographie sociale et électorale des espaces Périurbains Français (1968-2008). A travers les cas de trois Aires Urbaines Moyennes (Caen, Metz et Perpignan). Doctorat. Université de Caen Basse Normandie, Caen, France.

Stanley, B., 2015. The Polish Self-Defence party: from agrarian protest to the politics of populism, 1991-2007. Chapter 9. In: Strijker, D., Voerman, G. and Terluin, I.J. (eds.) Rural protest groups and populist political parties. Wageningen Academic Publishers, Wageningen, the Netherlands, pp 191-215.

Woods, M., 2003. Deconstructing rural protest: the emergence of a new social movement Journal of Rural Studies 19: 309-325.

Woods, M., 2015. Explaining rural protest: a comparative analysis. Chapter 2. In: Strijker, D., Voerman, G. and Terluin, I.J. (eds.) Rural protest groups and populist political parties. Wageningen Academic Publishers, Wageningen, the Netherlands, pp 35-62.

12 Rootless populists? The Dutch Pim Fortuyn List, the Freedom Party and others

P. Lucardie[*] and G. Voerman

Documentation Centre Dutch Political Parties (DNPP), Groningen University, Groningen, the Netherlands; a.p.m.lucardie@rug.nl

Abstract

It is possible to distinguish three waves of populism in the Netherlands. The first occurred between 1960 and 1990, involving parties with some populist elements but which did not articulate a populist ideology. Only one, the Farmers' Party, had roots in a rural protest movement, but it disappeared in the early 1980s. In the second wave in the 1990s, two parties that did articulate a populist ideology entered the scene: Liveable Netherlands and the Socialist Party (SP). While the former disintegrated within a decade, the latter has been quite successful and stable. In the third wave, after 2000, three right-wing populist parties were founded, List Pim Fortuyn, Proud of the Netherlands and the Freedom Party (PVV). While the first two disintegrated and disappeared within a few years, the third has been quite successful in electoral terms. None of these third-wave parties had roots in a social protest movement, whether urban or rural. In general, the parties were founded by political entrepreneurs, except for the SP, founded in 1971, which had a background in communism. However, the different rates of success can be attributed more to their ideological mix and organisation rather than their genesis. The SP shed its populist Marxist-Leninism and evolved into a left-wing social-democratic party with populist elements while building a very solid and disciplined organisation. The PVV has combined populism with nationalism, anti-Islamism and a conservative welfare statism. It lacks a formal organisation, however, which has created some problems in the past and might do so even more in the future.

12.1 Introduction

Populism is a diffuse term, with no scientific consensus about its meaning. In this contribution, populism is defined as a thin ideology focused on bridging a perceived gap between a relatively homogeneous people and an equally homogeneous elite, using

D. Strijker, G. Voerman and I.J. Terluin (eds.) Rural protest groups and populist political parties
DOI 10.3920/978-90-8686-807-0_12, © Wageningen Academic Publishers 2015

structural reforms such as referenda or the direct elections of political officials. The point of departure for these constitutional renewals is the notion of the people's sovereignty: it is argued that the popular will, or the will of the majority of the people, has to be decisive within the political system (cf. Mudde, 2007: pp. 23, 151-153). Populism is widely regarded as a thin or 'partial' ideology which, in political practice, generally makes its appearance in combination with a 'full' ideology such as socialism, liberalism or conservatism. Populism in pure form – without any ideological supplement – is rare and usually short-lived, as the fate of Liveable Netherlands (Leefbaar Nederland) illustrates (see below).

Populism considered in this ideological sense had been relatively weak in the Netherlands until the sudden rise of the Pim Fortuyn List (Lijst Pim Fortuyn, LPF), the Freedom Party (Partij voor de Vrijheid) and the short-lived Proud of the Netherlands movement (Trots op Nederland) at the beginning of the twenty-first century. None of these parties evolved from a protest movement, whether agrarian or urban, or from right-wing fringe parties. All three were founded by 'political entrepreneurs' who had been members of an established party: Pim Fortuyn had been a rank and file member of the Dutch Labour Party (Partij van de Arbeid) in the 1970s and 1980s; Geert Wilders had represented the conservative Liberal Party (Volkspartij voor Vrijheid en Democratie) in parliament for six years before he seceded from the party; while Rita Verdonk had been a Liberal Party's Minister and member of its parliamentary group for one year before she left the party.

In this paper we will describe the evolution of populism in the Netherlands after 2000, culminating in the formation of these three right-wing populist parties. In the first section we will sketch the institutional and historical context of Dutch populism. The following three sections deal with the history of Dutch populism, which we divide into three stages. In the first stage (roughly from 1960 to 1990), parties emerged with some but not all of the characteristics of populism as defined above: the Farmers' Party (Boerenpartij), Democrats 66 (Democraten 66, D66), the Centre Party (Centrumpartij, CP) and its successor, the Centre Democrats (Centrumdemocraten, CD). In the second stage, pure populism and left-wing populism gained influence at the local level – culminating in the foundation of the Liveable Netherlands party (Leefbaar Nederland) and the growth of the Socialist Party (Socialistische Partij, SP). After the turn of the century, right-wing or national populist parties – the LPF, Proud of the Netherlands and the Freedom Party – arose and began to attract substantial support. All things considered, the Dutch case seems to provide a perfect example of the many political forms that populism can take. Despite this ideological diversity, almost all of the populist parties mentioned here share one characteristic: they did not have agricultural roots. The only exception is the Farmers' Party, which will be mentioned here briefly but is described in more detail in Chapter 13 of this volume.

In the final section we will attempt to answer the question of why the LPF, Proud of the Netherlands and Liveable Netherlands disappeared from the scene within a few years, while the Freedom Party has survived quite well – at least until now – and the SP has developed into a stable and 'normal' party. In this regard, political scientists Sarah de Lange and David Art have compared the LPF and the Freedom Party and provide a plausible answer: 'Wilders made better choices than Fortuyn and his successors when it came to party organisation, recruitment, socialisation and training' (De Lange and Art, 2011: p. 1230). We agree with their 'agency-based approach', yet question some of their empirical conclusions.

12.2 Historical and institutional context

At first sight, the Netherlands would appear to have been a paradise for populist parties after the introduction of proportional representation and universal suffrage (in 1918 and 1922, respectively). The Dutch electoral system is an extreme example of proportional representation: seats in the lower house of parliament (Tweede Kamer) are distributed to parties according to the votes they receive throughout the country. The upper house (Eerste Kamer) is not elected directly by the people, and carries less weight than the lower house. As the lower house contains 150 seats, a party needs only 0.67% of the popular vote (in the parliamentary elections of 2012 nearly 63,000 votes) to obtain a seat. There is no legal threshold (as, for example, in Germany, Denmark and Austria). Obviously, this makes life easier for a new party. Between 1946 and 2012, 21 new parties gained entrance to parliament. Other institutional factors, such as the parliamentary system and the unitary character of the Dutch state, seem less relevant to the rise of new parties.

The electoral system contributes to the fragmentation of the party system, which in turn requires coalition governments. The tendency to form rather broad coalitions – beyond the minimal range – was reinforced by Dutch political culture and pillarisation, the segmentation of society along religious and political lines. Pillarisation implies close ties between political elites and the people in networks of ideological organisations – referred to as 'pillars' or columns ('zuilen' in Dutch). Almost every religious denomination and every social class produced, as it were, its own political party. For example, most Catholics and Calvinists voted for Catholic or Calvinist (confessional) parties, respectively (Lijphart, 1968: pp. 26-58). While pillarisation might lead to civil war, as the history of Lebanon has shown, this need not be the case if accommodation is pursued at the elite level, something deeply ingrained in Dutch culture even before pillarisation. The Dutch Republic (1581-1795) was in fact a confederation of provinces and municipalities where decisions were usually based on compromises reached after complicated negotiations – 'schikken en plooien', as the Dutch would say (Daalder, 1981; Randeraad and Wolffram, 2001). Although the three major confessional parties held a majority in parliament from 1918 to 1967, they often tried to include either conservative Liberals or – after the Second World War – Social Democrats (Labour) in their coalition governments. The consociational tradition was maintained when the confessional parties lost their majority.

While several studies suggest that consociationalism provides fertile ground for populist protest (Hakhverdian and Koop, 2007), pillarisation may have prevented the growth of populist protest movements. Given the control of pillarised organisations over social and political life, few people were able or willing to challenge them by joining protest parties. Dissatisfaction was mainly articulated within the pillars through pressure groups and occasionally through dissident organisations, including parties; for example, Catholic workers or Protestant farmers might support dissident Catholic or Protestant parties (Daalder, 1966; Vossen, 2003: pp. 99-108). However, the decline of the pillars after the 1960s created a growing potential electorate for new parties – populist or otherwise. A new and more 'post-materialist' generation not only questioned the consociational tradition, but also the authority of the elites – in political parties as well as in the churches, the universities, corporations and other institutions (see also Lucardie and Voerman, 2012: pp. 21-36). Paradoxically, the conservative Farmers' Party was the first party to take advantage of this new situation.

12.3 First generation of (pseudo-)populist parties

When the pillars started to crumble in the 1960s, a window of opportunity was opened. The bonds between parties, churches, trade unions and other civil society organisations loosened, and the link between parties and citizens was weakened. Voters switched between parties more frequently and began to vote for new parties more often. In the 1960s the Farmers' Party and D66 entered parliament, later followed by the CP and CD and many others. These parties were opposed to the political establishment, as are most new parties. However, could they also be said to be populist according to the definition used here?

The Farmers' Party: agrarian populists?

In 1958, a few farmers founded the Farmers' Party (Boerenpartij) to fight state intervention in agriculture as well as other areas (see also Chapter 13). The party combined conservative (if not reactionary) economic liberalism with a vague Christian moral conservatism, nationalism and a populist aversion to party politics (Vossen, 2005: pp. 250-255). Its leader, Hendrik Koekoek, achieved a certain popularity with his simplistic down-to-earth one-liners, his folksy style, regional accent and casual appearance (Vossen, 2005: pp. 258-260). He often appealed to common sense while criticising experts, and he asserted that he spoke on behalf of the 'common people' or 'the person in the street'. When confronted with statistics that seemed to contradict his argument, he said: 'Figures don't mean anything, I want to have proof!' (Bouwman, 1992: pp. 25-26).

In 1963, the Farmers' Party entered parliament with three seats, having obtained 2.1% of the popular vote. Four years later it reached its zenith with seven seats (4.8%), then decline set in. Koekoek retained his seat until 1981, but in the meantime he had alienated many supporters and party members through his authoritarian and inflexible leadership. The party organisation seems to have been very weak, and party programmes were always rather short and did not elaborate the party's ideology. The Farmers' Party usually called for lower taxes and more economic freedom, such as free trade for agricultural products. Already in 1971 the party criticized the European Economic Community (the predecessor of the EU) because it would prevent free competition between farmers and foster fraud. In its declaration of principles, it argued that a vibrant countryside was necessary to preserve the Dutch people in the future. In 1971 it also advocated lower salaries for ministers, no foreign workers and the introduction of a referendum (which was new) (Boerenpartij, 1971: p. 131). The last item was dropped from the 1977 programme, the others were retained (Boerenpartij, 1977). The Farmers' Party explicitly rejected popular sovereignty and argued instead for the recognition of God's sovereignty, which would be hard to reconcile with populism. Thus, while Koekoek's style was certainly populist, ideologically his party was not – at least not according to our definition.

Democrats 66: elitist populists?

Democrats 66 (Democraten 66, D66) has been called 'the Farmers' Party for intellectuals' by critics (Van der Land, 2003: p. 24). The Dutch historian Maarten van Rossem has described it as an example of 'decent, rather elitist, liberal populism' – which sounds like an oxymoron (Van Rossem, 2010: pp. 26-29). D66 was founded in 1966 by young intellectuals from different political backgrounds, united only by a common political project (Van der Land, 2003: pp. 19-34). This project entailed reforming the elitist and consociational Dutch political system in a radical democratic direction, through the direct election of the prime minister, provincial governors and mayors and possibly the introduction of a referendum. The party argued that proportional representation should be replaced by a single-member plurality (or 'first-past-the-post') system. As D66 did not view the political elite and the people as two homogeneous groups, it is rarely seen as populist.

Surprisingly, D66 won seven seats in the parliamentary elections of 1967, less than half a year after it had been founded. Its sudden success could be attributed to several factors: the charisma of its political leader, the eloquent journalist Hans van Mierlo; favourable publicity, partly arranged by Van Mierlo and his colleagues; and its political project and political alliances. In 1967, all three factors may have reinforced each other. The party was divided, however, over its strategy. A majority voted for a close alliance with Labour and the new Radical Party (Politieke Partij Radikalen) in 1971. In 1973 the alliance proved strong enough to form a government – although it still needed support from two Christian Democratic parties in parliament. Yet in 1973, D66 almost disbanded, after internal strife and dramatic losses at provincial elections. It recovered with a new leader and partly new issues, such as the environment. In the 1980s, it repeated this cycle: success in 1981, participation in government, followed by defeat, crisis and rebirth under a new leader – in fact, Van Mierlo again. In 1994 it broke all records, winning 24 seats and a central position in the new government – until the next crisis occurred with Van Mierlo's retirement and substantial electoral losses in 1998. Even so, the party continued to take part in government until 2002 and again from 2003 to 2006.

Its new leader from 2006 onwards, Alexander Pechtold, has become the main critic of national populism and a staunch defender of European integration and multiculturalism. As a Minister in 2005-2006, he criticised the 'dirty' aspects of Dutch politics – 'manipulation, power games, personal and partisan interests which affect the appointment of people and the decisions taken' – as well as the power of the political elite – the *regenten* – but his rhetoric cannot be considered Manichean or folksy (Dresselhuys, 2006: p. 64; Jansen, 2006). In 2014, the party, still led by Pechtold, was one of the largest in the polls. D66 has always cherished grass-roots democracy, but without abandoning formal structures and procedures. Although it may share some elements with populist parties, it does not qualify as populist in our opinion (or in the opinion of most other scholars).

Centre Party and Centre Democrats: nationalists with a populist touch

The Centre Party (Centrumpartij, CP) was founded in 1980 by former members of the Dutch People's Union (Nederlandse Volksunie), after the latter, a marginal party, had been taken over by national socialists (Van Donselaar, 1991: pp. 143-176). By identifying with the political centre the founders of the CP made clear that they rejected any association with the extreme right. In socioeconomic questions the party indeed advocated centrist, that is, conservative or liberal positions. Yet these questions were not central to the party's concern. The first aim of the CP was the 'preservation and development of Dutch culture', according to the party programme (Centrumpartij, 1982: p. 1). It argued that immigrants should be encouraged to return to their mother country. In addition, the party's second aim was to make the political system more democratic by introducing a referendum and a people's initiative. Hans Janmaat, soon to be elected chair of the party, justified this demand in populist terms, claiming that the established parties pursued their own interests rather than those of the Dutch people (Janmaat, 1980). His style, however, was not very populist or folksy. He had a degree in political science. In 1982 the CP won one seat in parliament, occupied by Janmaat, and a new chair was elected. Within two years, Janmaat and his successor quarrelled about the party's strategy as well as financial and personal matters. Janmaat subsequently left the CP and set up a new party, the Centre Democrats (Centrumdemocraten, CD). After his departure, the CP failed to close its ranks. Under a new name, CP'86, it became more radical and was eventually banned in 1998 because it promoted discrimination and public disorder (Lucardie, 1998: pp. 22-24).

The CD defended more liberal and conservative positions, similar to that of the early CP. However, its nationalism acquired a slightly more militant character. It maintained that only third-generation Dutch citizens should be allowed to hold public office or gain employment in the public service (CD, 1994: p. 8), while Dutch citizens should have privileged access to health insurance and social security (CD, 1994: pp. 23, 25). The CD advocated a more nationalist and self-sufficient agricultural policy, catering above all to Dutch needs (feeding the Dutch people). Populism remained a recurrent, albeit not very explicit theme: in the introduction to the manifesto of 1994, Janmaat asserted that established parties neglected the national interest of the Dutch population (CD, 1994: p. 3).

In 1989, Janmaat regained his seat in parliament, in 1994 the number of seats increased from one to three, and at local elections, the CD won 78 seats, mainly in the most urbanised western part of the country. However, by 1998, all seats were lost – except one local council seat. Many councillors failed to cope with their political duties, the hostility of the environment and the tensions within the party, while some turned out to have a criminal record (Lucardie, 1998: pp. 24-27). When Janmaat died in 2002, the ailing party died with him.

Conclusion

Returning to the question asked above: To what extent were the Farmers' Party, D66, the CP and CD populist? The first two parties clearly had populist characteristics, albeit in different ways. D66 rarely spoke in populist language and hardly, if ever, used the terms 'people' and 'elite' (although it opposed the 'regents'), but it was committed to the direct election of mayors and even of the prime minister, as well as a referendum. Thus, in the terminology used here, D66 might be better called radical-democratic than populist. The Farmers' Party, in contrast, condemned popular sovereignty and direct democracy, but attempted to mobilise the people using populist rhetoric.

In the 1980s, the CP and CD used populist language and articulated populist demands (for direct democracy). However, their populism remained clearly subordinate to nationalism: it was populist nationalism rather than national populism. After the Second World War, nationalism became negatively loaded in the Netherlands. Had this not been the case, Janmaat might have been more successful, especially in the 1990s when the country was governed by a coalition of D66 and the two parties representing the traditional political extremes, Labour and the Liberal Party. This 'purple combination' gave the impression that the established political parties were one of a kind and that political divisions did not matter any longer. In addition, the position of the major pillarised parties, the Christian Democrats (Christen-Democratisch Appel) and to a lesser extent Labour, was severely weakened by heavy election defeat in 1994. The CD, however, hardly benefited at all. Its criticism of immigration policy and multicultural society was shared by a growing number of voters, but the party was not able to break the taboo on nationalism. In addition, the same topics were also raised by Liberal Party leader Frits Bolkestein in a much more subtle way, with more electoral success.

Only after the turn of the century, when Bolkestein had been succeeded by the more left-liberal and 'multiculturalist' Hans Dijkstal as leader of the Liberal Party and the CD had disappeared, did the opportunity for a strong right-wing, national-populist movement arise. Before this, however, two other varieties of populism entered the scene, both benefiting from the Liberal-Labour coalition and the lack of real political opposition in the 1990s.

12.4 Populism in the 1990s

The latent discontent in the 1990s was not only exploited by the CD, but also by two senior-citizen parties – who in 1994 together won seven seats in parliament – and by the Socialist Party (SP), which also debuted in that year (with two seats). The senior-citizen parties did not seek to mobilise the people against the elite, but aspired to serve the interests of older people; therefore, they will not be discussed here. The SP, however, observed a distinction between the 'good' people and the immoral elite. It heavily criticised Labour for cooperating with the conservative Liberals. In addition, various local parties channelled dissatisfaction in their municipalities. In the late 1990s, two of them founded a party at the national level, Liveable Netherlands (Leefbaar Nederland). Their populism seems more straightforward and less ambiguous than that of the Farmers' Party, D66 and the CP/CD discussed above.

The Socialist Party: left-wing populists or left-wing social democrats?

The Socialist Party (Socialistische Partij, SP) was originally founded in 1971 as the Communist Party of the Netherlands (Marxist-Leninist), but changed its name in 1972. Very gradually it shed its Maoist and subsequently also its Leninist principles, but it retained a populist orientation to the Maoist 'mass line'. Like its shining example, Mao, the party wanted to listen to the opinion of the masses, to transform these into political demands and to mobilise the masses on the basis of these demands (Voerman,

1987, 2008: pp. 27-28). In 1983, for example, the party suggested that migrant workers should either assimilate into Dutch society or return home – a position perceived to be extremely right-wing at the time (Socialistische Partij, 1983). Although no longer opposed to the 'parliamentary road' to socialism, the SP remained critical of the Dutch political system in the 1980s and 1990s. It proposed that parliamentary candidates should be nominated not only by political parties but also by social organisations, neighbourhoods or enterprises (Socialistische Partij, 1989: p. 8).

The SP participated in parliamentary elections for the first time in 1977, obtaining only 0.3% of the vote and no seats. It had to wait until 1994 before it entered parliament, blaming Labour for having betrayed its voters. In 1994 it campaigned with the slogan 'Vote against, vote SP!' ('*Stem tegen, stem SP!*'). It castigated politicians who 'speak arrogantly about their intention to 'bring politics to the people'', instead of 'leaving political decisions to the people themselves' (Socialistische Partij, 1994: p. 16). The SP's programme reflected this populist outlook: the party called for the introduction of a referendum, popular initiative and the direct election of mayors and the head of state.

Since 1994 it increased its share of the electorate at almost every election, from 1.3% (two seats) to 16.6% (25 seats) in 2006. In 2012 the party received 9.7% (15 seats). In 1999 the SP redefined socialism in rather broad and diluted terms: human dignity, equality and solidarity (Socialistische Partij, 1999: p. 7). However, it continued to fight neo-liberalism, the privatisation of public enterprises and individualism. Its alternative was not so much socialisation and planning, but democratisation – in a way, a shift to social democracy in its original form (social and democratic). Its own view was that it now occupied the former position of the Labour Party. It still adhered to the introduction of a referendum and popular initiative, but now also explicitly wanted to protect and strengthen parliamentary democracy (Socialistische Partij, 1999: pp. 10-11). Regarding the position of farmers, the SP argued for a different European agricultural policy: less liberal and more democratic, less focused on subsidies and more on production controls. At the turn of the century, the SP started to strive for participation in government at the local, provincial and also national level. In its election campaigns, however, the party continued to use a populist style, with symbols such as a tomato (to be thrown at the elite) (Voerman, 2009).

From 1994 to 2008, a former factory worker, Jan Marijnissen, led the party on the ground as well as in parliament, with rather direct and folksy expressions and a sarcastic sense of humour. In 2008 he left the leadership of the party in parliament to Agnes Kant, who has a PhD in health science. She perceived a gap between the elite and the people, while calling for an alliance between the two – which could be seen as a rather soft variety of populism (Kant, 2009). By that time, the SP had ceased to demand the direct election of mayors and the head of state, as well as the introduction of a popular initiative. Kant resigned in 2010 after disappointing local elections. Her successor,

Emile Roemer, was a different character, more relaxed yet also folksy in his humour and style, and like Kant also referring now and then to the gap between the 'ordinary people' and the 'deaf' political elite.

The internal organisation of the SP has become more democratic over time. From a Leninist vanguard party, it evolved into a still fairly centralised social-democratic mass party, while retaining some populist elements (see also Lucardie and Voerman, 2012: pp. 37-70).

'Liveable' parties: parochial populists?

During the 1990s, several new local parties that had entered municipal councils considered the established parties and their local branches to be too centralised, too bureaucratic or too technocratic. Quite often the new parties opposed large-scale infrastructure projects while referring to *leefbaarheid*, 'liveability', a new word in the Dutch language, associated with quality of life, environmentalism and 'small is beautiful'. In 1999, the leaders of Leefbaar Utrecht and Leefbaar Hilversum decided to found a national party using the same adjective: Leefbaar Nederland (Liveable Netherlands). In June 2001, the new party held its founding congress. It elected a president with some political experience, Jan Nagel, who had been a leading activist in the left-wing of the Labour Party and a member of the upper house from 1977 to 1983.

In November 2001, a second congress discussed the party programme. It started with a clearly populist critique of the established parties, which had 'become part of a new aristocracy' (*regentenstand*) and reduced elections to a ritual legitimation of the system which distributed all political positions to the elite (Leefbaar Nederland, 2002: p. 312). The party called for a 'liveable Netherlands, where citizens have power (*zeggenschap*) as well as responsibilities' (Leefbaar Nederland, 2002: p. 312). To reduce the influence of political parties, Liveable Netherlands called for the direct election of the prime minister, mayors and provincial governors, a mixed electoral system, and a law allowing corrective (derogatory) referenda 'with high thresholds' (Leefbaar Nederland, 2002: p. 313). It wanted to reduce bureaucracy, not only in government but also in hospitals, schools and the army. The European agricultural policy with its subsidies had to be abolished; farmers should be offered the possibility of getting a job in nature management or in recreational businesses (Leefbaar Nederland, 2002: p. 315).

Liveable Netherlands did not want to side with the left or the right. However, the same party congress of November 2001 elected a leader who reacted more strongly against the left than the right: Pim Fortuyn. He was an intellectual who had belonged to the Marxist left-wing of the Labour Party, but shifted to the right in the 1980s and 1990s. In various publications he had not only criticised the purple coalition of the Liberals and Labour, but also all established parties, and especially the 'subsidy socialists' of Labour

(Fortuyn, 2002: pp. 106, 109, 119). He blamed the 'left-wing church' particularly, for what he considered to be a far too lenient Dutch immigration policy. His opinion on this issue and his critical comments on 'backward Islam' caused a conflict with the executive committee of Liveable Netherlands. In February 2002, Fortuyn was forced to leave the party and decided to present his own list of candidates at the parliamentary elections in May (see below). With a new and relatively unknown leader (without charisma), Liveable Netherlands won only 1.6% of the popular vote, still enough for two seats in the lower house of parliament. Internal squabbles did not improve the party image. At the snap elections of January 2003, Liveable Netherlands obtained only 0.4% and lost both seats. Four years later the party disbanded (see also Lucardie and Voerman, 2012: pp. 71-90).

Conclusion

In the 1990s the established parties in the Netherlands were faced for the first time with full-fledged populist parties. The breeding ground for populism was perfect, as the traditional opponents in Dutch politics, the conservative Liberals and Labour, had worked together for eight years, while the Christian Democratic opposition was very weak. The SP was very successful with its populist approach. Paradoxically, however, its series of electoral victories showed the limits of its populist electoral strategy. The stronger the party became in municipal councils, the more its rank and file expected the SP to participate in local government. After the turn of the century, the party also started to focus on taking governmental responsibility – which implied programmatic moderation and a less clear-cut populist orientation.

While the SP mixed socialism with populism, Liveable Netherlands seems to be one of the rare parties that did not combine populism with a different ideology, but propagated populism in its pure form. This might explain its relatively minor electoral success and rapid downfall. Experience abroad and in the Netherlands shows that the combination of populism with nationalism or socialism (the SP) is electorally more profitable. Liveable Netherlands' success in the polls at the beginning of 2002 was mainly due to the charisma of Fortuyn, combined with his nationalism and anti-Islamism. Fortuyn's break in February 2002 heralded the decline of Liveable Netherlands. The other factors that also contributed were internal divisions, organisational flaws, financial problems and the disappointing result of the elections of 2002.

12.5 Right-wing populism at the beginning of the twenty-first century

Leaving the tiny CP and CD aside it is remarkable that right-wing populism in the Netherlands appeared only after the turn of the century, much later than elsewhere in Western Europe. While in Belgium, Italy, France and elsewhere strong national populist parties had existed for decades, Dutch populism at first took other forms. However, when right-wing populism knocked on the door in the Netherlands, it caused an electoral landslide. The Pim Fortuyn List won 17.0% of the votes in 2002, and the Freedom Party 15.5% in 2010. Proud of the Netherlands gained 16.0% in the polls in the summer of 2008. However, it often proved easier to achieve short-term electoral success than to set up a sustainable party organisation and consolidate that success. The Freedom Party appears to be the most successful of the three parties, deciding not to accept members. However, the autocratic way the party was run also caused problems.

The Pim Fortuyn List: the right-wing populist breakthrough

After leaving Liveable Netherlands, Fortuyn set up a new party, the Pim Fortuyn List (Lijst Pim Fortuyn, LPF). Actually, it was not really a party. When elections were held on 15 May 2002, the LPF was only three months old and lacked a formal organisation, a formal party programme and also a leader: its charismatic founder was shot nine days before election day. After the election, Fortuyn was succeeded by Mat Herben, an amiable civil servant without charisma. Nevertheless, this curious quasi-party entered Dutch parliament with 26 seats (out of 150) and was subsequently invited to participate in a coalition government with two established parties, the Christian Democrats and the Liberal Party. One did not have to be a political scientist to predict a troublesome fate for this coalition. Indeed, it collapsed in October 2002 because the Christian Democrats and the Liberal Party felt it was impossible to govern with a party that was tearing itself to bits. Two LPF ministers were constantly in open conflict with each other, while the executive committee and the parliamentary party sided with one or the other, rather than trying to reconcile them. At the anticipated elections of January 2003, the LPF was reduced to one third of its previous size.

By then, the LPF did have a programme. It was inspired, as one might expect, by the ideas of Fortuyn – an eclectic but not necessarily incoherent mixture of liberalism, nationalism, communitarianism and populism (Lucardie and Voerman, 2002; Pels, 2003: pp. 16-21). According to Fortuyn, in a modern 'contract society' every citizen could be considered an entrepreneur, even if only selling his own labour: an 'entrepreneur of his own labour'. Thus, workers should negotiate pension plans and disability payments directly with their employers rather than depending on trade unions and civil servants to do so (Fortuyn, 1995). The LPF did not go quite as far in its platform, but suggested, for example, that collective labour contracts would no longer be binding for everyone (Lijst Pim Fortuyn, 2003: p. 383). According to the party, farmers and horticulturalists had to be freed from state regulation. A restructuring of the agrarian sector was considered necessary, which had to be implemented very carefully, without losing 'farmers' entrepreneurship' (Lijst Pim Fortuyn, 2002b: pp. 366-367).

Fortuyn was not only a liberal with respect to socioeconomic issues, but also in the moral realm. Equality between men and women, heterosexuals and homosexuals, was very important to him – an active and open homosexual. He believed that the state should not interfere in private affairs such as sexual behaviour, and considered that drugs should be legalised. This was also the position of the LPF (Lijst Pim Fortuyn, 2003: p. 371). However, in fighting crime, both Fortuyn and the LPF favoured a conservative rather than a liberal approach, emphasising repressive measures rather than prevention and social reform (Lijst Pim Fortuyn, 2003: pp. 370-371). The integration of immigrants and Islamic people had been Fortuyn's main concerns since the 1990s. These issues caused his break with Liveable Netherlands – and his growing popularity in urban areas where immigration and Islam were perceived as a problem.[1] Fortuyn considered Islam to be a 'completely backward agrarian ideology', not recognising universal values such as human rights, the separation of church and state, and the principle of individual responsibility. He warned about the 'Islamisation of our culture' and therefore favoured a very restrictive immigration policy and assimilation of immigrants to 'Dutch culture' – while also rejecting nationalism (Fortuyn, 2001: p. 105). In a penetrating analysis of Fortuyn's ideology, the social scientist Dick Pels dismisses this idea that he rejected nationalism and classifies Fortuyn (Pels, 2003: pp. 228-246) as an ethnic nationalist (*volksnationalist*). This may be exaggerated, as Fortuyn did not advocate cultural or ethnic homogeneity (see also Couwenberg, 2004: pp. 61-62). At any rate it seems fair to call Fortuyn at least a moderate or civic nationalist, even if he may have preferred the label 'patriot' himself.

This moderate nationalism chimes in with Fortuyn's – almost romantic – opposition to a federal Europe which would 'lack a soul' (Fortuyn, 1997). The LPF (2003: pp. 384-385) seemed to defend an even more moderate position with respect to European integration. Fortuyn agreed with Liveable Netherlands that bureaucracy as well as 'particracy' clash with democracy. Small is beautiful: small schools, hospitals and municipalities are able to govern themselves without bureaucrats and managers taking control. At the national level, it was considered desirable, but more difficult, to wrest control from the political managers and bureaucrats who controlled the political parties and all political positions. Even so, Fortuyn's greatest ambition was to restore democracy and take the power from the political elite (the 'caste' or the 'mandarins') and return it to 'the people in the country' (*de mensen in het land*) (Fortuyn, 2002: pp. 151, 184-186).[2] How? In the first place, public officials such as mayors and the prime minister should be elected directly by the people. In the second place, citizens should have direct contact with their

[1] There was a strong correlation between support for the LPF and the number of immigrants in a municipality – but the interpretation of aggregate data can always be contested (Orlebeke, 2002; Lemmens and Diederiks, 2003).

[2] In the party's constitution, its goal was defined as 'involve the citizens of the Netherlands and let them participate in public administration and social organisations, in short: give the country back to the people in the country' ('de burgers van Nederland intensief te betrekken bij en te laten participeren in allerhande vormen van openbaar bestuur en maatschappelijke organisaties, kortom: het land terug te geven aan de mensen in het land', Lijst Pim Fortuyn, 2002a).

representatives. In the third place, the representatives in parliament should not merely follow the party line but heed the interests of their constituents as well as their own convictions. Parliament and cabinet should become more independent of each other, and of political parties (Fortuyn, 2002: pp. 141-143). However, Fortuyn disapproved of a referendum or people's initiative: a politician should listen to his constituents but not shy away from his own responsibility. He should inspire public debate and exercise leadership (Fortuyn, 1993: pp. 131, 211). The LPF also favoured an elected prime minister and elected mayors, but did not reject the idea of a referendum (as Fortuyn had done). However, it preferred a consultative or non-binding referendum to a binding one (Lijst Pim Fortuyn, 2003: p. 379).

The LPF might have agreed on a programme by 2003, but it did not agree on strategy and leadership. It tried to set up a formal organisation but remained chaotic. Internal strife continued until the parliamentary elections of 2006. The voters did not seem to appreciate this, with only 0.2% of the electorate remaining loyal to the party – not enough to win a seat. Barely two years later the LPF dissolved (see also Lucardie and Voerman, 2012: pp. 91-136).

Geert Wilders: Fortuyn's successful successor?

The Freedom Party (Partij voor de Vrijheid) was founded in February 2006 by Geert Wilders, who had left the Liberal Party in 2004 but retained his seat in parliament. At the parliamentary elections of November 2006 the new party won nine seats. Actually, it is debatable whether the Freedom Party should really be considered to be a party. It has only one member, Wilders; even the party's MPs cannot join. Thus, Wilders makes all the decisions alone: he writes the party programmes, selects candidates and monopolises contact with the press, even if he may consult his other MPs (Staal and Stokmans, 2008). In his opinion, members of a party tend to serve their own interests and stand as a barrier between the parliamentary party and the electorate. Wilders' right-hand and party ideologist Martin Bosma supplied the theoretical justification for the memberless Freedom Party, in his view, 'the first modern political party in the Netherlands' and the 'model of the future'. Without members, the party 'would have a clear political line and a clear leadership': it would only be accountable to the electorate, not to 'that peculiar, unelected intermediate layer that sticks together at party meetings and in party bunkers' (Bosma, 2010: p. 30-31, 216). According to Bosma, the old parties were increasingly dominated by a small elite that kept the door shut on new ideas in society and tried to strengthen their own power. He regarded the Freedom Party as much more democratic than the old membership-based parties. Bosma argued that '[t]he memberless party means adding direct democracy to indirect (representative) democracy' (Bosma, 2010: p. 215; see also Vossen, 2013: pp. 179-184). In this way, his model fits seamlessly into the populist political programme of the Freedom Party.

In his usually short programmes and manifestos Wilders combines liberalism with nationalism. Freedom and independence for the individual as well as for the nation are his supreme values (Wilders, 2005: pp. 106-107). Both are threatened, in his opinion, by two forces: European integration and 'Islamisation', meaning the growing influence of Muslim immigrants. To counter the first threat, the Netherlands should withdraw from the European Union, unless the latter is reduced to an economic union with no supranational powers (later he would withdraw even this qualification). The Freedom Party also wanted to 'nationalise' agriculture and 'free the farmers from the claws of Europe' (Partij voor de Vrijheid, 2010: p. 353). To fight Islamisation, the Dutch government should refuse to accept any more immigrants from Muslim countries, stop the building of any more mosques and ban the Quran (Partij voor de Vrijheid, 2006: p. 409; Wilders, 2007).

With respect to moral questions, Wilders has remained a liberal, but in socioeconomic affairs he seems to be mixing liberalism increasingly with social-conservative if not social-democratic demands: still advocating lower taxes and lower state subsidies (particularly for cultural projects) but also opposing cuts in health care, social security and pensions, as well as the liberalisation of the labour market (see also Lucardie and Voerman, 2013: pp. 191-194). This mixture, which can also be found among other national-populist parties, could be described as welfare state conservatism or welfare state chauvinism, as immigrants are to be excluded from social benefits (at least until they have been assimilated).

Like other populist parties, the Freedom Party calls for direct elections of public officials such as mayors and the prime minister, and possibly even judges and police commissioners (Partij voor de Vrijheid, 2006: p. 409; Wilders, 2006). In his speeches, Wilders has used rather direct and aggressive 'Manichean' language – calling the Minister of Housing and Integration 'raving mad', for example (Stokmans and Valk, 2007).

In the lower house elections held in June 2010, the Freedom Party received 15.5% of the vote (24 seats). It lent support to the minority coalition of the Liberal Party and the Christian Democrats led by Mark Rutte until April 2012, when it failed to agree to a package of substantial cuts in government spending. At the same time it suffered internal tensions and the secession of a dozen representatives at all levels (local, provincial, national and European) (see also Lucardie and Voerman, 2012: pp. 151-186). In the national elections of May 2012 the Freedom Party lost about one third of its voters – the party obtained 10.1% (15 seats). In the European elections of 2014 it received 13.3%, almost 4% less than in 2009.

Rita Verdonk: Fortuyn's unsuccessful successor

In 2007 another liberal MP felt compelled to leave the Liberal Party while retaining her seat. Rita Verdonk, Minister for Integration and Immigration from 2003 to 2006, had almost won the leadership race in the party and had received more preferential votes at the lower house elections of November 2006 than her victorious rival Rutte. In October, she founded the movement, Proud of the Netherlands (Trots op Nederland). It was meant to be a 'movement for the rest of the Netherlands', those people who were no longer represented by the 'governing elite' (Trots op Nederland, 2007). She did not shun the label 'populist' (Duk and Sleutjes, 2008). Her style seemed firm, but not as aggressive as that of Wilders. She argued for direct elections of the prime minister and of mayors, as well as for a law on binding referenda in typical populist terms: 'measures to prevent the political elite from staying in the saddle' (Trots op Nederland, 2010: p. 18). Nationalism was implied in the name of the movement, as well as in its declared purpose to bring 'the character of the Netherlands' to the forefront (Trots op Nederland, 2008), although it remained rather mild in its party programme. The party manifesto was to be drafted by the supporters of the movement – a slow process that was not yet completed at the time the 2010 elections were called.[3] In the local elections of March 2010, the movement won about 60 seats, yet it did not obtain a seat in the lower house elections in June 2010. Initially, Verdonk did not want to build a party organisation, but changed her mind in 2009. A vicious conflict with her right-hand man, Ed Sinke, caused a crisis in the fledgling party, as well as considerable negative publicity and as a consequence the demise of the party (see also Lucardie and Voerman, 2012: pp. 137-150). In 2011 Verdonk retired and while the party won seven seats in local elections in 2014, it no longer seems to be active at the national level.

12.6 Comparison

In this section we will analyse and compare (in a rather sketchy and schematic fashion) the programmes and other ideological publications and the organisational structure of the five 'modern' populist parties studied here (Liveable Netherlands, LPF, Freedom Party, Proud of the Netherlands and SP). The results are summarised in Table 12.1 and 12.2.

Ideology

All five parties articulated populist ideas, but not always consistently and not at all times. The SP used populist rhetoric and symbols (a tomato to be thrown at the establishment) mainly in the 1990s, when the party also advocated direct elections

[3] The manifesto that appeared in 2010 focussed on 'core business' of the government and did not contain any ideas about agriculture.

Table 12.1. Ideological similarities and differences between Liveable Netherlands, LPF, Freedom Party, Proud of the Netherlands and SP.

Party	Populism	Nationalism and euroscepticism	Anti-Islamism	Socioeconomic position
Liveable Netherlands	strong	absent	absent	unclear → liberal
Pim Fortuyn List	moderate	moderate	moderate	liberal
Freedom Party	strong (increasingly)	strong (increasingly)	strong	liberal → social-conservative
Socialist Party	strong (1990s) → weak (after 2000)	weak	absent	socialist → social-democratic
Proud of the Netherlands	strong	weak	weak	liberal

Table 12.2. Organisational similarities and differences between Liveable Netherlands, LPF, Freedom Party, Proud of the Netherlands and SP.

Party	Time span	Membership	Leadership	Conflict management	Recruitment and training of candidates
Liveable Netherlands	1999-2007	low	confused	weak	external
Pim Fortuyn List	2002-2007	moderate	first charismatic (Fortuyn), then divided	weak	external and improvised
Freedom Party	2006-	very low (only one)	strong	not clear yet	external and thorough
Socialist Party	1971-	high	strong	repressive but effective	internal and thorough
Proud of the Netherlands	2008-	low	first strong, then divided	ineffective	external and improvised

of mayors (and of a president), as well as the introduction of a referendum law and people's initiative. By 2010, it still favoured the latter but not the former, arguing that mayors should be elected by the municipal councils rather than directly by the people (Socialistische Partij, 1994, 2010: pp. 217-218).[4] Nevertheless, populist rhetoric has not disappeared altogether. In its platform of 2012, for example, the party observed that 'the organisation of our country is too much in the hands of an ('old boys') network

[4] This demand was not included in the 2012 programme (Socialistische Partij, 2012).

of officials and managers who appoint each other to lucrative positions and part-time jobs, and award each other bonuses', therefore, citizens need a greater say regarding the organisation of society (Socialistische Partij, 2012: p. 180).

The programme of Liveable Netherlands in 2002 can be considered an example of pure populism – and not much else. Direct elections and referenda figured prominently (Leefbaar Nederland, 2002). Fortuyn agreed with the former but not with the latter, arguing that political leaders should 'incarnate' the people without being corrected all the time (Fortuyn, 1993: pp.131, 211; see also Lucardie and Voerman, 2012: pp. 108-110, 114). However, the main purpose of Fortuyn's LPF was to return power to the people (Fortuyn, 2002: pp. 151, 184-186).

Wilders, initially a parliamentarist rather than a populist, has increasingly used similar and even stronger rhetoric (see also Vossen, 2011). In his opinion, the multiculturalist elite has encouraged mass immigration and the Islamisation of the Netherlands (and of Europe) against the will of the people. In 2010 his party called for a referendum and direct elections of the prime minister, mayors and even senior judges and public prosecutors (Partij voor de Vrijheid, 2010: p. 319). In its programme of 2012, however, the Freedom Party only insisted on the direct election of the prime minister (and on the referendum) (Partij voor de Vrijheid, 2012: p. 109).

Verdonk's party did not go quite as far as the Freedom Party in 2010, but she did argue for more influence of citizens in the judicial process (via lay judges) and for more direct contact between voters and representatives, as well as direct elections of mayors and the prime minister, and a referendum (Trots op Nederland, 2010).

Populism was linked to nationalism by the LPF and the Freedom Party, but only weakly by Proud of the Netherlands and the SP and not at all by Liveable Netherlands. The SP has always criticised the EU, and opposed the constitutional treaty of 2005. In the 1980s it seemed opposed to mass immigration, but today it accepts the multicultural character of Dutch society. Fortuyn was critical of a 'soulless Europe' and concerned about Dutch identity (1997). Fear of 'Islamisation' led him to advocate a restrictive immigration policy and the enforced integration of immigrants (Fortuyn, 2002). Wilders was initially more moderate than Fortuyn, but became radicalised after his break with the Liberal Party, evolving into a fanatical anti-Islamist who wanted to ban the Quran, tax headscarves, close all Islamic schools and prohibit new mosques (Partij voor de Vrijheid, 2012: p. 113). Nevertheless, when the financial and economic crisis came to dominate the political agenda, he shifted his emphasis from Islam to Europe. When he broke with the coalition of the Christian Democrats and the Liberal Party in April 2012 and caused a crisis which would lead to elections in September of that year, the core issue seemed to be the 'diktat' of Brussels (the 3% ceiling for the budget deficit). According to Wilders, the Netherlands should replace the euro with the guilder and perhaps even leave the

EU. Verdonk did not agree, her party used nationalist rhetoric and favoured a stricter immigration policy, but was not explicitly anti-Islamist or anti-European.

The five parties held rather different positions in the socioeconomic sphere. The SP may have moderated its socialism and ceased to advocate the socialisation of the means of production, but it still defends egalitarian and interventionist policies and strongly opposes 'neoliberalism', privatisation and the liberalisation of the public sector. Liveable Netherlands did not have a clear socioeconomic platform in 2002, while it leaned towards a conservative-liberal position in 2003 (Lucardie and Voerman, 2012: p. 88). Fortuyn had developed his own brand of 'modernised liberalism' in the 1990s, mixed with a little social conservatism, while remaining adamantly liberal with respect to human rights and moral issues (Lucardie and Voerman, 2012: pp. 102-105, 110-115). Wilders started as a conservative and almost classic liberal, but gradually came to defend the welfare state against liberal reforms and cutbacks (such as raising the age for state pensions and reducing health-care spending, or liberalising labour relations) (Vossen, 2011; Lucardie and Voerman, 2012: pp. 174-175). His defence of the welfare state could be interpreted as almost social-democratic, but 'social-conservative' seems a better term here. Verdonk, however, remained a consistent conservative liberal, calling for a more modest state, a flat tax and welfare cuts (Lucardie and Voerman, 2012: pp. 146-147).

In summary, the five parties did not adhere to the same ideology or ideological mixture, as Table 12.1 indicates; they only agreed on populist reforms, and even there not completely. Of the parties involved, the LPF and Freedom Party show most similarities, even if the former seems more moderate than the latter. The five parties also varied in ideological coherence: quite high in the case of the SP, very low for Liveable Netherlands. Two parties experienced an ideological evolution: the SP becoming less populist (and also less socialist and more social democratic), and the Freedom Party becoming less liberal and more social conservative. The two parties concerned have the longest life spans of the five, which might also be related to their organisational structure.

Organisation

Populists often have a difficult relationship with political organisation. They often think that no intermediate or corporative bodies should exist between the government and the people, including no party organisation. Populist parties do exist, of course, but are often led by a charismatic leader who strives for direct contact with his supporters, without intermediaries. Charismatic leadership fares badly within bureaucratic and hierarchically organised parties where the decisions are made only after prolonged consultation through several committees or boards. In contrast, charismatic leaders prefer to make all the decisions themselves. Many populist parties call themselves a 'movement' rather than a 'party' and have a weak, mainly informal structure. Even if they have a formal membership, it may mean little: members have few rights and few duties, they may be allowed to attend mass meetings to cheer on the leader, but not to decide on the policy of the party. There is little room for intermediary bodies between the leader and their followers, such as party councils or congresses, because they would prevent the leader from directly understanding and implementing the will of the people. The question is whether or not this ideal-typical populist party model is found in the Netherlands.

The SP has evolved from a small and closed Marxist-Leninist cadre party to a more or less social-democratic mass party with about 44,000 members (in 2014). The party organisation seems highly structured, hierarchical and solid (Table 12.2). Leadership is unified and somewhat authoritarian – especially when the charismatic Marijnissen presided over both the parliamentary party and the party on the ground (Lucardie and Voerman, 2012: pp. 49-50). Internal conflicts have often been managed in a rather repressive way. Even so, members have substantial rights and duties and intermediary bodies do play a role. The party had obtained a large number of seats in municipal councils before winning seats at the national level. Parliamentary candidates could be easily recruited from MP's assistants, party officials and local councillors.

Liveable Netherlands was set up as a democratic 'amateur activist party' (on this concept, see Lucardie and Rihoux, 2008: pp. 6-7) that was decentralised and open – somewhat similar to D66. Members could vote at the party conference and select candidates at the regional level (Lucardie and Voerman, 2012: p. 78). However, in practice, the leadership intervened more than expected, internal conflicts escalated and they practically destroyed the party (Lucardie and Voerman, 2012: pp. 86-87). After the breach with Fortuyn, the party lacked conviction, let alone charismatic leadership, and membership remained modest (about 1,200 in 2002).

The LPF was the result of improvisation and never evolved into a stable structure (Lucardie and Voerman, 2012: pp. 116-117). As a charismatic leader, Fortuyn was not interested in a party organisation at all. Candidates had to be recruited externally in a rather improvised process. Internal conflicts erupted immediately at all levels after the violent death of its founder, within the party executive and the parliamentary group and between ministers (Lucardie and Voerman, 2012: pp. 127-132). The conflicts escalated beyond control and resulted in several party splits and declining membership – from a high of 4,000 in 2002 to less than 1000 a few years later, until the organisation was dissolved.

Wilders wanted to avoid the fate of the LPF and founded a party without members – the only members being Wilders and the Foundation Group Wilders, led by him (Lucardie and Voerman, 2012: pp. 156-157, 164-165). According to Dutch law, a foundation can be considered a legal person, hence the Freedom Party formally had two members, although in reality only one. Wilders drafted the party platform, selected and trained candidates – initially by himself, later with the help of a handful of assistants (usually also members of the parliamentary party). It was to be a 'virtual party' where parliamentarians would be accountable only to voters and not to rank and file party members (Bosma, 2010: pp. 214-217). Nevertheless, conflicts could not be avoided and dissidents at all levels left the party, including municipal councillors, provincial legislators, a member of the European Parliament and even senior members of the national parliament (Lucardie and Voerman, 2012: pp. 171-173, 179-185; Vossen, 2013: pp. 184-196). At this moment it is unclear whether the haemorrhaging has stopped and if future conflicts will be managed better.

Proud of the Netherlands also started as a virtual party, or movement without members, but local branches were soon set up. The party platform was to be written by the grass-roots membership, but in practice Verdonk and her assistants seem to have done most of the work and were also most involved in the selection of candidates. Membership remained modest (about 900) but conflicts within the leadership could not be controlled and led to a crisis and loss of support in the polls as well as in the parliamentary elections of 2010 (Lucardie and Voerman, 2012: pp. 144-145). After the departure of its leader, the party seemed to survive only in the periphery of the party system, with a few local councillors.

In summary, whereas the SP developed into a highly institutionalised and consolidated mass party – quite different from the ideal-typical populist party – the other four parties seemed unable to do so (Table 12.2). Liveable Netherlands, LPF and Proud of the Netherlands had a confused or divided leadership and were consequently weak or ineffective in solving conflicts. In addition, they had low membership figures, which means that they were poorly grounded in society and had limited resources. Contrary to the SP, these three parties were to a large extent dependent on their leaders. When the leader left, they either disintegrated within a few years (Liveable Netherlands, LPF) or only barely survived (Proud of the Netherlands).

The same risks are lurking for the Freedom Party. The party does not have members and is highly dependent on Wilders. His leadership is strong and the way he manages conflict is repressive – with a number of parliamentarians at the European, national, provincial and local levels leaving the party as a result. Whether the Freedom Party will succeed in becoming a stable 'virtual party' without formal members, and whether it would survive the departure of Wilders (whenever that might happen), remains uncertain but rather doubtful at this moment.

12.7 Conclusion

Of the five more or less populist parties studied here, only the SP seems an unambiguous success story in terms of survival and electoral performance. The only ambiguity here concerns its populist character after the turn of the century. Until 2012, the Freedom

Party was also quite successful, certainly in electoral terms, but its future seems somewhat less certain. The LPF, Liveable Netherlands and Proud of the Netherlands can be seen as failures in practically all respects. Thus, we may agree with De Lange and Art (2011) in concluding that agency is indeed an essential factor in explaining the success and survival of populist parties, but we might also add that agency may refer to organisation as well as ideology. The more successful parties seem to be more coherent and institutionalised organisations than the failures, but also ideologically more coherent.

In no case one can speak of agrarian populism. Agricultural policy and rural issues did not seem to concern Dutch populist parties a great deal. Generally only a few words are spent on the agricultural sector. In fact, even the Farmers' Party (which according to the definition we have applied here is not a populist party in ideological terms) devoted only a small part of its (rather short) platforms to agricultural policy. In most platforms of the populist parties involved here some rural conservatism can be found. Euroscepticism, however, is a theme common to all of them when they refer to agriculture, except for Liveable Netherlands.

Remarkably, none of the five parties had a real base in a social protest movement, whether rural or urban. Rural protests did occur in the Netherlands, even in the 1990s (see Chapter 3: Strijker and Terluin, 2015), and even gave rise to the foundation of two political parties, the Solidarity Farmers' Party and the General Democratic Party, but both failed to win any seats and disappeared within a few years.[5] One could argue that Liveable Netherlands had roots in local protest movements but they were political rather than social: socially heterogeneous groups of citizens protesting against local policies and local political elites. The LPF, Proud of the Netherlands and Freedom Party did not have any roots in social or political movements: the latter two emerged from an established political party, while the LPF was founded by a political entrepreneur. Only the SP has roots in society. It started as a Maoist splinter group arising from the Communist movement, with support from student activists – although it would be an exaggeration to argue that it was grounded in the student movement. During the 1970s and 1980s it developed active local branches and set up auxiliary organisations and mobilised protests against pollution, rent increases and other social problems. Thus, in a way, the SP might be regarded as the producer rather than the product of a social protest movement. In conclusion, a strong party organisation may be a necessary condition for the survival of a populist party – or any party – with or without roots in a social protest movement. To survive in the long term, populist parties need to overcome their ideologically motivated aversion to organisation – not an easy task to accomplish.

[5] Neither party could be considered a populist party in the ideological sense (Lucardie, 1995).

References

Boerenpartij, 1971. Urgentieprogram 1971. In: Goossen, P. (ed.) Parlement en kiezer 1970/1971. Martinus Nijhoff, The Hague, the Netherlands, p. 131.

Boerenpartij, 1977. Urgentieprogramma. In: Lipschits, I. (ed.) Verkiezingsprogramma's 1977. Sdu, The Hague, the Netherlands, p. 3.

Bosma, M., 2010. De schijn-élite van de valse munters. Drees, extreem rechts, de sixties, Groep Wilders en ik. Uitgeverij Bert Bakker, Amsterdam, the Netherlands, 376 pp.

Bouwman, R., 1992. Ross Perot in Nederland. Hervormd Nederland 48: 24-26.

Centrumdemocraten (CD), 1994. Oost west thuis best. Partijprogramma verkiezingen Tweede Kamer 1994. Centrumdemocraten, The Hague, the Netherlands, 34 pp.

Centrumpartij, 1982. Partijprogramma. Centrumpartij, Amsterdam, the Netherlands.

Couwenberg, S.W., 2004. Opstand der burgers: de Fortuyn-revolte en het demasqué van de oude politiek. Damon, Budel, the Netherlands, 256 pp.

Daalder, H., 1966. De kleine politieke partijen – een voorlopige poging tot inventarisatie. Acta Politica 1: 172-196.

Daalder, H., 1981. Consociationalism, Center and Periphery in the Netherlands. In: Torsvik, P. (ed.) Mobilization, center-periphery structures and nationbuilding. Universitetsförlaget, Oslo, Norway, pp. 181-240.

De Lange, S.L. and Art, D., 2011. Fortuyn versus Wilders: an agency-based approach to radical right party building. West European Politics 34(6): 1229-1249.

Dresselhuys, C., 2006. 'Ik wil geen schuwe aap zijn'. Minister Alexander Pechtold langs de feministische meetlat. Opzij 34(2): 64-67, 69.

Duk, W. and Sleutjes, L., 2008. 'Populist? Ik vind het een geuzennaam'. Interview met Rita Verdonk. Dagblad van het Noorden March 27, 2008.

Fortuyn, W.S.P., 1993. Aan het Volk van Nederland. De contractmaatschappij, een politiek-economische zedenschets. Contact, Amsterdam/Antwerpen, the Netherlands/Belgium, 215 pp.

Fortuyn, W.S.P., 1995. Uw baan staat op de tocht! De overlegeconomie voorbij. Bruna, Utrecht, the Netherlands, 78 pp.

Fortuyn, W.S.P., 1997. Zielloos Europa. Tegen een Europa van technocraten, bureaucratie, subsidies en onvermijdelijke fraude. Bruna, Utrecht, the Netherlands, 111 pp.

Fortuyn, W.S.P., 2001. De islamisering van onze cultuur. Nederlandse identiteit als fundament. Karakter Uitgevers/Speakers Academy, Uithoorn/Rotterdam, the Netherlands, 126 pp.

Fortuyn, W.S.P., 2002. De puinhopen van acht jaar paars. Een genadeloze analyse van de collectieve sector en aanbevelingen voor een krachtig herstelprogramma. Speakers Academy, Rotterdam, the Netherlands, 186 pp.

Hakhverdian, A. and Koop C., 2007. Consensus democracy and support for populist parties in Western Europe. Acta Politica 42(4): 400-420.

Janmaat, H., 1980. Een nieuwe lente, een nieuw geluid. Middenweg 1(1): 5-8.

Jansen, S., 2006. De regenten laten veel thema's liggen. Alexander Pechtold blikt blijmoedig op zes zetels. Dagblad van het Noorden November 9, 2006.

Kant, A., 2009. Niemand wacht op verheffing door Bos. NRC Handelsblad July 1, 2009.

Leefbaar Nederland, 2002. Leefbaar Nederland komt er nu aan! In: Van Holsteyn, J., Pellikaan, H. and Voerman, G. (eds.) Verkiezingsprogramma's. Verkiezingen van de Tweede Kamer 15 mei 2002 & 22 januari 2003. Rozenberg Publishers, Amsterdam, the Netherlands, pp. 311-317.

Lemmens, P. and Diederiks, J., 2003. Allochtonie en het electorale succes van de LPF. Facta 11: 1, 22-23.

Lijphart, A., 1968. The politics of accomodation. Pluralism and democracy in the Netherlands. University of California Press, Berkeley, USA, 231 pp.

Lijst Pim Fortuyn (LPF), 2002a. Statuten Lijst Pim Fortuyn. LPF, Rotterdam, the Netherlands.

Lijst Pim Fortuyn (LPF), 2002b. Zakelijk met een hart. Verkiezingsprogramma 2002. In: Van Holsteyn, J., Pellikaan, H. and Voerman, G. (eds.) Verkiezingsprogramma's. Verkiezingen van de Tweede Kamer 15 mei 2002 & 22 januari 2003. Rozenberg Publishers, Amsterdam, the Netherlands, pp. 368-391. Available at: http://tinyurl.com/pd4ake8.

Lijst Pim Fortuyn (LPF), 2003. Politiek is passie. Verkiezingsprogramma 2003. In: Van Holsteyn, J., Pellikaan, H. and Voerman, G. (eds.) Verkiezingsprogramma's. Verkiezingen van de Tweede Kamer 15 mei 2002 & 22 januari 2003. Rozenberg Publishers, Amsterdam, the Netherlands, pp. 361-367. Available at: http://tinyurl.com/pw2t5r6.

Lucardie, A.P.M. and Rihoux, B., 2008. From amateur-activist to professional-electoral parties? On the organizational transformation of green parties in Western Democracies. In: Frankland, E.G., Lucardie, A.P.M. and Rihoux, B. (eds.) Green parties in transition. The End of Grass-roots Democracy? Ashgate, Farnham, UK, pp. 3-16.

Lucardie, A.P.M. and Voerman, G., 2002. Het gedachtegoed van Pim Fortuyn: liberaal patriot of nationaal populist? Socialisme & Democratie 59(4): 32-42.

Lucardie, A.P.M. and Voerman, G., 2012. Populisten in de polder. Boom, Amsterdam, the Netherlands, 240 pp.

Lucardie, A.P.M. and Voerman, G., 2013. Geert Wilders and the Party of Freedom in the Netherlands: A Political Entrepreneur in the Polder. In: Grabow, K. and Hartleb, F. (eds.) Exposing the demagogues. right-wing and national populist parties in Europe. Centre for European Studies/Konrad Adenauer Stiftung, Brussels/Berlin, Belgium/Germany, pp. 187-203.

Lucardie, A.P.M., 1995. Binnenkomers en buitenstaanders. Een onderzoek naar partijen die in 1994 hun entree in de Tweede Kamer trachtten te maken. In: Voerman, G. (ed.) Jaarboek 1994 Documentatiecentrum Nederlandse Politieke Partijen. Documentatiecentrum Nederlandse Politieke Partijen, Groningen, the Netherlands, pp. 123-148.

Lucardie, A.P.M., 1998. Een geschiedenis van vijftien jaar centrumstroming. In: Van Holsteyn, J. van and Mudde, C. (eds.) Extreem-rechts in Nederland. Sdu, The Hague, the Netherlands, pp. 17-30.

Mudde, C., 2007. Populist radical right parties in Europe. Cambridge University Press, Cambridge, UK, 404 pp.

Orlebeke, J.F., 2002. Het succes van de Lijst Pim Fortuyn. Facta 10(7): 10-11.

Partij voor de Vrijheid (PVV), 2006. Verkiezingspamflet. In: Pellikaan, H., Van Holsteyn, J. and Voerman, G. (eds.) Verkiezing van de Tweede Kamer der Staten Generaal 22 november 2006. Verkiezingsprogramma's. Rozenberg Publishers, Amsterdam, the Netherlands, pp. 407-410.

Partij voor de Vrijheid (PVV), 2010. De agenda van hoop en optimisme. Een tijd om te kiezen: PVV 2010-2015. In: Pellikaan, H., Van Holsteyn, J. and Voerman, G. (eds.), Verkiezing van de Tweede Kamer der Staten-Generaal 9 juni 2010. Verkiezingsprogramma's. Rozenberg Publishers, Amsterdam, the Netherlands, pp. 301-359.

Partij voor de Vrijheid (PVV), 2012. Hún Brussel, óns Nederland. Verkiezingsprogramma 2012-2017. In: Pellikaan, H., Lubbe, R. and Voerman, G. (eds.) Verkiezing van de Tweede Kamer 12 september 2012. Verkiezingsprogramma's. Rozenberg Publishers, Amsterdam, the Netherlands, pp. 93-122.

Pels, D., 2003. De geest van Pim. Het gedachtegoed van een politieke dandy. Anthos, Amsterdam, the Netherlands, 316 pp.

Randeraad, N. and Wolffram, D.J., 2001. Constraints on clientelism: the Dutch path to modern politics, 1848-1917. In: Piattoni, S. (ed.) Clientelism, interests, and democratic representation. The European experience in historical and comparative perspective. Cambridge University Press, Cambridge, UK, pp. 101-121.

Socialistiese Partij, 1983. Gastarbeid en kapitaal. Socialistiese Partij, Rotterdam, the Netherlands.

Socialistiese Partij, 1989. Een maatschappij voor mensen. Handvest 2000. Socialistiese partij, Rotterdam, the Netherlands.

Socialistische Partij (SP), 1994. Stem tegen, stem SP. Verkiezingsprogramma Socialistische Partij. Socialistische Partij, Rotterdam, the Netherlands.

Socialistische Partij (SP), 1999. Heel de mens. Kernvisie van de SP, zoals vastgesteld door het congres op 18 december 1999. Available at: http://tinyurl.com/mbjsrgr.

Socialistische Partij (SP), 2010. Een beter Nederland, voor minder geld. Verkiezingsprogramma SP 2011-2015. In: Pellikaan, H., Van Holsteyn, J. and Voerman, G. (eds.) Verkiezing van de Tweede Kamer der Staten-Generaal 9 juni 2010. Verkiezingsprogramma's. Rozenberg Publishers, Amsterdam, the Netherlands, pp. 207-255.

Socialistische Partij (SP), 2012. Nieuw vertrouwen. Verkiezingsprogramma SP 2013-2017. In: Pellikaan, H., Lubbe, R. and Voerman, G. (eds.) Verkiezing van de Tweede Kamer 12 september 2012. Verkiezingsprogramma's. Rozenberg Publishers, Amsterdam, the Netherlands, pp. 173-212.

Staal, H. and Stokmans, D., 2008. 'We staan pas aan het begin van ons avontuur'. Geert Wilders en zijn PVV na 'Fitna'. NRC Handelsblad May 24, 2008.

Stokmans, D. and Valk, G., 2007. Wilders 'gijzelt' Kamer in islamdebat. NRC Handelsblad September 7, 2007.

Strijker, D. and Terluin, I.J., 2015. Rural protest groups in the Netherlands. Chapter 3. In: Strijker, D., Voerman, G. and Terluin, I.J. (eds.) Rural protest groups and populist political parties. Wageningen Academic Publishers, Wageningen, the Netherlands, pp 63-77.

Trots op Nederland, 2007. Trots op Nederland – de nieuwe beweging van Rita Verdonk.

Trots op Nederland, 2008. De visie van Trots op Nederland.

Trots op Nederland, 2010. Partijprogramma 'Vertrouwen en handhaven'.

Van der Land, M., 2003. Tussen ideaal en illusie. De geschiedenis van D66, 1966-2003. Sdu, The Hague, the Netherlands, 512 pp.

Van Donselaar, J., 1991. Fout na de oorlog. Fascistische en racistische organisaties in Nederland 1950-1990. Uitgeverij Bert Bakker, Amsterdam, the Netherlands, 249 pp.

Van Rossem, M., 2010. Waarom is de burger boos? Over hedendaags populisme. Nieuw Amsterdam, Amsterdam, the Netherlands, 126 pp.

Voerman, G., 1987. De 'Rode Jehova's'. Een geschiedenis van de Socialistiese Partij. In: Koole, R.A. (ed.) Jaarboek 1986 Documentatiecentrum Nederlandse Politieke Partijen. Documentatiecentrum Nederlandse Politieke Partijen, Groningen, the Netherlands, pp. 124-150.

Voerman, G., 2008. The disappearance of communism in the Netherlands. In: Backes, U. and Moreau, P. (eds.) Communist and post-communist parties in Europe. Vandenhoeck & Ruppert, Göttingen, Germany, pp. 15-38.

Voerman, G., 2009. Van Mao tot marketing. Over het populisme van de SP. Socialisme & Democratie 66(9): 26-32.

Vossen, K., 2003. Vrij vissen in het Vondelpark. Kleine politieke partijen in Nederland 1918-1940. Wereldbibliotheek, Amsterdam, the Netherlands, 319 pp.

Vossen, K., 2005. De andere jaren zestig. De opkomst van de Boerenpartij. In: Voerman, G. (ed.) Jaarboek 2004 Documentatiecentrum Nederlandse Politieke Partijen. Documentatiecentrum Nederlandse Politieke Partijen, Groningen, the Netherlands, pp. 245-266.

Vossen, K., 2011. Een nieuw groot verhaal? Over de ideologie van LPF en PVV. In: Van Baalen, C., Goslinga, H., Van Kessel, A., Van Merriënboer, J., Ramakers, J. and Turpijn, J. (eds.) Jaarboek Parlementaire Geschiedenis 2011. Waar visie ontbreekt, komt het volk om. Boom Uitgevers, Amsterdam, the Netherlands, pp. 77-87.

Vossen, K., 2013. Rondom Wilders. Portret van de PVV. Boom Uitgevers, Amsterdam, the Netherlands, 312 pp.

Wilders, G. 2005. Kies voor vrijheid: een eerlijk antwoord. Groep Wilders, The Hague, the Netherlands, 132 pp.

Wilders, G., 2006. Klare wijn. Available at: http://tinyurl.com/n7w7mk7.

Wilders, G., 2007. Genoeg is genoeg: verbied de Koran. De Volkskrant August 8, 2007.

13 Agrarian parties in the Netherlands: the Plattelandersbond and the Boerenpartij

K. Vossen

Political Science, Radboud Universiteit Nijmegen, the Netherlands; kpssvossen@hotmail.com

Abstract

Throughout most of the twentieth century, agrarian parties have been a common phenomenon in European party systems. In Scandinavian countries and Switzerland, agrarian parties developed into enduring and relevant parties, while in other countries, such as France, Belgium and Germany, agrarian parties were flash-in-the-pan parties. The Dutch agrarian parties are a somewhat unusual case compared to many other countries. In terms of electoral strength and influence, they were rather irrelevant, although the pre-war Plattelandersbond and the post-war Boerenpartij both managed to hold at least one seat in parliament for nearly twenty years. Between 1963 and 1967, the Boerenpartij even served as the main populist party in the Netherlands. This contribution will analyse the similarities and differences between the two parties, as well as the political opportunity structure related to each of the parties in the Netherlands.

13.1 Introduction

Windmills, wooden shoes, the polder landscape and of course Dutch cheese: many of the traditional images of the Netherlands have to do with agriculture. This may seem odd as the Netherlands is one of the most urbanised countries in the world. These images, however, originate from the nineteenth century, in which the Netherlands was indeed predominantly a nation of farmers and fishermen. In 1947, 20% of the Dutch population was still employed in the agricultural sector.

The soil for farmer parties in the Netherlands appeared to be very fertile not only in terms of socioeconomic structures but also in terms of the electoral system. In 1918 an electoral system of proportional representation was introduced which was very favourable for new parties. There were virtually no barriers to participation in elections, with new parties

D. Strijker, G. Voerman and I.J. Terluin (eds.) **Rural protest groups and populist political parties**
DOI 10.3920/978-90-8686-807-0_13, © Wageningen Academic Publishers 2015

requiring only a list of 25 signatures to participate. The electoral threshold was also quite low: initially 0.5% of the vote was enough to gain a seat in parliament.

Thus, one might expect that the political opportunity structure was very favourable to an agrarian political party. Between 1918 and 2012, about ten such parties, established to advocate the interests of farmers, participated in national elections. However, none of them developed into a powerful and enduring agrarian party, and despite the low electoral threshold, only two parties managed to enter parliament: the Plattelandersbond (Countryside League) in 1918 (from 1933 under a new name) and the Boerenpartij (Farmers' Party) in 1963.

This contribution will thus focus on these two parties, asking why they failed to establish themselves as serious and enduring parties? What was their ideology and strategy? Could both parties be perceived as populist parties? In accordance with most scholars, I will consider populism to be a thin-centred ideology which is based on the perception of politics as entailing an irreducible conflict between two homogeneous and antagonistic groups – a virtuous people and a malicious elite. The aspiration of populism is to build a polity in which the will of the virtuous people prevails. The denunciation of the elite as the incarnation of evil and the glorification of the people may be considered the basic hallmarks of populism. The core of populism entails some other closely connected features which reinforce and highlight the core concept. These 'flavour enhancers' are an inclination towards conspiracy theories, a popular and folksy style, a strong voluntarism, and a preference for both plebiscite-based democracy and charismatic leadership (Albertazzi and McDonnell, 2008; Mudde, 2004; Taggart, 2000; Vossen, 2010).

13.2 The Plattelandersbond

The Plattelandersbond was founded in February 1917 and can be considered the first serious agrarian party in the Netherlands. The timing of its establishment was certainly no coincidence, as the Dutch agricultural sector was in turmoil during the First World War (Moeyes, 2011). Although the Netherlands remained neutral, the country experienced grave economic problems due to the war at sea. In order to regulate food supply in the cities, the government heavily intervened in the agricultural sector. Many farmers felt that they had made a sacrifice for which they were not given enough credit by the 'city-folk' (*stadsen*). At the same time, due to the new electoral system, there seemed to be an opportunity for an agrarian party. In the previous electoral majority system, which was based on districts, the chance of success of a distinct agrarian party was less likely. Although in some rural districts there had been candidates who presented themselves as advocates of rural interests (Van Klinken, 2003), these candidates were all more or less associated with one of the established parties in the Netherlands, that is, one of the liberal parties, the Catholic or orthodox Protestant parties or the social-democratic party.

However, in 1917 there were signs that the old political constellation was about to change, as some long-standing political issues that had caused much opposition were resolved in the constitutional amendment of that year. Universal male suffrage (in 1919 also for women) was introduced (one of the demands of the social democrats and liberals) and religious schools were funded by the state in the same manner as public schools (a Catholic and orthodox Protestant demand). As these old, highly divisive issues were now resolved, many thought that a new era in politics would be ushered in, with a new kind of politician focused on concrete economic matters. In addition to the Plattelandersbond, a whole plethora of new parties participated in the national elections for the Lower House in 1918 (Loots, 2004; Vossen, 2003).

With a rather brief manifesto – which consisted of some concrete proposals to improve the situation in the countryside – and with a number of unknown candidates, the Plattelandersbond managed to reach the electoral threshold, thus becoming the first agrarian party in the Netherlands to enter parliament (Table 13.1). Here, Frederik Bos, the party's only MP, joined a new parliamentary group that consisted of various new parties, all demanding a sort of 'new politics' in which economic matters would replace old ideological issues. This so-called 'neutral' parliamentary group was primarily considered liberal, and in 1921 most of its members joined a new liberal merger party, the Vrijheidsbond (Freedom League), which was established to end the division of the various liberal groups.

Table 13.1. Electoral results for agrarian parties in the Netherlands, 1919-1937 (%).[1]

	1918	1922	1925	1929	1933	1937
Plattelandersbond (Countryside League)[2]	0.7	1.6	2.1	1.0	1.3	0.2
Plattelandspartij (Countryside Party)	-	0.4	-	-	-	-
Nationale Boerenbond (National Farmers' League)	-	-	-	0.1	-	-
RK Boerenpartij (Catholic Farmers' Party)	-	-	-	0.1	-	-
List Meertens	-	-	-	-	0.2	-
List Wolfswinkel	-	-	-	-	0.1	-
Actie Bouwman (Action Bouwman)	-	-	-	-	-	0.6
Total	0.7	2.0	2.1	1.2	1.6	0.8

[1] Available at: http://tinyurl.com/pezfreg
[2] In 1933 and 1937 the Plattelandersbond participated in the elections as the Nationale Boeren, Tuinders en Middenstandspartij (NBTM Party). In 1925 the Plattelandersbond participated with two opposing lists: List Braat and List De Boer.

The Plattelandersbond, however, did not join the ranks of this new party, due to its new MP, the obstinate Arend Braat. In 1919 he had replaced Bos, who, according to contemporary accounts, was suffering from homesickness. In the following years, Braat dominated the image and strategy of the Plattelandersbond and became one of the most notorious Dutch MPs of his time. Braat transformed the Plattelandersbond from a vaguely liberal interest party into an agrarian protest party with some strong populist characteristics. In parliament, Braat attracted attention with his roughly textured speeches, in which he made wild accusations against his fellow MPs. According to Braat, they were all 'city-folk' who despised farmers and wanted to destroy the countryside. 'Hundreds of thousands of farmers' were, according to Braat, 'ready to destroy Amsterdam, Rotterdam and The Hague', should there be a leftist revolution like those in Russia or Germany. Speaking and dressing like a boorish farmer, Braat became the 'crudest person ever to enter Dutch parliament', a man who brought rural wisdom and manners into Dutch politics (Vossen, 2003: pp. 118-120). Due to this anti-elitist stance, which was flavoured with conspiracy theories and a boorish style, nobody in The Hague seemed to take him seriously. However, Braat gained some popularity among a small segment of the farming population. Especially in the isolated and scarcely populated province of Drenthe in the north of the Netherlands, the Plattelandersbond in fact gained considerable support, where in some small peasant villages the party won more than 40% of the vote.[1] It was largely as a result of these votes from Drenthe that the Plattelandersbond obtained a second parliamentary seat in 1922.

Despite this modest electoral success, Braat's populist style caused internal conflict. A group of party members wanted to transform the Plattelandersbond into a more policy-directed, decent agrarian party with achievable goals and a more respectable attitude, which would contribute to the education and integration of their constituency. These members were mostly the more educated countrysiders (agricultural engineers, veterinarians or schoolteachers at rural schools), who regarded themselves as a vanguard for agrarian emancipation (IJzerman, 1925). They considered the Scandinavian and Swiss agrarian parties their role models, while Braat in contrast was increasingly approaching early fascist movements in the Netherlands. The clash between populists and moderates continued well into the early 1930s and led to various splits in the party. Eventually, the more moderate and 'decent' faction seems to have won the power struggle. In 1932, at the height of the economic crisis, the Plattelandersbond changed its name to Nationale Boeren, Tuinders en Middenstandspartij (NBTM Party) and adopted a new, more elaborate party platform. Braat was replaced by Cornelis Vervoorn, a more respectable gentleman farmer. Vervoorn earned more respect from his fellow MPs but lost most of the constituency of the Plattelandersbond: in 1937, the NBTM Party vote dropped from 1.3% to a mere 0.2% and it lost its only seat (Table 13.1). The party never recovered from this electoral catastrophe and was disbanded soon afterwards.

[1] For example, the towns of Vries (46.2 %), Rolde (40.8 %), Ruinen (30.9 %), Anloo (30.5 %), in 1925.

The most obvious explanation for the decline of the Plattelandersbond lies in the rise of the Nationaal-Socialistische Beweging (NSB: National Socialist Movement) in the 1930s. Founded by the engineer Anton Mussert at the end of 1931, the NSB became the most successful fascist party in the Netherlands, obtaining nearly 8% of the vote in the 1935 provincial elections. Various electoral studies have shown that even in 1935, large parts of the constituency of the NBTM Party had defected to the NSB (Kooij, 1982; Von der Dunk, 1982). In the small villages in Drenthe, the NSB now became the strongest party. However, the failure of the Plattelandersbond to establish itself as a powerful agrarian party had causes which went much deeper. Although the notion that Dutch society was 'pillarised' in this period (i.e. segmented along religious lines) has been criticised and increasingly questioned by historians, it does help to explain the fate of the Plattelandersbond. The party was not able to attract Catholic and orthodox Protestants voters, who seemed almost hermetically locked up and isolated in their pillars. From the early twentieth century onwards, Catholic and orthodox Protestant farmers had their own interest organisations, which were tightly connected with the congenial political parties. While the Plattelandersbond did attempt to attract Catholic and orthodox Protestant farmers with two lists of candidates, it was without success. As a result, the party saw its potential constituency narrowed down to a rather small segment of non-religious or non-orthodox Protestant farmers.

Of course, the agricultural crisis of the 1930s also affected Catholic and orthodox Protestant farmers. Eventually, the dissatisfaction among these farmers was also expressed via pillarised structures. Catholic peasants in the southeast of the Netherlands felt more attracted to a Catholic protest party such as the Bouwman Action (Actie Bouwman). Between 1935 and 1939, this party, led by tree-nursery employee Alfons Bouwman, obtained striking results in Betuwe and East Brabant (around 30 to 40% in some villages). Outside this region Bouwman's electoral attraction strongly diminished, which according to some contemporary sources also had to do with the fact that he used to speak a dialect which was barely understood outside this specific region (Van Velthoven, 1983). It may well be that this problem also hindered the farmer Driek Meertens, who in the early 1930s attracted a following among the Catholic farmers in the south of Limburg. In the national elections of 1933 Meertens gained around 50% of the vote in some of the remote villages in the most southern part of this province. Since both Meertens and Bouwman received close to zero votes outside their own regions, they were only minor parties at the national level and never won seats in parliament. There is some evidence, however, that the parties of Meertens and Bouwman served as stepping stones to the NSB for many Catholic farmers, who later switched to the NSB, albeit only temporarily. Afraid of a mass defection of its believers to the NSB, in 1936 the Catholic Church openly condemned the NSB and forbade Catholics to become members of the party. Thus, faced with a dangerous new party, the Catholic pillar showed its sheer strength. As a result, in the elections of 1937 the NSB lost more than half of its constituency in the Catholic provinces.

Orthodox Protestant farmers also expressed their discontent through pillarised channels. Those who disagreed with the harsh economic policies of the orthodox Protestant prime minister Hendrik Colijn could vote for smaller Protestant parties such as the Staatkundig Gereformeerde Partij (SGP; Political Reformed Party) and the Christelijke Nationale Actie (CNA; Christian National Action), which during these years articulated the interests of farmers and fishermen. Hendrik van Houten, the leader of a union for small peasants in Friesland, became the leading candidate of the Christelijk-Democratische Unie (CDU; Christian Democratic Union), a more progressive Protestant party (Langeveld, 1988). Inspired by the example of the Catholic Church, the Gereformeerde Kerken in Nederland (Reformed Church) attempted to prevent the defection of orthodox Protestant farmers to the NSB as well as the CDU.

The vicissitudes of the Plattelandersbond and other new parties such as the NSB demonstrate the constraints of pillarised democracy in the interwar years. Dutch society was strongly segmented, and as a result it was difficult to establish a new party for all religious denominations. Discontent was also expressed in accord with the pillarised structure: Catholic farmers went to Catholic protest parties, Protestant farmers to Protestant ones. In the case of agrarian parties, regional differences also constrained their constituency. Parties which attempted to 'break through' the pillars, such as the agrarian parties, but also the NSB and even the social democrats, were confronted with fierce opposition from the churches. As a result new parties had to focus on the non-pillarised or less pillarised segment of society.

13.3 The Boerenpartij

After the Second World War it took until 1958 before a new farmers' party, the Boerenpartij, was founded. The post-war restoration and even reinforcement of the system of pillarisation once again made it difficult for new parties to gain a foothold. Moreover, after 1945, agrarian parties were associated with collaboration and fascism. Various farmer leaders of the 1930s, such as Van Houten, Braat and Bouwman, had shown sympathy for fascism and/or had collaborated with the German occupying forces, and were excluded from the political scene well into the 1950s (Van Donselaar, 1991). Also, in the 1950s, the modernisation of the Dutch agricultural sector resulted in a decrease in the number of farmers. The share of the population working in this sector dropped from 19.8% in 1947 to 12.5% in 1960 and 7.0% by 1970.

In 1954, a new corporatist organisation was established – the Agricultural Board (Landbouwschap), which was a council governed by public law and consisting mainly of delegates from the various pillarised agricultural organisations. This situation did cause some discontent among farmers around the country. In 1946, an organisation had been set up to protest against governmental interference in agriculture, the

Landelijke Vereniging voor Bedrijfs Vrijheid in de Landbouw (National Association for Freedom of Enterprise in Agriculture, BVL) (Terluin, 2012). Hendrik Koekoek, or 'Boer Koekoek', as he was usually called, became the figurehead of this post-war farmers' movement which was nicknamed 'the Free Farmers'. The Free Farmers managed to gain a lot of publicity because of their fierce protests against taxes introduced by the Agricultural Board. Inspired by the successful French protest party (Union de Défense des Commerçants et Artisans) led by Pierre Poujade, in 1958 Koekoek founded a political party, the Boerenpartij, which would operate as the political branch of the Free Farmers movement.

In the national elections of 1963 the Boerenpartij, somewhat surprisingly, managed to enter parliament with three seats. In the following years Koekoek became a high-profile national figure. Like his prewar predecessor Braat, Koekoek was an extraordinary figure in Dutch politics because of his boorish style, strong regional accent and his love of peasant wisdom. In his speeches in parliament he used anecdotes from farming life to highlight his arguments. Rather than looking for compromises he preferred taunting his opponents. It is almost needless to say that Koekoek, like Braat, was not taken very seriously by the political establishment. Unlike Braat, however, Koekoek managed to gain some popularity beyond his own (relatively stable) constituency of farmers. In particular, readers of the most popular daily newspaper *De Telegraaf* took a liking to him: 'One of the few politicians who is talking our language', as one of the readers stated in a survey (Vossen, 2005). As a result, the Boerenpartij managed to obtain almost 7% of the vote in the provincial and municipal elections in 1966 (Table 13.2). Since the NSB in 1935, no new party had been so successful.

What made the result even more unusual was the fact that Koekoek, a Protestant, also attracted a lot of voters in the cities and in Catholic areas. In the Catholic provinces of Brabant and Limburg, the success of the Boerenpartij was a clear sign that the Catholic farmer organisations were losing control over their members. In big, secular cities such

Table 13.2. Electoral results Boerenpartij, 1959-1982 (%).[1]

	1959	1963	(1966)[2]	1967	1971	1972	1977	1981	1982
Boerenpartij (Farmers' Party)[3]	0.7	2.1	6.7	4.8	1.1	1.9	0.8	0.2	0.3

[1] Available at: http://tinyurl.com/pezfreg

[2] The result in 1966 was achieved in provincial elections.

[3] Excluded here are the many splinter parties, such as Binding Rechts and List Voogd, as they were not clearly agrarian parties.

as Amsterdam and The Hague, but also in Catholic cities such as Nijmegen and Breda, the Boerenpartij obtained around 10% of the vote. Unfortunately, data on the political and social characteristics of the urban electorate are all but unavailable (Kuypers *et al.*, 1967). In electoral terms, it was not the relatively secular province of Drenthe but the Catholic province of Brabant that became the party's stronghold.

However, the success of the Boerenpartij proved to be short-lived. By 1966, the party was already troubled by internal conflict and scandal. Journalists discovered that various new candidates had collaborated with the Germans during the Second World War. *De Telegraaf* withdrew its support for the party when Koekoek appeared reluctant to openly condemn these candidates. Moreover, the leadership of Koekoek was also questioned by various members within the Boerenpartij. Koekoek was used to leading the party as if it were his personal possession.[2] However, some more intellectual members had attempted to formulate the ideological principles of the party, with the goal of transforming the Boerenpartij into a more decent and 'normal' conservative and/or Christian party which could influence governmental policies (Van den Bos, 2007). The orthodox Protestant lawyer A. Zeegers even wrote a programme of principles in which the Boerenpartij rejected the people's sovereignty. The status of this programme is dubious as shortly after its publication Zeegers was expelled by Koekoek.

Although Koekoek had always been keen to attract a more conservative following, he did not show much interest in attempts to transform the party, or in any ideological debate. He preferred to surround himself with old stalwarts, mostly from the Free Farmers movement, and distrusted most of the new candidates. Various candidates were expelled, or left the party themselves. In 1968, four of the seven members of the parliamentary group broke away to form the conservative party Binding Rechts. As a result of all the internal conflicts the Boerenpartij vote dropped from 4.8% in 1967 to 1.1% in 1971 (Table 13.2). In the 1972 national elections the party made a modest comeback, with 1.9% of the vote, but again this success marked the start of more internal conflict. Nevertheless, thanks to his old and loyal constituency, Koekoek managed to stay in parliament until 1981.

Considering the political dilettantism of the Boerenpartij, combined with its target group, which was clearly shrinking, the question is perhaps not so much why this party did not succeed but why it managed to obtain so many votes in the first place. Due to a lack of data on the electorate, however, it is difficult to answer this question. Clearly, because of his specific image as a telegenic antihero, Koekoek fitted well into the new media frames. As a result he could serve as a relative harmless mouthpiece for people who felt uncomfortable with the rise of new-left issues and movements in Dutch politics, but also with the continuing hegemony of the old pillarised elites.

[2] Based on the documentary by OVT Radio broadcast, 'Het Spoor Terug', 24 February 2008; available at http://tinyurl.com/oqpksqc.

Characteristic of the latter is Koekoek's demand for the abolition of the pillarised control of Dutch broadcasting services. In fact, Koekoek's rise also symbolised the growing desire for more personal freedom. Considered from this perspective, it is no surprise that the younger generation was overrepresented within the electorate of the Boerenpartij (Kuypers *et al.*, 1967: p. 40). Koekoek was more in tune with the times than is often assumed, as the Dutch pillar system was clearly crumbling in the 1960s, and many voters were longing for a new kind of rightist party which, on the one hand, was less patronising and less elitist and, on the other hand, more conservative and less inclined to accommodate the new left than the established parties on the right. For a short while Koekoek satisfied this need. Whereas the new left sought to emancipate common citizens, Koekoek championed a strong belief in the common people, with their common taste and their justified need for more personal freedom.

In the 1970s, however, he lost his unique selling point, as politicians such as the Liberal party leader Hans Wiegel and Christian Democratic party leader Dries van Agt also began to counterattack the condescending new left from a popular and conservative perspective. The Boerenpartij probably also suffered from the abolition of the legal obligation to vote (or more precisely, to turn out at the polling station) in 1970. Nonetheless, the party clearly benefited from the depillarisation of the farming population. While the Plattelandersbond was forced to focus on non-pillarised groups, the Boerenpartij also attracted Catholic and orthodox Protestant farmers. In electoral terms, however, these groups had simply become too small to serve as the basis for a durable party. After 1981, the Boerenpartij only managed to survive in some small rural municipalities.

13.4 Dutch agrarian parties: a conclusion

The Dutch agrarian parties are a somewhat distinct case compared to many other countries. In Scandinavian countries and Switzerland, for example, agrarian parties developed into enduring and relevant parties, while in other countries, such as France, Belgium and Germany, agrarian parties were somewhat flash-in-the-pan. In the Netherlands, the agrarian parties were rather irrelevant – in terms of influence and electoral strength – but the Plattelandersbond as well as the Boerenpartij nevertheless managed to hold at least one seat in parliament for nearly twenty years.

Both parties were highly personalised, which also contributed to their organisational weakness. For the outside world, they were above all the vehicles of their leaders 'Boer Braat' and 'Boer Koekoek'. Both party leaders displayed a populist style, which was intended to highlight the cultural division between the political elite of condescending city people and the authentic, shrewd farmer. The often disdainful reactions of fellow parliamentarians and the media confirmed this image. With some success, Koekoek managed to extend this image to a more general populist opposition between the old elites and the 'little man'. Both Braat and Koekoek met some opposition within their own ranks because of their authoritarian styles of leadership and their vulgar antagonistic political style. Some opponents aimed to establish a more decent and normal political party which could achieve some practical results. These attempts eventually failed, which came as no surprise.

For most of their voters, the Plattelandersbond and the Boerenpartij above all served as protest parties through which they could express their feelings of discontent and lack of respect. For a more practical defence of their own interests, farmers could still rely on the confessional (and later Christian democratic) parties and the affiliated (pillarised) farmers' organisations well into the 1990s. In the Netherlands, these parties were firm defenders of agricultural interests. In the twentieth century, the Minister of Agriculture was usually a confessional politician who cooperated closely with the parliamentary spokespersons on agricultural affairs. In fact, the 'Green Front' – as the agrarian lobby inside and outside parliament is referred to – did make a distinct agrarian party superfluous (Andeweg and Irwin, 2004: p. 142). The only option for an independent agrarian party therefore seemed to be full-fledged protest. Eventually, however, the number of farmers in the Netherlands became too small to support a distinct farmers' party, and their number will only diminish further in the future. It is estimated that in the period to 2035, some 24,000 agricultural enterprises will be terminated (Gies *et al.*, 2014). Thus, the time for an agrarian party in Dutch politics seems to be over forever.

References

Albertazzi, D. and McDonnell, D., 2008. Twenty-first century populism. The spectre of Western European democracy. Palgrave Macmillan, New York, NY, USA, 288 pp.

Andeweg, R.B. and Irwin, G.A., 2004. Governance and politics in the Netherlands. Palgrave Macmillan, Basingstoke, UK, 328 pp.

Gies, T.J.A., Nieuwenhuizen, W. and Smidt, R.A., 2014. Vrijkomende agrarische bebouwing in het landelijk gebied. InnovatieNetwerk, Utrecht, the Netherlands, 57 pp.

IJzerman, A.W., 1925. Braat, de vader des plattelands? 'Ontwikkeling', Amsterdam, the Netherlands, 15 pp.

Kooij, G.A., 1982. Het echec van een volkse beweging. Nazificatie en denazificatie in Nederland. Hes, Utrecht, the Netherlands, 359 pp.

Kuypers, G., 1967. De Nederlandse kiezers in 1967. Enkele eerste resultaten van een landelijke enquête. Agon Elsevier, Amsterdam/Brussel, the Netherlands/Belgium, 80 pp.

Langeveld, H., 1988. Protestants en progressief. De Christelijk-Democratische Unie. Sdu, Den Haag, the Netherlands, 560 pp.

Loots, J., 2004. Voor het volk, van het volk. Van districtenstelsel naar evenredige vertegenwoordiging. Wereldbibliotheek, Amsterdam, the Netherlands, 287 pp.

Moeyes, P., 2011. Buiten schot. Nederland tijdens de Eerste Wereldoorlog. De Arbeiderspers, Amsterdam, the Netherlands, 459 pp.
Mudde, C., 2004. The Populist Zeitgeist. Government and Opposition 39: 542-563.
Taggart, P., 2000. Populism. Open University Press, Buckingham, UK, 142 pp.
Terluin, I.J., 2012. 100 Jaar boer Koekoek. Faculteit der Ruimtelijke Wetenschappen, Groningen, the Netherlands, 160 pp.
Van den Bos, M., 2007. Tegen! Een geschiedenis van de Boerenpartij. Afstudeerscriptie Universiteit van Amsterdam. Universiteit van Amsterdam, Amsterdam, the Netherlands, 99 pp.
Van Donselaar, J., 1991. Fout na de oorlog. Fascistische en racistische organisaties in Nederland 1950-1990. Uitgeverij Bert Bakker, Amsterdam, the Netherlands, 249 pp.
Van Klinken, G., 2003. Actieve burgers. Nederlanders en hun politieke partijen 1870-1918. Wereldbibliotheek, Amsterdam, the Netherlands, 686 pp.
Van Velthoven, J.C., 1983. De Actie-Bouwman. Spiegel Historiael 12: 653-660.
Von der Dunk, H.W., 1982. In de schaduw van de depressie. De NSB en de verkiezingen in de jaren dertig. Sijthof, Alphen aan den Rijn, the Netherlands, 222 pp.
Vossen, K., 2003. Vrij vissen in het Vondelpark. Kleine politieke partijen in Nederland 1918-1940. Wereldbibliotheek, Amsterdam, the Netherlands, 319 pp.
Vossen, K., 2005. De andere jaren zestig. De opkomst van de Boerenpartij. In: Voerman, G. (ed.) Jaarboek 2004 Documentatiecentrum Nederlandse Politieke Partijen. Documentatiecentrum Nederlandse Politieke Partijen, Groningen, the Netherlands, pp. 245-266.
Vossen, K., 2010. Populism in the Netherlands after Fortuyn: Rita Verdonk and Geert Wilders Compared. Perspectives on European Politics and Society 11: 22-38.

Index

A

B

C

D

E

F

G

H

I

J

K

L

M

N

O

P

R

S

T

U

V

W

Z

www.ingramcontent.com/pod-product-compliance
Lightning Source LLC
Chambersburg PA
CBHW081125180926
51611CB00004B/54

* 9 7 8 9 0 8 6 8 6 2 5 9 7 *